WASTED WORLD

WASTED WORLD

HOW OUR CONSUMPTION CHALLENGES THE PLANET

ROB HENGEVELD

THE UNIVERSITY OF CHICAGO PRESS
CHICAGO AND LONDON

ROB HENGEVELD is affiliated with the Faculty of Earth
and Life Sciences, Department of Animal Ecology at Vrije
Universiteit, Amsterdam, where he was an honorary
professor, and with the Centre for Ecosystem Studies at
Wageningen UR.

The University of Chicago Press, Chicago 60637
The University of Chicago Press, Ltd., London
© 2012 by The University of Chicago
All rights reserved. Published 2012.
Printed in the United States of America

21 20 19 18 17 16 15 14 13 12   1 2 3 4 5

ISBN-13: 978-0-226-32699-3 (cloth)

ISBN-10: 0-226-32699-3 (cloth)

Library of Congress Cataloging-in-Publication Data

Hengeveld, Rob.
   Unlivable Earth : how our consumption challenges the
planet / Rob Hengeveld.
       p. cm.
   Includes bibliographical references and index.
     ISBN-13: 978-0-226-32699-3 (cloth : alk. paper)
     ISBN-10: 0-226-32699-3 (cloth : alk. paper)   1. Nature—
Effect of human beings on.   2. Population—
Environmental aspects.   3. Waste products—
Environmental aspects.   4. Waste minimization.   I. Title.
GF75.H45 2012
304.2′8—dc23
                                        2011025255

⊗ This paper meets the requirements of ANSI/NISO
Z39.48-1992 (Permanence of Paper).

To Claire, who asked me to write this book, and to Eke, Sytse, and Ilona, who have to live with it

CONTENTS

INTRODUCTION

WHAT IS THIS BOOK ABOUT?

This book is about problems principally arising from too many people living on Earth. It is not optimistic in its contents or in its conclusions. As a warning, it has its dark side, telling of the threat to the future the present contains. It is bleak in its descriptions of depleting and wasting our resources; of famines and widespread disease; of uprooting age-old social relationships, customs, and civilizations; of desperate wars being waged for the remaining resources. But it does have a lighter side: there is hope—if we humans take the right measures. We need to reduce our numbers drastically, and soon. Though this is difficult, it is doable if we not only know but also understand what's happening. This book aims to give you that understanding, the insight you need.

One set of problems concerns the exhaustion of resources, and another the generation of waste. This book explains not only how these problems arise but also how they interconnect and what to do about them. Of course, if they arose solely from there being too many people, taking countermeasures, however difficult this would be, would still be relatively simple. But matters are more complicated than the number of people getting too large. Problems also arise from the high standard of living experienced by only the rich part of the world's population, a quality that has to be spread more evenly across the poor part as well. Yet other problems result from the way we use our energy and material resources; their utilization should be organized less wastefully.

People often opt to analyze and solve problems one at a time, but apart from the advantage of keeping things simple, this has disadvantages when the problems are interconnected. For example, most books on global trends in the environment concentrate on climate warming, on salinization, or imminent food or energy shortages, but hardly any of them give a comprehensive overview of the various trends. In this book, I want to give such an overview by discussing a whole suite of problems and showing how they are interconnected, either directly or because they share a common cause: the huge number of people on Earth. Leaving out certain problems will affect our understanding and thus reduce the efficacy of any measures we take to deal with them.

In fact, rather than being a closing entry to balance the budget, the environment is basic to demographic and economic processes on a large scale. The world's demographic makeup is the result of the socioeconomic system we have built over the millenia and also its driving force. The system provides our food, basic materials, housing, and clothing; facilitates the flow of food and materials; and makes our lifestyle comfortable. Not only has it grown, it has also diverged from processes normally found in nature. As such, it remains a special form of ecology: the ecology of humans. We have to look at all these issues in their totality and in their historical context. We need to work from an overall view.

While working on these problems, I came to realize not only that many of the processes now going on in our global environment are interconnected but also that they all lead to the same result: a shamefully wasted world. These are processes that may threaten the future of humanity. And worst of all, the threats seem to be converging. Cataloguing a whole suite of interconnected, coinciding problems in our earthly environment can easily give a bleak picture of the future. This is certainly not my intention, though it's inevitable, given the nature and urgency of the problem.

Increasingly, we are understanding more about the world around us. And our growing understanding enables us to increase our mastery over our world, with the result that major innovations are spreading across the world improving the living conditions for millions. Although this has its price, the price tag is up to us. If we don't take the right measures, the price will be enormous: the world and any possibility of further progress will go to waste. We will waste our future and all we have achieved so far.

HOW ARE THINGS GOING WRONG?

All the resources on Earth could not possibly sustain life in all its abundance and diversity for a single year, let alone for the almost four billion years that life has already existed. Instead, the waste generated by one organism always becomes the food of another, this food results in the waste of the next organism, and so on, until the first organism finds the nutrients again in one form or another as its food. The continual reuse of resources is essential if life is to be sustained for any length of time. Life has developed not only the variety and diversity of species we see around us but also the cycles for long-term sustainability. These cycles have turned resources into waste, waste into resources, these resources again into waste, and again and again for billions of years and, hopefully will do so for another four billion years, after which conditions in the solar system will have deteriorated too much for life on Earth to keep going any longer. We are now living roughly halfway along this path, at life's zenith.

Within these broad cycles of material use, each organism feeds on another one. Even plants do this by living on gases expelled into the air by other organisms, such as animals, and from other nutrients left as excrement in the soil or released when organisms die and decay. Using sunlight as their energy source, plant life is the ultimate motor for keeping the nutrient cycles going. Or, to put it another way, the energy from the sun keeps those cycles in perpetual motion.

Humans, together with all other organisms in their environment, are part of these global nutrient cycles. For most of their existence on Earth, human beings have lived as part of the great cycles of nature, eating some species and being eaten by others. They never stood apart. However, gradually, humans began to dominate as a species, using more than their share of resources and generating much waste. Humans also found other sources of energy and nutrients, could cope with many diseases and parasites, and were seldom eaten by other species. The number of humans grew and grew. Humans began to change their environment by clearing the forests, irrigating the soil, and exhausting its nutrients. The present deterioration of the natural environment is therefore not new—it is age-old. The relatively recent huge growth in the human population has accelerated the rates of resource use and waste generation to an excessive degree. These excessively high rates strained the cycles until they finally broke down—a process that took only a couple of centuries to happen.

So what exactly has been happening during the last couple of centuries? What went wrong, and can we still restore the ancient biological processes of nutrient recycling, and thereby halt the decline? What needs to be done?

The driving force behind the deterioration of the environment is simply that the human population has gotten out of hand. Natural processes could not supply the growing numbers with enough energy and nutrient sources or turn our waste back into nutrients fast enough. In the 1970s, the human population numbered roughly 2.5 billion. Now, only 35 years later, it is just over 7 billion. That's an average increase of some 1.3 billion people every decade, which means that each year there are an extra 130 million humans to be fed, housed, and clothed. And every year, there are an extra 130 million people generating more waste. Although the rate of population growth was higher during the 1970s, the total number of people has continued to increase, putting a strain on the great nutrient cycles of life.

In the 1970s, some scientists still thought that Earth might be able to support a human population of 30 billion; some estimates even put Earth's carrying capacity at between 10^{16} and 10^{18} million people, which was 100 million times more than the alarming 10 billion that people anticipated would be reached during the next few decades. Therefore, there seemed to be a comfortable safety margin, so nothing needed to be done. And indeed nothing was done for three decades. Since the beginning of this century, however, the estimated maximum population has been adjusted downward to 9–10 billion, which means that right now we're only a worrying 2.0–3.0 billion people away from reaching the limit of what our planet can support. Given the present rate of growth of the human population, we have very little time left (certainly much less than the 30-year window envisaged in the 1970s) to slow down or reverse the trend. Because of our optimism and negligence we have certainly wasted valuable time. Yet many people continue to hope that the human population will stabilize at this level of 9–10 billion people without external stresses, such as shortages of food, drinking water, or energy or pollution of arable land.

Throughout history, however, resource use has intensified and grown at even higher rates than the population. Each of us is now using more resources than our parents did—and more than we ourselves did only one or two years ago. And each of us is generating more waste than our parents ever did. Those in rich countries use more equipment and appliances at home and have one or more cars per family—and the cars are larger. And, partly because of the growth of megacities, people eat food from other parts of the world, which has to be transported over ever-increasing distances and has to be conserved and stored for ever-longer periods. Furthermore, more and more people are going on vacation once or twice a year to destinations further and further away from home. And so on. The result is that our resources are being depleted at accelerating rates by each of us, and each of us is generating increasing amounts of waste. Moreover, in order to grow and manufacture more products, we use energy resources whose generation requires itself ever more energy, and inevitably produces waste. The effects of these waste products are already being felt: the atmosphere and the oceans are warming, with the result that the polar ice caps and mountain glaciers are melting. The main waste product here is carbon dioxide produced by our cars and industries and by the concrete we use when building houses, offices, factories, and roads. And finally, what makes all this worse is that we will soon run up against the limits of our energy sources, and thus the limits of our food supply, as well as of our supply of raw materials.

The carbon dioxide produced by cars and industries is polluting the atmosphere and thereby causing the atmosphere to warm up via the greenhouse effect. This indirect effect of pollution by a waste product affects us by altering our living conditions (a hotter and drier climate, or colder and wetter conditions—depending on which part of the world we live in); and worse, it also affects the living conditions of the crops on which we depend for our food. In parts of the world, conditions may become too hot or dry for the survival and growth of some crop species. Food crops adapted to cool conditions will be unable to grow or will wilt because the soil from which they extract their water dries up. Moreover, valuable topsoil can easily been blown away from dry soils, leaving behind infertile subsoil, hardpans, or crusts of salt form in which plants cannot germinate and grow. Thus, climate warming in the world's traditional food baskets means that sooner or later we will face severe shortages, because many of our food crops will be unable to cope with the changed conditions.

Warmer conditions also affect organisms that recycle our waste, as well as those that cause human disease or diseases of livestock or crops. Other species around us forming links in the great nutrient cycles on Earth are unable to keep pace with our consumption and waste generation. Thus, unknowingly, we have even begun to affect the organisms that break down our waste. In short, we are using too much, we are destroying too much, and less and less of our excess waste is being recycled.

Only a century ago or less, most people depended mainly on natural products as resources for their way of life. The waste these products left could be taken up again in the natural cycles of all nutrients. But now, in most societies there are cars, industries, MP3 players, computers, and so on—all of which require energy and material resources, and all of which make waste. We're not only creating more waste, the waste we are producing has become widely different too. And what is worse, often, this different waste cannot be recycled through digestion by bacteria, plants, fungi, or animals. It remains undigested and pollutes, poisons, or chokes these animals, fungi, plants, and bacteria.

Because there are no organisms on Earth that can use much of our waste as their food, we are not just straining and breaking the biospheric nutrient cycles, we are bypassing them. Ultimately, we will be unable to eat other species further down in the cycle, because our resources have turned into unusable waste. The plants and animals on which we rely for our energy and food are dying out or becoming toxic because of the toxicity of our waste. Our resources are being exhausted, and our waste is beginning to pollute our environment and food on a large scale.

So, we are polluting our agricultural land and turning other land into salty desert. We are turning mountains into deep pits by mining for metal or coal and are lowering the groundwater level over vast tracts of surrounding land. And we are forcing species to shift, extend, or reduce the geographic area they inhabit. We are turning some species into weeds or pests and causing others beneficial to us to die out. We are wasting ever-larger parts of Earth— for ourselves and for thousands of other life-forms around us: species that feed us, that recycle our waste, and that used to clothe our environment and make it comfortable for us to live in. Unless we take countermeasures, our planet will become uninhabitable for us and all the other species on which we depend. We are browning our blue and green Earth.

The exhaustion of our material and energy resources and the saturation of the environment with our waste could have occurred far apart in time, but they will more or less coincide. This is partly because these processes are related: they all result from the causes they have in common—excessive population growth and the increasing resource use and generation of waste. This increase in resource use and waste production is not spread out equally among all humans; it applies only to the richer part of our global population. Because our attitudes have to change within only a couple of decades, we should address the prime motor of our resource utilization and waste generation: our own numbers and the rate at which our population is growing. The measures we must take concern the numbers in which we can sustain our presence on Earth and assure a certain quality of life for each of us: a quality we also—too often quite unjustifiably—wish to improve. If our elected leaders do not act soon, conditions beyond our capacity will take over, causing immeasurably greater difficulties for most of us.

WHAT SORTS OF PROBLEMS CAN WE EXPECT?

At the beginning of the present millennium, the world population was 6.5 billion, all needing to be fed and clothed. This feeding and clothing depends on many systems: agricultural and industrial, educational and health, administrative and judicial, communication and transport. There are systems for finding, mining, and processing mineral resources, sewage systems, and so on. Imagine a megacity with millions of people living cheek-to-jowl, all totally dependent on food from elsewhere. Imagine the various flows of transport needed to feed them, supply their factories with raw materials, and remove waste. The transport networks extend across continents and oceans. What would happen if they broke down? Imagine a world in which medical and pharmaceutical services no longer function—a world without hospitals and local family doctors. How long could we be cut off from electricity or water? A day? A couple of days? What would happen?

We are increasingly dependent on interactions between people, and our growing population implies increasing numbers of such interactions, such as those of the service industries. Moreover, all these small subsystems that coordinate activities to keep the larger system running need to be coordinated via a hierarchy of superimposed systems: local, regional, state, national, and international. There are separate hierarchies for each field of human interaction: business, health care, agriculture, and so forth. All these coordinating activities require energy, material, and huge numbers of people. As our numbers increase, the numbers and size of our coordinating organizations increase even more, so that we need ever-larger amounts of a limited and rapidly shrinking supply of energy and raw materials.

Apart from all the people living on Earth today or in the near future, we also ought to think about all the people who will live long after we are dead. Most predictions stop at the year 2050, or at best at 2100, but this is clearly too short a time-span; it is merely a few generations, a time our children and grandchildren will see. As well as thinking about our current use of our limited resources and our current generation of waste, we must think about those in the distant future. At the least, we need to set up an efficient, large-scale recycling industry. (Yet, recycling itself depends on the use of much energy, and it will never be perfect.)

Even now, our reserves of oil and gas are running out. Their global supply will begin to decrease seriously as soon as 2015–2030. We have known they would run out since the mid-1960s and early 1970s, when many people all over the world first became concerned about the limits to growth. Reluctantly, scientists, industries, and politicians are only now beginning to invest in research to find alternative energy sources, concentrating on biofuel or hydrogen. Growing plants for biofuel, however, costs fertilizers made from natural gas and reduces the area for growing food at precisely the time when our food production needs to increase even further to feed our growing population. At the same time, the available agricultural land is being reduced by erosion and pollution, and crop production is stagnating because of climate change. This is aggravating the current problems of undernourishment or large-scale famine in parts of the world, and will do so even more in the near future when we need to have 70 percent more food than at present.

There are alternative sources of energy. Like oil and gas, hydrogen can be burned, releasing energy that can be used. It can be produced on a large scale. Solar energy is also an option, as are wind and water power. However, with oil, we not only possess a large supply of energy, but also a large resource for manufacturing widely different products. Some—fertilizers, herbicides, and pesticides—we use when growing, processing, and storing food; others are used to manufacture pharmaceuticals, clothes, plastics, nylon, rubber, building materials, machinery, computers, roads, footballs, and toothbrushes. Almost anything. We also use oil as a source of energy so we can transport food and manufactured goods from one place to another, usually over long, ever-growing distances across the world, supplying the millions in our megacities. So even if we replace our fossil fuels with hydrogen, we will still have many problems to solve.

And this is only part of the problem. As we use up our energy and material resources, we cannot avoid generating waste—both energy and material. Waste production is tightly connected to all our activities; it is the flip side of the coin of resource utilization. It's impossible to use any resource without producing waste. This waste is polluting our environment: the land, the rivers, the seas and oceans, the air. And although we can reduce the generation of waste by using cleaner and more efficient processes, the real problem of pollution is that we can reduce it only to a limited extent; we cannot stop it altogether. Every system—be it a bacterium, a human being, a society, or a mowing machine—produces waste. Moreover, the more our world population grows, the more waste we produce, no matter the measures we take to reduce the production of carbon dioxide and all other waste. Growing populations use more and more of their resources, thus exhausting these resources more quickly and generating more waste. The growth of the human population, of the interactions among its members, and the rapid increase in waste generation are directly connected with our resource utilization.

In recent decades, much more has happened than this. We are also exhausting our farmland, soil fertility, and groundwater. And we are exhausting the large stocks of fish in the rivers, seas, and oceans. We are polluting arable land with fertilizers and herbicides or with plastic sheeting to prevent weeds or evaporation. When soil dries out, dust bowls can arise over vast areas, and wind and runoff remove the precious topsoil. Moreover, we are taking up more and more of the fertile land to build towns or construct highways. We are mining groundwater for crop irrigation and drinking water, thereby depriving other plants and animals. Our human, household, and industrial waste is polluting surface and groundwater, threatening our supplies of water for drinking and irrigation. We are wasting our living space by intensifying its use. We are also polluting the air we breathe: in some cities, people wear face masks as a matter of course, and in China alone, air pollution results in the death of about two million people each year. Our world is wasting away not only in terms of certain raw materials and because of generations of waste, but also because of the deterioration of our food-producing land, the water we drink, and the air we breathe.

There is one pivotal process on which all others depend: our own reproduction—not our own, personal reproduction but that of all humans together. This overall reproduction is expressed not by parental interest and love, but by the numbers of humans added every year to the existing total. Our numbers are spiraling out of control, with the result that our resources are being depleted faster and our waste is polluting and destroying our environment at matching rates. The main cause of this acceleration is our growing numbers.

However, the exhaustion of our resources and generation of ever-more waste threaten our growing numbers. If we do not limit our reproduction, our numbers will soon be checked by external factors in some way or another. But if we are to avoid the human population being checked by a violent stampede for what remains when one or more of our resources run out, or by some kind of pollution, it is vital that we reduce population growth. If we continue reproducing at the present rates, we will unavoidably get into all sorts of difficulties and disasters, which will begin by reducing our quality of life, and then our life expectancy. However difficult for personal, moral, or religious reasons it will be to check our own reproduction, we cannot avoid taking drastic measures, and we should do this sooner rather than later. Population control by famine, drought, genocide, or wars fought over the remaining resources is worse than are the difficulties caused by changing our attitudes and customs ourselves.

The measures to be taken not only determine the future of our immediate progeny—our sons and daughters and grandchildren—more importantly, they also concern the long-term survival of humans on Earth. As measures for a long-term sustenance of humanity, they should be different from anything we have done so far. They should stretch our imagination and capacities. It is essential that we develop and take our measures soon; we no longer have any choice.

Of course, it is not our numbers alone that produce problems—it is also the demands each of us make and the growth of those demands. Measures to curb and reduce those demands will be difficult, but not that drastic— certainly not compared with changing our reproductive habits. As such, measures against overuse of resources must be the first priority. They should preferably be taken when other measures are just beginning to be implemented.

It makes no sense to take measures against each individual development without taking account of how they interconnect. We all know that the amount of resources we use depends on how many people are using them, as well as how much each person uses. So, reducing only the number of people, or only the amount used per person, is not enough (reducing the number of people might increase the amount used per person, for example; conversely, if personal consumption is reduced, the world will support more people).

However difficult it will be to make decisions like these, it is much more difficult to see how we should tackle, for example, the recycling of our waste: How much can we actually recycle, at what rate, and at what energy cost? Moreover, given a certain amount of an alternative nonfossil energy source, would this rate ultimately determine the number of people on Earth? If so, how many people would this be? Is this recycling rate the only factor determining the maximum number of people in future? To what extent is it unavoidable that we will still exhaust our resources, despite our recycling efforts? At what rate would we lose some resource during the recycling process anyway? And what would that mean for our long-term existence? What measures can we still take, how much is still in our own hands just because so many developments are interconnected? Can we press some master button, or do we have to press several at the same time in a coordinated way? Unlike any other species on Earth, humanity has opted for a certain path, but will this mean that in the end our numbers will dwindle just because we have chosen this path?

This book covers the entire process of using, wasting, and recycling energy, materials, and spatial resources. It is about how we are wasting our living conditions and what is still left to do something about. I particularly want to make you understand what is going on, how various processes interconnect, and how they enhance each other. We need to understand what is happening so we can take the correct and most effective and—above all—the most humane measures, however inhumane they may now seem.

PART ONE
NATURAL PROCESSES

In part 1 of this book, I describe processes in biological systems, such as bacteria, plants or animals, but only insofar as this is relevant to the future of our world. It is useful to know about these natural processes for two reasons. First, we should know these physical, chemical, and biological processes because we as humans depend on biological organisms for our food. In our future society, matters may be arranged similarly: mimicking green plants, some chemical procedures will split water into hydrogen and oxygen with the help of solar energy. We will then burn the hydrogen again; that is, we will bind it again to the oxygen released earlier. Thereby, we will obtain the energy to go for a jog; run our machines, equipment, and cars; make new chemical compounds; and drive our programs to recycle the waste we unavoidably make. In this way, the study of natural processes and their origin teaches us how we can persist on Earth in a complex human society. And, second, it is important to know how, in their evolution, organisms have become organized internally and in relation to each other by continually adapting to stressful situations. This organization may suggest how we should organize human society in future too. We can learn by example.

1 THE NATURE OF LIFE:
MAKING WASTE

Feeding adds matter and energy to an organism. Yet, after some time, an organism gets hungry and begins to feed again. Having to feed again means that energy and matter have been used somehow; they are used to sustain the organism's life, and the byproduct is waste expelled from the organism.

In very broad terms, this is what happens to all organisms all the time. Organisms big and small, simple and complex. Plants and animals, bacteria, and molds. Life is a never-ending stream of energy and matter—a stream entering a system as energy-rich food, doing work inside, and leaving it again as energy-poor waste. Actually, this stream has already flowed uninterruptedly for almost four billion years. Before they die and hand over the flame to their offspring, all organisms feed and turn food into waste, which they expel. This continues for generation after generation of organisms. Their waste, and eventually their bodies, become food for yet other organisms. And so on, and so on. Indefinitely. This food can be water or minerals from the soil, plants or animals, or gases from the air. The same perpetual cycle applies to waste: the water vapor we breathe out, the minerals returning to the soil, the oxygen plants release into water or air—or the dead elephant decaying away, over the months returning to the soil and air all its constituents assembled and used over the years. Except, of course, its energy, which it has continually dissipated throughout its life, first into the environment, and from there eventually into outer space—from which it originally reached us as solar energy.

In one way or another, the result is a never-ending stream of energy and matter flowing through several kinds of organism: from food to plant to animal to waste. Put crudely, an organism is an organized flow of energy and matter, ending up as waste. Or, more succinctly, it is a mechanism processing resources into waste. The energy and matter together form this mechanism, which both transfers and stores them for some time. This extremely intricate mechanism is what we know as an organism. And together, many organisms using each other's waste as their resource, as food, form a mechanism recycling mineral resources. These cycles are driven by energy that at present comes from the sun.

In slightly more detail, what happens is more complicated: the food has often first been stored within the organism and is used to power various life functions. Apart from storing energy and matter, organisms also grow, becoming larger and larger. Obviously, energy and matter are needed just to enable an organism to become larger (that is, an organism needs food so it can grow). Energy and matter are also used to maintain the organism. Like most things, all organisms wear out and decay; unavoidably, something goes wrong in the very complex and intricate processes within the organism's body so that its parts must constantly be repaired if they are to maintain their original function. These ongoing repairs are powered by the energy and matter in the organism's food. This maintenance is usually insufficient to restore all functions adequately, however, and ultimately the organism dies.

Yet another way matter and energy are used is for reproduction. Parents reproduce themselves in new, young organisms—their offspring; these new organisms eventually take the place of their parents when the latter wear out and die. Finally, some types of organisms, most of them animals (but never plants and molds), are mobile. This means that to feed or reproduce, for example, they can move from one place to another, using some of the energy which they have ingested.

Having lost energy and material in these various ways, the organism has to replace them. That is why animals get hungry and begin to search for food. Plants also get hungry but they use their food mainly for storage, maintenance, and growth, and at some point in their life cycle also for reproduction, that is, for making flowers and seeds. Plants also get thirsty: they evaporate water with their leaves, and also use part of it for building up their body. They drink water from the soil using their roots. The difference between plants and animals is that it is customary that we never talk about hunger or thirst, eating or drinking of plants, which we do in connection with animals. But in fact, they are doing the same.

Although for some time the energy and matter form part of the organism for all these functions, on its death the organism itself becomes waste, forming food for other organisms, such as bacteria or fungi, which break it down. Thus, the dead elephant mentioned earlier is first used as food by vultures and hyenas, and later what little they leave is consumed by bacteria and fungi. Eventually, all these organisms expel their waste into the environment, where it becomes food for plants again, and so on in an endless cycle of resource use and waste production. The cycle is like a snake biting its own tail, forming a circle. But if the snake didn't bite its tail and was instead a more or less straight line, the system would be exhaustive: the waste would remain unused forever, never to become food again. This straight line represents the process of resource depletion, and, at the same time, the piling up of waste, which pollutes the environment. By contrast, the two processes of resource depletion and waste production are connected to each other and are consequences of each other. The one does not go without the other. They are yin and yang. Resources enter the processing system and leave it as waste, which turns into food. Without recycling, resources deplete and waste accumulates.

A different pattern holds for the energy use of plants. They derive their energy directly from sunlight. Using this energy, a plant can build up complex chemical compounds from gases in the air and from certain minerals and water in the soil. The plant uses these compounds to build the cells of its body. Thus, the plant's food consists of solar energy, plus chemical compounds and water from the air and soil. All our energy ultimately comes from the sun and returns back to space. It keeps the great cycles of life processes turning, although overall it is a noncyclic, linear process of energy degradation. By depleting our resources and leaving our waste unused, we similarly follow a noncyclic, linear, exhaustive type of process, not only of energy but also of materials.

The question is, of course, why this difference between linear and cyclic processes? Why are cyclic processes typical for biological systems? Is there a reason nature follows that particular strategy in all organisms, and at all levels, from within organisms (the chemical level) to between organisms (the ecological level)? And why do we apply the simpler, linear process? Is it simply because we don't adjust the amounts of the production of resources and waste to each other? If anything, one might have expected the opposite to be the case: our sophisticated human society, with all its deliberate planning, might have developed the more complex cyclic type of processes. *Homo sapiens* is the only knowing species. With our greater intelligence, we would easily be able to master the complexity of cyclic processes, individually and with all their interactions. This would have been impossible for organisms of all those other species that are much less sophisticated than we are. Surely we are cleverer than a bacterium, aren't we?

But, disturbingly, with regard to recycling, the reverse is true. So, why? You may even wonder why those more complex cyclic processes occur in nature anyway. And, again, are they bound to biological systems and do they happen under all conditions, or are they only found under particular, restrictive conditions under which those systems are found? Moreover, if, one day we found ourselves living under similar, restrictive conditions, would we then need to develop similar systems of cyclic processing—for example, when either resource depletion or waste production become pressing, or perhaps both become a threat to our sustainability, our existence on Earth? How rigorous and rigid ought these cyclic systems be in order to guarantee our sustainability as a species and a society? Also, would they impose restrictions on our behavior, on our resource use, or even on the number of humans occurring on Earth? Would those restrictions be more or less stringent than those found in the nonhuman living world?

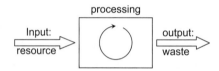

Let's depict the two types of processes in the form of a diagram. The figure above shows the stretched-out (linear) exhaustive system. It consists of a processing box with an input arrow on the left and an output arrow on the right. The input arrow can represent anything: food, gasoline, heat, garbage, wood chips—whatever. Similarly, the processing box can also represent anything: you yourself digesting your food; your automobile burning the gasoline to make heat, which moves a piston in the engine and ultimately makes the automobile move; and so on. The output at the right can be a product or mere waste. Thus, the product you made can also be something you really wanted to make one day, like a chair or a nicely polished gemstone for your loved one. Or it could be a task you wanted to do, like pollarding a willow tree in a meadow after the winter cold. Or it can be combustion heat—energy—for driving your heavy removal truck uphill, or it can be some wanted chemical product, like sunscreen. Similarly, however, waste as a processing output can also consist of some poison polluting the groundwater or heat lost somewhere in the environment (warming up a nearby river, for example).

Obviously, the amount of matter found on either side of the box must be the same, because we can neither make nor destroy matter. The same amount of matter that comes in at the left of figure, comes out at the right. In fact, the same applies to energy, as we cannot destroy energy either. But there is a difference: we lose unusable energy; it becomes useless heat. As a result of the processing activities inside the box, energy is said to degrade. Plants input solar energy into their chemical system, storing it temporarily, but in the end, they lose it as degraded, unusable heat. And when they are eaten by some animal, it obtains the stored energy, which it decomposes, releasing the energy as heat. That is no recycling either—it is extending the linear process the plants followed. And the same happens when we eat plant or animal food or when we burn oil, natural gas, or coal as fossil remains of ancient algae and plants. So, in all systems, there is an overall loss of energy in the form of heat. We degrade energy to a lower form by using it, but it is still there at the end of the process, and in the same amount—but in a useless form. We can release energy by burning wood or oil in stoves or in turbines or by digesting starch in the cells of our bodies. Or we can extract it from the heat radiated by the sun. Therefore, similar to matter, neither can we make energy ourselves, nor can we destroy it. And here's the difference between energy and matter: we can recycle matter, but we cannot recycle energy. If we want to recycle energy, we would first have to upgrade it again from its degraded form, which is impossible. Matter can be recycled, whereas energy always follows this linear, down-slope path of resource depletion toward waste production. Moreover, the compounds to the right of the box are overall more stable than the compounds to the left. And this has something to do with the overall loss of heat, which in turn has to do with the need to add new energy in order to be able to recycle these stable compounds.

The most general law in physics tells us just this: wherever you look in the universe—at large galactic processes or at small ones within or among atoms—there is always an increase in the amount of waste energy, heat accompanied by a rundown of material organization into disorderly waste. The running of organizations, be they individual organisms, sports clubs, governmental organizations, or societies and civilizations, indicates that energy degrades during this process, and at some point leaves as heat. We can't escape this law; we can only hope to slow down our own processes of waste production by being economical. And that's exactly what cyclic systems do. They minimize the amount of material used by recycling it internally. And when some material happens to be thrown away as waste, out of the cell, or out of the body as feces, some of it can still be taken up and used by some other organism—that's interorganism recycling. This is exactly what has been going on in nature since life began. And it goes on because the resources on Earth are finite, so recycling is imperative.

Energy too is limited. Living systems must have run out of usable chemical energy quite soon after they came into existence. Degraded energy cannot be upgraded again, and waste matter cannot be taken back into use without inputting the energy that was lost when orderly matter became disorderly waste matter. Fortunately, there is an external source of energy available to life on Earth: the energy from our nearest star, the sun. Every day for billions of years, the sun has supplied us with huge amounts of new energy that can be used to convert waste into a reusable form and allow new elements and new processes to be incorporated into new chemical "machinery" added to the existing machinery but using more energy.

So, we—along with plants, animals, bacteria, and fungi—are all driven by solar energy. But where does the sun itself get its energy? Some sixty years ago, we learned that in the core of the sun, the same fusion processes as in the operation of the hydrogen bomb have been going on for billions of years; the sun, therefore, can be considered one enormously large hydrogen bomb. It's a fusion bomb because two atoms of the hydrogen it contains fuse into one larger chemical element, helium, and this, in turn, fuses into yet other, even larger elements. In a sense, the whole universe appears to contain an unbelievably large number of gigantic hydrogen bombs: the stars. Some of them explode with incredible force, whereas others simply fade away when their hydrogen runs out. Some of the vast amount of energy produced by our sun reaches us here on Earth as light and heat. It is the process of hydrogen fusing into helium that has warmed Earth for all 4.5 billion years of its exis tence. And the sun will continue to warm Earth for about the same length of time in the future.

It seems that this kind of process occurs throughout the known universe: at first, there was only hydrogen, the element with which all other processes in the universe began. The history of the present state of the universe is written in terms of hydrogen. All chemical elements on Earth—those that make up our bodies and all things and organisms around us—form only a tiny fraction of all the hydrogen that exists. They are completely overshadowed by the enormous amount of hydrogen surrounding us in the universe. On this extremely large scale—as large as the distance covered by light during some fifteen billion years—it's almost all hydrogen around us, for as far you can see. The solar system, Earth, and we ourselves within it are the great exception.

And hydrogen also proves to be crucial for life. Before life began on Earth, there must have been plenty of hydrogen molecules available in the environment. Life began using these molecules because of the relative ease with which hydrogen can form and break chemical bonds. My personal theory is that most of the available hydrogen was bound first by selenium, and then by sulfur and phosphorus—and of course, still later by oxygen. Hydrogen and oxygen together form the compound water (H_2O), which literally fills up whole oceans. Other elements, those able to bind with hydrogen and also able to break up chemical bonds quite easily again, began to cycle hydrogen around and around. However, over time, they unavoidably bound more and more of the free hydrogen molecules. The result was that in order to obtain or release hydrogen from the first elements, other elements that could pull harder on hydrogen and retain it for longer now entered the processes of life. Eventually, carbon was thus included in life processes, in its turn taking a share of the available hydrogen. At the same time, the reaction products of these hard-pulling elements also became less likely to break up; the elements that joined in later pull so hard that they are less reactive, and therefore their molecules are more stable. So, the flip side of the evolutionary coin is that compounds made by recent living organisms are less reactive than they must have been in the most ancient life forms. But retaining the hydrogen longer also meant storing some binding energy longer; in particular, it was the carbon compounds that specialized in energy storage (think of starch in plants or of animal fats and oils).

Yet this meant that plants and animals would also produce nonrecyclable biological waste, that is, nonreactive stable compounds that other organisms could not break down. Also, the organisms became more complex, each of them using a greater number of the available elements. They tended to keep using the former elements in the old mechanisms, which were always kept operating, and they used new elements for their newly patched-up mechanisms. And the complexity grew ever-more rapidly, because any new mechanism had to be supported itself by one or more other ones, and these, in turn, by yet other mechanisms. The more complex forms were larger than the simpler ones and therefore contained more of the available hydrogen. Easily available hydrogen was running out.

Obviously, therefore, the scarcer free hydrogen became, the stronger the elements had to pull to get their share. The initial elements could not start by pulling hard themselves, because their reaction products would be too stable, but they could be substituted by other ones with more pulling capacity. This is why selenium will have been substituted by sulfur, which then formed further compounds with the help of phosphorus. Sulfur may have been the first element to be substituted for selenium, because their chemical properties are similar; it was particularly because of its stronger pulling force that it soon came to replace selenium. Then its other properties could also be incorporated into the processes. Later, and for the same reason, sulfur itself was replaced by oxygen, which again has similar properties, but which has even more pulling power than sulfur. On the other hand, as already mentioned, the molecules these elements form are less and less reactive and so were more stable, requiring more energy to break their bonds in the great recycling processes of life. And if the bonds remained intact, the reaction product remained inactive and, hence, useless in the processes of life. These products became the chemical waste of life. Thus, life became ever-more energy-hungry, ever-more complex, and ever-more wasteful.

At some later stage, it became possible to form carbohydrates from carbon dioxide and water. At this stage, however, the electron-pulling capacity of the newly introduced element carbon was not enough to obtain sufficient hydrogen directly from the environment. Another process happening in special structures found in the membranes surrounding the cells was added to the existing processes, namely, the splitting of, first, hydrogen sulfide (H_2S), and then water (H_2O) by light energy from the sun. Hydrogen was now obtained from an inexhaustible source—water—with the help of an equally inexhaustible source of energy—that obtained from the radiation of the sun. This process generated the hydrogen needed, rather than using up the hydrogen naturally available in the environment. The direct result was that oxygen was released as waste into the water of the oceans, from where it escaped into the atmosphere as a gas. With the hydrogen available from water, and by a biological mechanism using the inexhaustible source of solar energy, the early structures and processes could now form sugars, a special kind of chemical compound for storing energy. When these sugars were broken down, their hydrogen became free and reacted with the oxygen again, forming water. This reaction released the energy that initially came from the sun and was used in the splitting process, making it available for use in further reactions. The sugars thus developed into temporary "storehouses" for solar energy. This energy could be released in times of energy shortage, for example, during the night when no solar energy can be captured. The first sugars, however, were still nonreactive because of the very strong binding capacity of carbon, and so they were expelled as useless waste into the environment. Only later did a set of suitable enzymes evolve to break them down, molecules specifically facilitating certain chemical reactions, speeding them up or making them possible.

Throughout the process, reduction processes—the binding with hydrogen, or more precisely, the binding with its components (protons and electrons)—remained of central importance in the cells. Reduction involved binding elements and chemical compounds with hydrogen, eventually with the help of energy from the sun. Throughout the history of life, the ancient life processes depending on reduction were therefore maintained with the help of solar energy. Moreover, this remained possible because enough hydrogen was available in the water within the cells in spite of the fact that gradually the supply of hydrogen in the environment outside these cells had diminished.

In a way, plants on the one hand and fungi and animals on the other are complementary life-forms: plants take up carbon dioxide and water as part of their food from the air and the soil and expel oxygen as a waste gas back into the air. Carbon dioxide forms their gaseous food, and water, their liquid food ("food" here being taken broadly as material taken up by some living system). Meanwhile, plants make carbohydrates, such as the starch in wheat, rice, and potatoes, as chemical energy that is stored in their cells. In contrast, fungi and animals, including you and me, take in those plant carbohydrates by eating and digesting—burning—them. For this to happen, they take up the oxygen the plants expel and use it as a feeding gas, and they expel carbon dioxide as a waste product. Oxygen as a gas is actually an essential part of animal "food." Therefore, we use as our food the starch of plants (directly, as in the case of potatoes and rice, or processed, as in bread and pasta), we breathe in oxygen, take in water by drinking our coffee, soft drinks, or beer, and expel carbon dioxide and water as waste products. Meanwhile, we dissipate heat from our body that has been generated by all the chemical reactions that have taken place. This heat is the energy we release by breaking down chemical compounds from the plants, binding the hydrogen from the hydrocarbons with the oxygen the plants expelled earlier. The plants extracted it from the solar radiation and stored it as starch in their cells, which is the main constituent of our bread, pasta, and so forth. And before releasing this energy this way, we use it for walking and talking, for maintaining our bodies, or for begetting and educating our children.

Of course, this isn't even the end of this amazing story. Biology has more in store. The animal-like cells, as I've already explained, produce carbon dioxide as waste products by breaking down the sugars. They burn them in some way, using oxygen. This is similar to what our cars do. But our cars use gasoline instead of sugars, and this gasoline was formed long ago—tens of millions of years ago, in fact—from the same sugars. Similar to today's algae and plants, those ancient plant-like cells needed carbon dioxide to build up their sugars, thereby producing oxygen. So, in this respect, these primordial animals and plants complemented each other perfectly. Thus, taken together, both carbon dioxide and oxygen were produced as waste products, but were also needed as chemical resources. And hydrogen, going backward and forward between those two chemicals, was the great intermediary. Talk about recycling in nature!

Primordial plant and animal life-forms became mutually interdependent, supplying each other with food via their waste and protecting each other against the deadly accumulation of their waste gases. Thus, they began to recycle oxygen and carbon dioxide, which from then on were found in the atmosphere and the oceans as waste products and resources. These recycling processes form very large cycles, present throughout all parts of Earth where life can be found—called the biosphere. They are therefore known as biospheric nutrient cycles, the word "nutrient" having to do with to nourish, to feed. It will be obvious that there are more than these two types of nutrient cycles. Other crucial cycles are the nitrogen cycle, sulfur cycle, phosphorus cycle, iron cycle, and water cycle. In all evolutionary stages and at all levels, recycling became basic to life.

Plants and animals therefore perfectly complement each other, each using the other's waste as food. The flow of their matter forms a cycle running from plants to animals and back again. Thus, matter is recycled again and again for long periods of time in the more or less closed cycle of nutrients formed by plants and animals. Matter has flowed through this cycle for millions and millions of years. The carbon dioxide in your sugar has gone through the stomachs and bodies of billions and billions of previous animals and plants. That's the chemical reincarnation of life.

The fact that this cycle is more or less closed means that, in principle, hardly any matter gets wasted or lost. What does get lost eventually forms thick geological layers, strata of limestone, oil, gas, and coal. However, what is continually and irretrievably lost in a few steps from plants to grazing animals, and from these to their predators and parasites, and ultimately to the decaying bacteria and fungi, is energy. Plants receive the energy they need from the sun and transfer this to animals, fungi, and bacteria, which eventually release it into the air as heat—useless, spent energy. Energy cannot be recycled forever; as I've already said, it degrades in the process, getting lost in the great material cycles of life. It is only the matter that goes around in those cycles. And it is energy that drives them by degrading into worthless heat.

The large carbon cycle in the biosphere is not really perfect; part of the matter, which still contains energy, always leaks away into the environment. For example, some plants and animals are not eaten or do not decay completely; instead, they are preserved in the form of peat, or they fossilize in sand that has blown over them. Their material waste, therefore, is not recycled but rather accumulates, resulting in the thick, continent-wide layers of peat under the permafrost of Canada and Siberia, or in vast underground fields of coal, oil, and gas, for example.

The oil or gasoline you use for driving your car is no less than ancient fossil material from which you extract the necessary energy to get moving. But some fossil material cannot be used as a source of energy or minerals for plants or animals. For example, most of the oil and gas found deep in the earth cannot be used as fuel unless subjected to a very special set of geological conditions and processes. And oil can also be locked up in asphalt lakes in which long-ago large animals, such as mammoths and saber-toothed tigers, could become trapped. Moreover, peat, coal, oil, and gas cannot be taken up by plants and animals themselves; they can be decomposed by fungi or molds, which transform them into carbon dioxide, or by bacteria, which turn them into another gaseous waste, methane. Ultimately, methane reacts with oxygen in the atmosphere, forming carbon dioxide and water, which in turn can be taken up by plants as part of their food. Shells and bones that accumulate to form huge strata of chalk and limestone cannot be used as energy sources, either. The chalk and inorganic limestone cannot very easily been taken up by organisms as a mineral, calcium. And plant and animal remains in topsoil take many—often thousands—of years to decompose before plants can take them up.

We can access only a small part of the ancient organic waste products and burn them for energy. When they are burned, the carbon dioxide and water vapor that are formed are released, warming the air as greenhouse gases. Methane, a gas produced by cattle and from peat by certain bacteria, is also a greenhouse gas. Greenhouse gases all have the same effect on the energy budget of the atmosphere: they raise its temperature. How this happens is simple. The relatively high-energy radiation coming from the sun loses some of its energy by warming the soil and ocean water of Earth, but another part bounces back into space. After warming the soil and water, the first part leaves Earth as low-energy, infrared radiation. That part, the infrared radiation, can be absorbed by the molecules of the greenhouse gases, which become more energy-rich, that is, warmer. All those individual, energy-rich molecules together form a warmer atmosphere. In turn, this warmer atmosphere warms up the terrestrial environment and the oceans. The total process is known as climate warming.

The carbon compounds that are not burned but, instead, are separated from the fossil material before burning are used in industries for making nylon, plastics, fertilizers, herbicides, pharmaceuticals, asphalt and so on—all essentials of our modern society. Most of these products, however, cannot be broken down easily by biological organisms, if at all. Many plastics known or advertised as decomposable can be decomposed only to a certain point; the small particles still remain in large, invisible quantities in the environment, often choking microorganisms. Therefore, they remain as human waste.

As we turn ancient, natural waste into present-day human waste, our own personal and societal waste is gradually heaping up, as pollutants in the soil and in the water, and as greenhouse gases in the air. This is happening because the world population is growing, and the technology used for enhancing our quality of life is expanding. Moreover, our products are not intended to be recycled but, instead, to be kept. We don't make a staircase such that it can be recycled, either by natural processes or by our own. It's intended to be used as long as possible. The difficulties created by these kinds of nondecomposable waste will peak at more or less the same time, as most of these wastes are the result of the utilization of fossil fuels. And the peak will roughly coincide with yet other problems arising from too many humans on Earth.

How are we going to cope with this? Finding another energy source is feasible, but without finding a new supply of material at the same time, it won't be an option. And merely reducing the amount of carbon dioxide released as human waste into the atmosphere won't help us out either. Dwindling energy supplies and increased carbon dioxide emissions are epiphenomena of the growth of our global population.

Instead of myopically focusing on finding new energy sources and reducing carbon dioxide emissions, we should take two steps. One is to reduce production, in the sense both of our own reproduction—that is, the number of humans—and in the sense of reducing consumption and industrial and household production—the average produced by us and for us, in terms of products and waste. Production is the prime cause of both problems. These two process components form the forces driving our resource exhaustion and waste production; they are processes whose rates we can influence to some extent, which eventually could bring the relief we are looking for. And the second essential step forward in finding a solution to approaching global disasters to humanity and the rest of the natural world is to replace our linear, exhaustive system of resource utilization and waste production with a rigorously maintained cyclic one, keeping all the matter we need in circulation. There are no alternatives.

2 NATURE GOES IN CYCLES

How exactly does chemical and biological recycling work? Let us for a moment consider chemical equilibriums. When you put two chemical compounds into a flask together, they either do or do not react. If they do, they may react either explosively or slowly. In both cases, at some point they stop reacting altogether when one or both of the reactants are used up. Explosive reactions end very soon, whereas slow reactions end after longer periods. The figure below shows these various possibilities in a very simple diagram. The arrow pointing to the right in the first figure shows the process when the two compounds react: they are used up totally and replaced by the reaction product, the new compound(s).

A ——————————→ B

Conversely, when the two initial compounds refuse to react, the arrow points in the opposite direction (below).

A ←—————————— B

And somewhere in-between these two possibilities there is yet another process, represented by the arrow with heads at both ends.

A ←—————————→ B

This indicates that the reactions go both ways at the same time, forming a new compound at a certain rate but also continually breaking it down again into its two initial components. And, on average, the rate at which the new compound is formed balances the rate of breakdown.

Two of these three possibilities are interesting for us at this moment; the second actually is irrelevant for our purposes, since nothing happens in that case. As I've already said, in other respects the failure to react can be interesting, as in the case of waste products and many pollutants. In fact, this is no more than the extreme case of reactions in the first possibility. This kind of reaction is interesting because it indicates the existence of a stable equilibrium in which ultimately nothing happens anymore. At that point in the process, a nonreactive compound has been formed, either locking up some useful energy or depleting it altogether. If no other organism can digest this compound, it is lost forever as a resource for any other organism. That is, no other organism can break it down and release its energy or matter for further use. Thus, the resource concerned shrinks and eventually runs out. In some cases, only a large amount of energy can break the bonds of such a stable compound, making the matter available again for further use. This happens, for example, with carbon dioxide, water, and simple nitrogen compounds such as the metabolic products of animals. These can be recycled only if there is abundant solar energy available to power the process in which hydrogen or some molecular groups can be attached to them. Thus, plants bring carbon dioxide into circulation again, and some bacteria can do this with certain nitrogen compounds. For this to happen there must be not only a superabundance of energy, but also many specialized molecules—proteins—to reduce the amount of energy required. Therefore, because they all produce stable end products (that is, waste), living systems have developed several mechanisms to bring these stable waste products back into circulation again, and these prevent them from dying out early because of resource depletion. After all, the natural end of all matter and energy is the stability of chemical waste and degraded energy—heat. The biological mechanisms that developed over eons prevented this from happening; wasted matter could be taken up again into the production process, but only with the help of a large amount of energy coming from an external source—the sun.

The last possibility, where the reactions going both ways, is biologically the most interesting from the viewpoint of biological recycling. This is because long ago, life developed reaction cycles in whole systems of interdependent reactions—recycling processes—in which elements and molecules are continually returned to the production process. As I mentioned, the forward and backward reactions balance each other: as many bonds are formed as are broken down again at the expense of energy. On average, the same chemical compounds are always found, forming a mixture: the initial reactants together with the reaction products. Thus, contrary to a reaction leading to a stable equilibrium in which nothing happens anymore after a forward reaction has taken place and in which a backward reaction is impossible, the equilibrium is a dynamic one in which forward and backward reactions happen side by side. This may suggest that nothing happens; it seems that the reactions have reached some end point, but in fact they have reached an equilibrium in which the processes of building up average out those of chemical breakdown. But below the calm surface of this average result is a dynamic world in which lots is going on continually. We therefore call this equilibrium a *dynamic* equilibrium: it's not a still, dead world, but a bustling one because of all the forward and backward reactions that are going on nonstop.

Because the equilibrium is dynamic, if any of the newly formed compounds is used up in another reaction, it is immediately and continually replenished by its renewed production, reestablishing the equilibrium value. Biologically, this is highly important, because this new compound can be tapped as long as there are sufficient resources available for the replenishing reactions to occur.

If the reaction products are continually being removed by C, the system remains out of equilibrium and keeps producing the compound for as long as it is being tapped for another reaction. This is the essence of the nonequilibrium state in which living systems permanently exist: the possibility of tapping a reaction product guarantees the continuity of the flow of matter and energy.

Most intriguingly, in the almost four billion years of life on Earth, chemical equilibrium has never been reached in living systems, despite the fact that many reactions last only a few seconds or even much less than that—one millionth of a second, for example.

The next step in our reasoning is a seemingly simple one. Imagine not only the existence of one reaction mechanism connecting the reactants at the left in figure 2c with the reaction products at the right. In a way, this shows a primitive "cycle" of two chemical components emerging. Now, we can also insert yet another compound into one of the reactions: the backward reaction, for example (as the top figure on p. 18 shows). As above, compound B is being tapped by C, but C is in turn being tapped by A. The result is a primitive cycle of three components. In fact, this remarkable representation contains the essence of a biological cycle: a dynamic, nonequilibrium reaction in which reactants and products continually swap roles. Life could thus have started up from one dynamic reaction equilibrium, building up stepwise to an elaborate set of complex, mutually interacting cycles of more dynamic equilibriums.

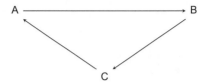

Of course, overall, energy is continuously being lost as heat in these reactions, and during all this time, this loss has to be compensated for by a nonstop inflow of energy (below).

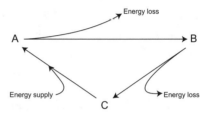

In biology, we can say that the cycles are driven by external energy, that is, by energy originating from other reactions outside the system—from the environment or from the sun. This nonstop stream of external energy drives all cycles of life. A stream that enters the system, drives it, and leaves it again as degraded energy; and the chemical compounds mostly remain behind, in principle cycling around and around forever, being built up, broken down, built up, broken down, and so on in perpetuity. Within organisms, certain compounds within these cycles are continually tapped, resulting in a permanent flow of matter and energy, because this maintains a chemical disequilibrium—one that has existed for almost four billion years. But between organisms, these compounds are brought back into circulation again, which gives an overall equilibrium in which hardly any matter is lost. How much material would life had to have used up if it hadn't developed such an elaborate system of cycles—internal, biochemical cycles and external, biospheric ones? Would all the material in the entire planet have been sufficient? But life only scratches an extremely thin film of Earth's surface. Living bacteria have been found in rocks at a depth of three kilometers, while the radius of Earth is six thousand kilometers.

Thus, from early on, under the stresses of finite resources and the pollution of the environment by biological waste, the waste of the one life-form became the resource of another. This happened first of all within cells, then between them, then within them again, and then between individual cells and their multicellular successors: the animals and plants around us. And the cycles were all driven by energy external to each of these levels, and soon originated from outside the immediate surroundings of the earthly environment—from the sun. And the energy dissipates from Earth back again into outer space as degraded energy, as heat, never to return again as usable energy. This kind of organization of flows of energy and matter is essential for the persistence of life; without recycling systems, life would either soon have run out of its supplies, or it would have been smothered in its own waste.

And now think for a moment of what we as *Homo sapiens*—the only knowing and understanding species—are doing. How long can *we* survive by applying a linear flow of matter? Or are we, in fact, the only stupid ones?

But, of course, biological systems are not perfect either. Even the earliest life-forms, however primitive and tiny they were, must have sucked up hydrogen and electrons from their immediate surroundings to live, so that, locally, at a minute scale, the environment became depleted of electrons. Compounds with the weakest bonds must have formed the stuff of life. Their bonds could not only be readily formed, they could also be broken relatively easily, so that atoms and parts of molecules could be kept in circulation perpetually rather than becoming fixed in stable, nonreactive—and therefore unusable—molecules of biochemical waste. Stronger electron pullers and donors could also break up the waste of earlier systems. Step by step, ever-stronger electron pullers and donors were included into the living structures, as the easily available electrons in the mineral resources of Earth were used up, and as waste accumulated in which these electrons were bound. Being stronger, these strong electron pullers were able to extract the much scarcer electrons from the environment, inserting them into the system to form molecular bonds. In these stronger, more stable bonds, they could also retain those electrons longer. In contrast, metals that easily donated or exchanged their electrons were increasingly being utilized by developing early life-forms. Sometimes, stronger pullers and donors replaced the former, weaker ones in the same reaction mechanisms.

Thus, the overall evolutionary trend was toward using elements that form increasingly stronger bonds—that is, toward lesser reactive, more stable molecules. Such molecules, however, often constitute the waste of living systems as well. So because their waste products are nonreactive, living systems produce more and more chemical waste. This applies to each organism individually, but its waste serves as food for another organism, which means that taken together the nutrients form large, biospherical cycles all driven by energy. The components—the species to which the individal organisms belong—are interchangeable, whereas the flow of nutrients and energy persists. Therefore, the species do not necessarily keep each other in numerical balance, as is often presumed, nor do they necessarily adjust a specific carrying capacity that they together establish. Species come and go in space, rise and decline over time, or evolve and die out independently of each other over the ages. As we will see, their numbers can grow and then crash without any regulatory control. And despite part of the nutrients is scraped off from the biospherical cycles, forming thick geological strata and mountains of biological waste, the grand flow of material and energy kept running uninterruptedly over billions of years, even intensifying over the eons.

We can also look at the problem in a different, more general way, now contrasting human resource use with the biological use described above. One point that I've mentioned several times is important in this respect: the energy flow is continuous and linear, whereas the flow of nutrients forms cycles. I have shown why we can say that the energy stream shaped the molecules, the chemical systems within cells, and therefore the organism. If we combine these two observations, we might say that actually our food is not about the nutrients themselves but about the energy the nutrients contain. The chemicals, the nutrients, are carriers of the energy. They go around in cycles as long as possible, so that as little as possible is needed to keep the flow of energy going. And they are conservative energy users; the chemicals are kept within the system as long as possible. Thus, the energy flow is primary, and the flow of chemical nutrients is secondary, subordinate to it.

This contrasts with our use of chemicals and energy; when we want to build something, we use some energy to achieve that end, and produce some unusable waste, which, in principle, is chemically stable. This puts our own processing schemes in a completely different perspective: all our efforts are directed toward material resource use and, consequently, toward the production of material waste. We want the energy and don't bother about the chemical waste. In contrast to what happens in biological processes, in all our efforts we use energy as a means to an end, the end being a nondegradable structure. We would not dream of building a skyscraper or an airplane—or even a chair—from easily degradable materials. This means that the destruction of such structures requires energy, often the maximum amount possible. In contrast, in a biological process, the continuation of an energy stream requires molecular structures and processes releasing as little energy as possible and which, moreover, fit into some material recycling scheme.

Therefore, the very stable compounds we make later require huge amounts of energy in order to be reincorporated into a recycling scheme (below). So in this way we lose both materials and energy, whereas biological processes are very economical with both. Cradle-to-cradle schemes usually omit this aspect of the energetics of the process, concentrating on its material side.

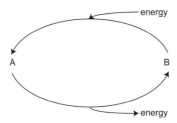

Exactly the same processes of resource depletion and waste production as those seen in the biological world have been going on in the last few centuries of human history—only much worse. As a result, we are now on course to exhaust the fossil fuels that supply us with energy and many petrochemical materials, such as plastics, pharmaceuticals, paints, and herbicides, without which our society cannot function. And by using up the minerals, the soils, and the forests, we have produced a vast amount of waste in the form of garbage and pollution of all sorts and sizes in our immediate environment and the oceans, and of gases such as carbon dioxide in the air.

Although adaptation has often taken millions and millions of years to happen, life has been able to adapt by evolving new life-forms, always utilizing new resources and reusing natural waste. And we can't.

PART TWO

ONGOING PROCESSES IN THE HUMAN POPULATION

The origination and maintenance of any system requires energy, biological or societal. In order to feed, our most basic activity to which all others boil down, we need a sufficient agricultural system. This system itself depends on a large supply of energy for pumping up water, for the production of fertilizers and other chemicals, as a trading and a transport system, as a financial system for making payments, for keeping the value of money under control, and so on. All these systems individually require energy and material, as do the necessary interactions between individual people. Moreover, they show a dynamic of their own—a dynamic beyond the control of humans.

I

POPULATION GROWTH AND ITS LIMITATIONS

Populations tend to grow, and they do this at the expense of their environment by exhausting their resources and by polluting them with their waste products. Eventually, shortfalling resources and pollution can result in the population crashing to low numbers—or to their total extinction.

In chapters 3–14, I describe a number of these processes. In the chapters in the section "The Growing Problem of Mankind," I concentrate on population growth itself and put this in the context of responses that are meant to cure the effects of the higher numbers but that both stimulate further growth, often even enhancing the rates of numerical growth. These responses are improvements in agricultural production and the beginning and expansion of industrial production. Recently, both processes have come together and have also fused with financial systems forming agribusiness systems, even at the corporate or the state level.

The chapters in the section "Exhausting and Wasting Our Resources" describe direct exhaustion and waste, whereas the following section, "Exhausting and Wasting Our Environment," does the same for our environment on which we more loosely depend. Making this difference does not mean that the latter processes are unimportant, but they can operate over longer periods of time. For example, consequences of a shortage of energy make themselves felt as soon as its sources finally run out, whereas air pollution and its consequence, climate warming, builds up over decades or centuries. Still, at present, climate warming is of major and growing concern and can destroy large parts of Earth so that the persistence of human life and of life in general, if not the whole Earth, is at stake. And deforestation, though covering centuries or millennia, can speed up this process of climate change.

A

The Growing Problem of Mankind

A pivotal factor for the future of humankind is the number of humans present on Earth; as this number is already too high, any countermeasures against future threats should concern this number in the first place. As it is pivotal to anything else, we will concentrate in this section on population numbers.

3 POPULATION GROWTH AND AGRICULTURAL PRODUCTION

Imagine, you go into your bathroom one morning and see some water on the floor. And you find a tiny little hole in one of the pipes. Not too large, but still, it has to be fixed. Luckily, you know a handyman, and yes, he can come over immediately. Unfortunately, he doesn't have a piece of replacement pipe with him, but he'll think of something. After some time, you go and see what he has done so far, and you find a real mess: water all over the place, spouting out of big holes, the man splashing around desperately—at the very moment you come in, he's sawing into one of the few remaining parts of your pipes. What he has done was take out a piece of good pipe to replace the part with the tiny hole with it. But this left him with a bigger hole in the other part of the pipe. He'd repeated this procedure several times, each time taking out a bigger piece of good pipe from some other part to make good the hole he'd made himself in the first place. He was fixing a hole by creating a bigger one.

A similarly disastrous repair job has been going on during the last couple of centuries of human history, or, perhaps, throughout our whole history. And, worse, in our case, the holes in the pipes kept on growing while the handyman worked. Two growth processes, one on top of the other: the growth of the size of pipes needed as replacements, plus the growth of the holes. A plumber's worst nightmare. The handyman must have been desperate indeed. And so should you.

You can find what has been happening in any book on the history of agriculture. As always, when you take something out of a bowl, say, less remains in the bowl. So, when you harvest your crop of wheat over a good number of years, the soil becomes exhausted, and the next harvests are somewhat smaller. The remedy is to put back some nutrients into the soil by manuring. But the livestock need to be fed in order to produce the dung you need, and so you have to add on some grazing land to your plot of arable land. Each animal, however, needs more land than one member of your family. This land, in turn, also becomes exhausted over the years, because you are removing the nutrients from it to put them on your arable land. One of the next measures you can take is to leave the land lying fallow for one or more years, but this means that you'll harvest less. So to harvest the same amount of food you and your family need, you have to extend your arable and grazing land.

And then there's another problem: only some of your livestock can be kept over the winter, because all that time they have to be fed on hay, which requires more land to grow it in a favorable season. So, in the fall, some livestock have to be slaughtered. At some point in history, farmers found they could plant their fallow land with fodder crops (beets, clover, and grass, for example) so that the livestock didn't need to be slaughtered in the fall. Actually, more animals can now be kept grazing on the former fallow land, producing more meat, as well as more manure for the arable field. And this means that more people can be fed. But all these extra people also have to work the land and have to be fed, and so on. You're filling one hole but creating another.

The system adopted is a growth system, and we are still applying it. We need both population growth as well as material and energy growth to survive. Production systems will never be fully circular but will always contain one or more linear elements. And the longer and more we have grown, the faster we have to grow.

This is what, in broad lines, has happened during human history: growth out of necessity—growth generating new growth—with the result of an ever-growing amount of resources being used, and waste being made. And worse than this: just like the holes in your bathroom pipes appeared to be growing themselves, the human population has been growing too, one growth process on top of the other. In the end, growth out of growth out of growth of necessity, *ad infinitum.*

That's what's been going on especially during the last thousand years of European history. At present, we are living in only one stage of this long process of continued, ever-accelerating growth, proceeding in spurts with each new revolution in society, spurred by the whip of our demographics— one that is in turn more and more dependent on new sources of energy and material, and on new technologies that have arisen during successive revolutions. To understand the condition we are in at present, and certainly in the near future, we need to understand the long runway of our past during this extraordinary millennium.

The last couple of centuries in particular have seen two types of really spectacular expressions of growth of the human population, absolutely revolutionary ones: an agricultural revolution and the Industrial Revolution. Although, in a sense, one could also consider them as one single weird response to population growth, which actually worsens it. The industrial revolution was an answer to the shortcomings of the agricultural one, and yet it further aggravated the problems. These growth processes enhanced each other in complex ways, always generating newer growth processes, each operating on top of the other ones and enhancing the dire effects of the others even more, enhancing their combined effects on the state of Earth by using more resources, by wasting more, and, in the end, by producing more growth necessitating further growth. Together, these growth processes account for many of the problems we are facing now and will face in the future: resource exhaustion and the associated production of waste. Is there a real plumber around?

What happened during those two centuries, whether we consider this process as two revolutions or as one, is that a new source of energy was added—had to be added—to the traditional ones. And this additional amount of energy may have made all the difference. During the first agricultural revolution ten thousand years ago in the Neolithic, additional energy had already done this: the new amount of available food allowed people to organize themselves and to build up their societies and towns and make new tools to assist them. Like any processing unit, people need energy, energy encapsulated in, or carried by matter—their food. With more food available, they could process more, and, as humans, produce and organize more. People themselves are flows of energy and matter, and with more energy and matter, people could do more, organize themselves and each other, and more people could live together. Thus, the size of a society as well represents a certain flow of energy, a flow additional to that of the sum total of all the individual people. Hence, the rise of large communities with all their waterworks, stores, transport systems, houses, palaces, and temples, all their armies to be maintained itself required the use of increasingly more material and energy. Organizations, their origin and maintenance, and all the activities happening within it represent energy and, like organisms, require a continuous flow of energy and material. In fact, it has been speculated that much earlier, the very origin of humankind as a different, entirely new phenomenon in nature, came from the invention of the cooking of food, which saved much energy and time previously spent on chewing and digesting. This gave more leisure time and allowed more people to live together and communicate and organize. The processing machine ran more efficiently. The Agro-Industrial Revolution of the eighteenth and nineteenth centuries, then, is not the second but the third major invention, pushing humankind up the ladder of civilization, allowing an even easier way of living—as well as pushing them up the ladder of population size.

For millennia, the energy source in agricultural production was essentially human, supplemented by animal power. Humans drained and irrigated their land, hauling buckets of water from one level to the next. Plows could in principle be pulled by animals, but in many cases, the poor pulled them themselves, as contemporary drawings testify. From very early on, boats for transport could be driven by wind, but here again, humans often pulled the oars, or men or horses towed vessels upstream. Similarly, on land, people or animals had to pull or push carts, usually along muddy and potholed dirt roads. Humans, too, sometimes by the hundreds of thousands, dug canals and other irrigation systems. Traders, and often whole populations or tribes, walked long distances, many thousands of kilometers, occasionally migrating over Europe or riding on horseback westward from Mongolia into Europe. In Roman times, they built long, straight roads across Europe, paved with worked cobbles and lined with strongholds and fortresses. Or they built long defensive walls, as they did in Britain and the Chinese along their northern border. In the Middle Ages, human labor was facilitated by pulleys, but the energy source for agricultural production, as well as for town and road building, remained essentially human.

The consequence of all this physical hardship was that almost throughout human history, life expectancy was short—normally up to only some thirty-odd years—and the rates of birth and mortality were high. And being undernourished and exhausted, people were more susceptible to disease, particularly in times of climate instability or food shortage; during the frequent epidemics and famines, they died by the thousands, if not by the millions. But not only the quantity of food was important, particularly when it ran short; it was often monotonous. The poor in particular, who formed the majority at that time, could not afford a more diverse and healthy diet. And how would they grow it? At which local market could they buy or sell their new product? How far away was there a market? Did they have the time, energy, and the means of transport to get there?

These questions weren't only asked many times over during the last couple of centuries, but also throughout the last millennium, beginning around AD 1000, the time of the well-known Romanesque period of architecture in Europe. In an earlier demographic and agricultural revolution happening between 700 and 1100, the expanding and urbanizing Muslim world introduced many new crops into southern Europe from India, the Far East, and Africa. Then, coinciding with—or justifying—the Crusades, there were intense commercial connections among the Scandinavian countries in the north, the Mediterranean countries in the south, and the highly developed Muslim world in the east. At that time, the first significant wave of European urbanization and agricultural reorganization and mechanization occurred, necessitated by intensified demographic growth and expansion. The Low Countries saw a time of their greatest land reclamation. Over time, the Viking civilization in the north was succeeded by the Hanseatic League connecting various parts of northwestern Europe, whereas during the fifteenth century, the commercial interests of Spain and Portugal spread westward to South America, southward along the African coast, and eastward to India and China. Later, the Dutch took over the trade to the Baltic countries and to the Levant in the eastern Mediterranean, and then to North and South America and the Far East, acquiring control over the Arabian spice and cloth trades. During all these centuries, Europe experienced waves of demographic growth and urbanization; those of the architectural Romanesque period, for example, was followed at first by the early and then the late Gothic periods, merging into those of the Renaissance and the Baroque. Those times, too, saw large-scale deforestation and the growth of commerce in food and wood across Europe, first fed by Arabic silver, and then by the silver and gold from South America. That whole period was a time of a rapidly broadening mercantilist colonialism, during which the first financial and banking systems were developed.

From 1560 to 1690, a horticultural revolution took place in the urbanizing Low Countries, followed by a similar revolution in England, with the result that a more varied diet came within reach of a larger part of the population. And large, wind-driven machines, the well-known, much romanticized Dutch windmills, took over many of the heavy tasks previously done by humans. Thus, the early Dutch phase of industrialization mostly operated on wind energy. This development toward an early industrial society was also possible because much of the work to be done was organized differently: internally, there was a shift from a partly local subsistence society to one largely based on industry and trade. In Holland, canals and lakes were used and extended for intense and regular traffic by boat, connecting towns with each other and with areas of food production. For greater agricultural production, new land was reclaimed, too—first in the Low Countries, then in England—and windmills drained marshes and fens for which the English hired experienced Dutch engineers. Later, this Dutch revolution in energy use, organization, commerce, and food production was taken over completely by England, where coal was soon introduced as a new, richer energy source. Externally, these were also times of revolts and wars for independence and colonial supremacy, the best-known of which started the eighty-year war of independence between the Low Countries and Spain, during which Holland also took over some colonies and trade routes from Spain.

With a larger surface area for food production and trade, a more diverse diet because of horticultural improvements, and, above all, another energy source to do the hard work of pumping, sawing, hammering, grinding, and heaving, the frequency of famines declined in England, as they did in Holland before. Yet, from 1690 until the start of the agricultural revolution in England in the mid-1700s, food production slowed down considerably. Instead of exporting food, England had to import it. By then, industrial colonialism had replaced mercantilist colonialism, generating a triangular trade mainly spanning the Atlantic Ocean and extending along the southern rim of Asia. Slaves were brought from Africa to the cotton and sugarcane fields in North and South America; cotton and sugar were shipped from there to England, where the cotton was manufactured and the sugar used; and from there, the cloth and finished products were sold in India and Africa. The profits thus gained allowed England to build up its industry and agricultural mechanization, its system for importing food and other products, its sanitary and medical systems, and, hence, the number of its people, which rapidly spread out across the world. This emigration released their country from problems of severe shortage of food and space, safeguarded its income from trade, and sowed the seed for the present stage of commercial and industrial globalization and of further, enhanced population growth.

It has been suggested that waves of high mortality (50 percent or more) due to recurrent famines and deadly disease are typical for agricultural societies and ended only after industrialization set in. Therefore, whenever and wherever industrialization set in, it always had the same result: better average living conditions, lower and more regular mortality rates, and, more rapid population growth. After England, the same developments followed in France, Germany, Italy, and Scandinavia, although often with a long lag time.

During the 1970s, when concerns were first expressed about the limitations of Earth's resources, one concern was obviously the cause—the growing world population using those resources. Paul Ehrlich, an ecologist, initially emphasized particularly the growing number of humans. The reasoning Ehrlich followed was that the replacement rate for each individual in the population—as measured for females only—should be equal to 1 if the population is to be stable, that is when it does not grow. This means that, on average, each female will be replaced by one new female in the next generation. If this rate of growth exceeds 1 (say, on average, to 1.2), the population increases; if it is smaller, it decreases. Moreover, when it is larger than 1, the increase in the total number of people will not follow a straight, upward line, but one that bends upwards more and more steeply, meaning that, over time, the total number of people will grow faster and faster. For example, with a replacement rate of 2, if two girls of a family survive and start new families themselves, the average number of female grandchildren becomes 4, that of the female great-grandchildren 8, after which it becomes 16, and so on. Another way of measuring the replacement rate is to relate it to both husband and wife, which means, roughly, that they have to be replaced by 2 children, independent of sex. As the mortality of men is slightly higher than that of women, and mortality in underdeveloped countries is also higher than that in developed countries, the average number of children per family becomes 2.33 for the entire world. Therefore, the replacement fertility rate of some underdeveloped African countries with high mortality rates amounts to 3.35, whereas it is 2.06 for European countries, showing that this rate decreases as conditions improve. In 2008, the global fertility rate had already fallen to 2.6, which means that we have almost compensated for the global average of between 5 and 6 children per family in the early 1950s and that in many countries the population is already decreasing.

Barry Commoner, another biologist, but one also active in politics, recognized this growth of the number of humans as a problem, but also allowed for the growth in resource use per head in the human population—a growth process happening on top of Ehrlich's increase in the number of heads. People were no longer living under or on the breadline, but often well above it. The ever-accelerating increase in the number of humans Ehrlich described also came to apply to the amount of food consumed by each person. And the better they were fed, the greater their numbers became, because the rates of reproduction grew higher and those of mortality lower. The combined effect of the accelerating rates of increase operating in sync is that the total increase in demand for food soon accelerates incredibly fast: two superimposed faster-growing processes are speeding up and enhancing each other. However, as we have seen, these two growth components directly pertaining to the people themselves are not the only ones, as the organization fulfilling their demands itself requires energy and material. These demands are growing even faster so that at present they far outweigh the first two components.

Apart from their growing demands, populations themselves grew fast—at first, only locally and in conjunction with their agricultural production, but later, the world population as a whole—so that more energy was needed to fuel this growth. Preparing fields, growing greater amounts of food and fodder, harvesting and trading them, and so on, all costs energy. Each person represents a daily stream of energy of a particular strength, as does the organizational infrastructure needed to keep this stream going. Step by step, as societies grew, they also became more organized with regard to the production of food, and hence, society itself became more organized. However, as food production and society were growing and becoming more organized, ever-more material and energy were needed per head to maintain society. This is because the organization itself requires not only manpower but also much material and energy. The more people, and the more complex the organization, the more energy and material are expended to build it up and to maintain it. In its initial stages, the various aspects of the organization were still straightforward: the more cows, the more grazing land, the more people tending the cows. But quickly, some people were needed to coordinate the various actions and people involved, followed by another layer of people coordinating the first-stage coordinators, and so on. The need for these organizational cross connections means that the number of people in the organization itself grows faster than the number of people actually being organized; the organizers themselves have to be organized. Thus, the larger the population, the faster the number of possible interactions in the organization grows. Similar to the number of people, society as an infrastructure between people itself grows exponentially. Each interaction in this organization costs energy and material, and the number of interactions grows faster than the number of humans. And this amount of energy and material grows faster and faster, just like the number of humans does, but now growing per individual in the population. That is, the larger the population, the even faster the growth in the amount of resources used and the amount of waste produced. At the moment, the percentage of people needed to organize the upkeep of our hundreds of millions in Western society is about 98 percent. And that is not enough: they are all aided by machines of all sorts to do part—if not most—of their work and all running on nonhuman energy.

Even in ancient Mesopotamia and Egypt, a class of people was needed to organize food transport and storage, organize irrigation, make laws and to solve conflicts, defend the territory, and so on. Wagons were needed for transport and facilities were needed for storage (not only for crops, but also for clay, papyruses, weapons, etc.); defensive walls were built, and so on. Over the seasons, stored food had to be checked continually for damage done by insects or rot. And toward the end of the year and in times of food shortage, the stored food had to be distributed or eked out among the population, which again required legislation, administration, and transport, among other things. The construction and maintenance of a network of ditches and canals forming the irrigation system also entailed extra costs; a single family would not have needed such an elaborate control system. Soon, all this extra effort cost a large part of the labor force available and, hence, of the potential for food production, plus all the extra resources for making and maintaining the wagons and weapons and building the store houses, as well as the houses and palaces of the organizing class, and so forth. Primitive forms of industries arose, technically supporting all these activities; the activities were soon concentrated in the first towns, where the administration was also to be found. In other words, the resource use per head of the population greatly increased and diversified with the growth in the size of the population, and urbanization followed suit.

But once the organization began to grow and diversify, costing an ever-greater part of the resources, an amount on top of that required for the subsistence production of before, the organization also began a life of itself. It began to have different needs of its own and itself needed organizing. The latter development spawned a similar one in turn, and so on. Apart from this, certain people, groups of people, and societies and nations demand more than others, particularly when they are developing a more complex organization, which complicates the picture. The point here, however, is that a large world population utilizes far more material and energy than the sum total of production units would have done if based solely on individual subsistence. And the growth in utilization of resources itself increases: it grows faster and faster per head of the population.

History gives the nod to both Ehrlich and to Commoner: the growth in the number of heads required the production of more food, which required better organization, which was more expensive in terms of material and energy demand—that is, it required a growing resource utilization per head of the population. On top of all these necessary expenses in terms of material and energy is the extra cost of luxury goods. These goods take ever-increasing amounts of extra material and energy. And remember, each bit of resource used enters as waste in the budget of life.

The world population at its present size and as it currently has to be organized requires an enormous amount of material and energy. It will be very hard—if not completely impossible—to reduce this amount, as it is tied to the growth in population. Moreover, this overutilization is set to continue growing at a rate higher than the global rate of population increase. It will not be canceled out by the currently decreasing rate at which we are replacing our numbers, so therefore, we must reduce our numbers still more if we are to reduce overutilization and waste production by the infrastructure. Though commendable, achieving reductions per head of the population by improving the insulation of our houses, using public transport, or eating less meat is not enough.

In retrospect, the growth rate of the population had reached a historical high—2.19—in the 1960s. In Ehrlich's terms, this means that each woman was replaced by 2.19 women in the following generation, on a global average. At this rate, the entire world population would have roughly doubled in only 32 years. Imagine if this growth rate had lasted until now! However, since that time, the overall growth of the global population has slowed down somewhat; the rate is currently about 1.17, which makes the doubling time slightly longer, about 60 years. This is still quite short: doubling the population within the lifetime of a Westerner. Because population growth has slowed down, although possibly independently and in spite of Ehrlich's warnings, the 3.7 billion humans of 1970 did not grow to over 7.4 billion by 2008 but instead reached 6.6 billion. But this is not yet the end of the growth of our numbers.

In fact, our global growth rate still exceeds 1, and the number of people in the world is therefore still growing. Some countries, such as France, are still actively stimulating the growth of their population; when I began writing this book, it proudly announced that its reproduction rate had exceeded 1. Governments often advocate population growth, because a country needs a certain percentage of young people to keep it dynamic and to feed the elderly, whose numbers are increasing due to improved life expectancy, thanks to improvements in living standards. Usually, these improvements themselves cost extra money and, therefore, extra people and resources, not to mention more administration. Growth because of growth—a self-accelerating process.

But how can the world feed such numbers? How can it keep feeding them? "Well," some say, "so far, it's been able to feed us, hasn't it, and surely, our numbers could become even higher, say, 10 billion. After all, every year, plants produce another crop, don't they? Just sow or plant some more, and that's it. So why worry?" According to this reasoning, this annual growth rate may therefore go on for a long time, year after year, a new, slightly larger crop. Moreover, if the gradual decline in population growth over the last decades continues, growth will stop at around 2050, having reached a total of about 9.5 billion humans on Earth, which—all being equal—is a number that can be sustained. But some experts think that this stabilization level is optimistic as the population will continue to grow slightly after 2050, and estimate the population will be 11 billion by the year 2100, a year many of our grandchildren will see.

So, by simply growing some more plants to feed on, we may escape the worst, a worldwide famine. What a relief. Yes, but each year, we have to keep fertilizing the soil for more plants to grow, and we have to supply them with more water. More people, more plants, more fertilizer, and more water. All being equal. But what if everything does not remain equal? What if there's a bad season, a run of bad seasons, or if the climate or soil deteriorates? What if our mineral resources run short or run out? Phosphate, an essential component in our fertilizers, for example, will be exhausted within the present century, yet our food production has to be no less than 70 percent more by 2050. If our numbers have indeed stabilized by then, our demands are expected to keep increasing, and our population needs to persist well after 2050, the time of our children. Yet, we cannot replace phosphorus with any other element. What about our supply of fossil fuels? We certainly need to take the finite amount of these resources on Earth into account. Another resource, freshwater, is already running out in large parts of the world, and most of the water is taken up by the plants before harvest. Already, some of the largest rivers on Earth no longer reach the sea or the ocean. The first discussions and quarrels about water have already begun, and national defense organizations are making their first preparations. The ancient origins of words like "rivalry," in fact, stem from using the same stream, *rivus*. So, our worries are still real. Pumping up water from aquifers is an option, although this costs energy, and the costs of energy are rocketing. And the costs of acquiring larger and larger pumps to reach deeper and deeper aquifers are already prohibitive for many farmers. In Mexico and Uganda, to name but two countries, one farmer after another has pulled out of farming because it is too expensive to buy and run more powerful pumps. And where regular water is becoming in short supply, salts are accumulating in the topsoil during irrigation, in that way making it impossible for plants to grow. This salt cannot be washed away easily, if at all. We know this from the ancient Mesopotamians, who eventually had to leave their fields for this reason. And the same happened much later to the Aztecs, as it happens at present in the American Southwest, in the Australian Murray-Darling Basin, in large parts of India, around the former Aral Sea in southern Russia, and in China.

I have concentrated on the fact that our use of resources is growing, together with the increasing number of heads in the population. However,

Commoner's reasoning makes the picture even more dire, as he takes the total resource use per head into account. Over the years, each of us has been using more and more and still more, ever more. So, our per capita resource utilization is growing faster and faster. It may therefore soon be far worse than Ehrlich expected, and worse than what our crude population calculations indicate. Ehrlich advocated that strict demographic rules should be applied to keep control over the numbers of people on Earth. But it is not the mere numbers that count. People's standard of living, their social conditions, and their societal organization also have their demands. More and more people want to eat meat, but each kilogram of beef costs seven or eight kilograms of grain to produce. And more and more people not only want to drive cars as a luxury, but they also need them in and between our growing cities as part of the societal infrastructure. Commoner therefore advocated not only controlling the number of humans but also imposing rules to keep control over the total resource utilization by each of us. But we cannot even control the numbers of people, let alone their resource use per head. Overpopulation has to be considered in respect to a rapidly growing number of people, as well as in respect to concomitant too rapid resource use by each of us. And our resource utilization is not only of renewable resources—the crops themselves—but also of nonrenewable resources the plants use (such as fertilizers) and that we as humans need, apart from those in our food.

This increasing use of limited, nonrenewable resources has to be considered over all the years we want humankind to prolong its existence on Earth. But for how long: two thousand years more? Or ten thousand? Have we thought about that? Seen in such a long-term perspective, are we sure that we have enough of these nonrenewable resources on Earth? Just to feed all those billions of humans, year after year, irrespective of supplying their other basic needs, transport of food, housing, clothing, computing, organization, and so on, and all these for such a length of time. With regard to nonrenewable, nonrecyclable resources, we need to multiply the annual resource use of ten billion users by the number of years we want to prolong our existence on Earth. Over ten years, ten billion people use the same as one hundred billion people would use in one year, and over the hundred years of the present century, this amounts to the use of one thousand people in one year. We need to do the same for the production of waste. If recycling of materials could alleviate our situation, how perfect do our recycling processes need to be to make life sustainable in the longer term? The longer the time, the less we may allow to go down the drain as waste. Can we put these questions in the perspective of our history to get any feeling of their weight? What has happened to us already, what have we achieved so that we can at least roughly know what to expect? Or at least roughly know what to do?

The first agricultural revolution, the Neolithic revolution, which happened in the Middle East some nine thousand to ten thousand years ago, had already led to an intensification of land use. In Mesopotamia, the area between the rivers Euphrates and Tigris, and in Egypt along the Nile, extensive water-works were designed, crops were grown, and silos were built in order to pre-serve the harvest for later use. Food production and preservation were thus intensified and mechanized as much as the organization and the tools of the day allowed. In that way, the land could produce a large amount of surplus food so that a considerable part of the people no longer needed to live on the land but could begin to live in towns. Here, they kept shops, worked in administration and other forms of organization, or they found their way into extensive religious institutions with hundreds or tens of thousands of priests for which during a brief period enormous pyramids and ziggurats were built. Many people were also involved in international trade, which began to take shape at that time, thereby disposing of surpluses in yet another way. In ancient cities, moreover, people could also be occupied with manufactur-ing (agricultural and other tools, for example) or they could develop their cultural activities, start up science, and build their houses, palaces, temples, ships, and defence works. Or they fought large wars in distant lands, partly to defend, and partly to extend their territory. With regard to food production, the large numbers of builders, transporters, administrators, lawyers, soldiers, and priests formed the nonproductive part of the society. Yet they were pro-ductive in terms of their contribution to building and maintaining roads and transport systems for bringing food from the countryside into the towns, as well as bringing material resources like flint, wood, and precious metals to the towns, often from great distances. But despite generating surpluses, these regions still became overpopulated, although occasionally, at times of climate change, famines decimated the population. And, as mentioned, over time, their fields became saline, that is, they became encrusted with salt so that their crops could not grow there. As always, one easy solution was to expand the territory of the country or emigrate, which led to mass move-ments and, hence, to the inevitable wars.

At the end of the eighteenth century, as had happened already so often in human history since Egyptian and Mesopotamian times, people in sev-eral countries in Europe became concerned about their growing numbers. In France, the great mathematician Pierre-Simon Laplace (known as the French Newton) began to study demography, and his friend, the chemist

Antoine-Laurent Lavoisier, experimented on measures of land improvement on his estate near Paris. Lavoisier realized that, over time, minerals may fall short, and that therefore his land had to be fertilized in a particular way. In fact, France lagged far behind the Low Countries, where the first experiments had begun a couple of centuries earlier, and Britain, where innovations had also been made in agriculture, and the land enclosure movement had followed. Now, France followed suit; gradually, crop diversification and rotation also replaced the dependence of the French on growing oats, barley, and, for the rich, wheat. As in other countries, waterlogged land had to be reclaimed by large-scale drainage of fens and marshes. Later, northern Germany and Italy followed suit too. In that way, agricultural production could be improved, intensified, and extended, allowing greater numbers of people to live on the same surface area while providing a larger surface area for inhabitants. In France, large numbers of people also lived high up on the mountain slopes, deforesting them up to the last tree and thus losing much fertile topsoil through the resulting erosion. In Italy, the effect of land drainage in the valleys can still be seen very easily, where the older parts of the ancient cities perch high up on widely scattered hill tops, whereas millions of present-day Italians now live on the former marshland in the valleys in between. These valleys contain the fields of fertile arable land, as well as the numerous industries that are currently the backbone of Italy's economy. These land reforms, improvements, and expansions formed the core of the second agricultural revolution, thousands of years after the first. However, they stimulated population growth and its supporting infrastructure and technology even more than the previous revolution, despite the fact that all the measures were taken to relieve the problems of the growing numbers of people. The measures did not solve those problems but ended up by aggravating them. The measures improved the living conditions in various ways, thereby increasing the reproduction rate, not by increasing the fertility rate but by lowering the rate of mortality.

There are parallels between the first, Neolithic agricultural revolution in the Middle East, the second agricultural revolution in the Muslim world, and the one during and after the eighteenth century in Europe. The second revolution similarly concerned a large-scale mechanization and eventually even a complete industrialization and urbanization of the countryside. These trends set in on a large scale during the nineteenth century and sped up considerably during the twentieth, particularly after the Second World War. As a result, very large parts of the population now depend for their food on a small and dwindling number of farmers, either in their own country or abroad.

Food production, especially if it is in other countries, requires an enormous increase in the transportation system, as well as in other aspects of the infrastructure, such as harbors, retail markets, financing systems, road networks, and distribution systems.

The result has been an ever-accelerating rate of urbanization and migration, not only from the rural areas into the cities, but also by founding and populating new land and cities abroad. After the expansionist Roman Empire, there was population expansion and emigration, first from the world of Islam, then from Scandinavia. Vikings spread westward to Iceland, Greenland, and North America, eastward to Ukraine and Poland, and southward to England, France, and southern Italy. Then, during the Middle Ages, Central Europe began to expand eastward into the Baltic countries, Poland, Czechoslovakia, and Russia. During the nineteenth and twentieth centuries, emigration from all parts of Europe now took on global proportions: people emigrated to the United States, Canada, South America, South Africa, Australia, and New Zealand. Within Russia, Siberia was populated, and within China, people moved to Mongolia and the central and western provinces. Of course, right from their beginning, these founding populations began to grow, expand, and seed out themselves. This resulted in settlements encroaching into the new, still pristine surroundings, accompanied by agricultural and industrial development. Less than a century ago, the enormous, ancient forests around Chicago were cut for their timber on an unbelievably large scale. And elsewhere in North America, the vast prairies were turned into fields.

At present, there is hardly any inhabitable space left in the world for emigration—let alone expansion—in order to solve problems of population growth elsewhere. Many countries throughout the world now have laws restricting emigration. Recently, South Africa and Europe have legislated to keep out immigrants from surrounding countries, and the United States is building extensive electric fences on the Mexican border, and India is erecting a three-thousand-kilometer-long fence along the border with Bangladesh. Migration has also become internal as a population contracts into rapidly growing cities of unbelievable size. Although attempted all over the globe, building denser cities with higher buildings will not be an option in the long term. The world is being saturated not only by the spectacular building and by the rapid and extensive outgrowth of towns and megacities but also by the increasing areas used to grow food and for industrial production, mining, and road building, as well as by the increasing, inevitable waste disposal resulting from any kind of resource use. We have to admit that in terms of human habitation, plus the production of food and industrial products,

Earth is becoming saturated. In terms of predictions of the resource and land use in the future relative to what resources and land remain at present, it is already very, very overpopulated.

The problem is that with respect to the present availability of resources and waste production it still seems unsaturated and therefore underpopulated. Given the continued growth and presumed level of stabilization of the world population in relation to the trends in resource depletion and waste production, however, we are already overshooting the habitability of Earth. This means that, as seen on a longer term into the future, further than our present existence, we are greatly overpopulating the planet.

The migration and spatial concentration of people, however, was not enough. Food production still had to be increased faster in order to feed the ever growing numbers. Therefore, measures for efficient land use were also taken. As mentioned, one important measure was to reclaim shallow parts of the sea or waterlogged countryside. The new land could be taken into production or settled. The newly created land not only attracted people from elsewhere but—as already mentioned for Italy—reproduction rose even more in response to the newly created opportunities, whereas mortality continued to decline. Agricultural production was improved first by liming, then by deep plowing, and later by using more effective chemicals to increase soil fertility or to replenish nutrients lost. Finally, production could be increased by combating crop diseases, harmful insects, nematodes, and other pests. Yet, other measures included the leveling of the countryside and the enlargement of the fields, thus physically making the land more uniform and more efficient to work. And when the environmental conditions could not be improved any further, the yields of the crops and livestock themselves were optimized relative to those field conditions: their genetic constitution was improved and made uniform to optimize the match between the plants' increasing demand for minerals and water and what the land supplied. This genetic improvement of crop and livestock culminated in the creation of large, centralized seed factories and sperm banks. This optimization of genetic traits, unfortunately, led to an enormous loss of water and nutrient resources as well as to the loss of the vast genetic variation in old varieties and breeds. The same happens as a result of breeding trees and fish. Biodiversity is declining as are the riches of Earth in general.

In the days of Lavoisier, just before 1800, many other people were already thinking about the same problems surrounding population growth. The challenge was how to curtail or stop this growth. A couple of decades after

Lavoisier's death, the French mathematician Pierre François Verhulst formulated an equation for population growth, according to which the rate of population growth starts off by increasing, then gradually slows down and eventually ceases, after which population size remains constant. He assumed that external conditions were constant, and therefore the decrease in growth rate was brought about by mechanisms internal to the population. However, Verhulst did not indicate which mechanisms would be involved: mathematically, a great variety of mechanisms are possible.

Other scholars in Europe, particularly in France and England, had more specific ideas about the main mechanisms operating. In England, Adam Smith and Thomas Robert Malthus thought that reproduction remained as it always had been, but that the larger the population grows, the more mortality would increase, particularly among the poor, the less fit individuals in a population. The dominant idea of the time was that poverty was a self-imposed condition, and therefore, the poor didn't need to be helped. Their higher mortality rates would thus relieve the stress in society resulting from too high densities and irresponsible living. Mortality through food scarcity among the poor would thus both account for and justify Verhulst's constant population size. At present, demographers think, instead, that the high fertility rate will decrease in some way, thus adjusting to the presently low rate of mortality.

The system of maximal agricultural production is determined not only by uniformly optimal conditions but also by conditions that remain constant. In fact, as already mentioned, Verhulst's model assumed these conditions would be constant. Yet, although humans can largely optimize conditions, and this uniformly in a field or an agricultural region, we cannot optimize them over time. Short-term seasonal and year-to-year fluctuations in weather or longer-term changes in climate from decade to decade remain, and these can always radically destabilize the population-agriculture system, as they have always done in the past. This often happened during the late Middle Ages, particularly in sensitive, marginal areas, such as in Ireland, Scotland, or Scandinavia, or slightly earlier in Greenland and Iceland, or later during the Little Ice Age in Europe from the sixteenth to well into the nineteenth century, just to mention a few examples. In those northern parts of the world, Greenland and Iceland in particular, harsh food and living conditions once led to a population of dwarfed and crippled individuals, whereas later and further south, Napoleon's Great Army was still one of men on average only four feet nine inches tall. Just as in the European Alps, farming—and farming villages—high in the mountains of Scotland were abandoned because of falling temperatures; the fields became overgrown and the villages fell into ruin. Such events went on regularly in Europe until less than a hundred and fifty years ago, when millions were still dying from hunger in places like Belgium and Finland (1867–1868). This also happens to people in the Sahel region of the Sahara or in other agricultural subsistence systems in Africa under the warming and drying conditions of the present. Weather and climate variability can thus throw large parts of the world population into jeopardy, particularly in the driest parts, where at present almost half a billion of us are already in acute danger. Also, between 1846 and 1848, the potato blight, a fungal disease, destroyed the staple food of the poor in Ireland, causing a famine from which about a million people died and which forced another million to emigrate to North America. Thousands of other Irish migrated to England to work in the factories there.

Other factors interfering with a smooth demographic pattern of growth and stabilization are earthquakes, cyclones, drought, large-scale inundations, and mudflows, the effects of which are reported frequently in our news broadcasts and newspapers. Other processes, such as desertification, salination, erosion, climate change, and soil and water pollution, don't lead smoothly to stabilization around some balanced level, but instead to the ruination of productive agricultural areas.

Genetic and environmental diversity were both deliberately greatly reduced, destabilizing the plant-environment system. Now, as soon as a new disease breaks out, crops have no resistance to it, and as the large areas of monocrops favor the spread of disease, crops can be destroyed over large parts of a continent, sometimes within months. Diseases and pests, including species accidentally introduced from afar, have become widespread, and their eradication is costly: hundreds of billions of dollars a year within the United States alone. The response is to keep on producing new varieties and breeds industrially. This industrial production of new genetic lines, in turn, makes the system as a whole very vulnerable to small changes in production and to the political willingness of seed-producing countries to sell new varieties. Poor farmers are increasingly unable to buy those industrial seeds. Former increases in crop production have turned into decreases.

Therefore, in many parts of the world, we are reaching or overshooting the limits of agricultural production for present-day conditions. Britain, for example, consumes six times more than its own land mass would allow. This means that in the human population, food production is becoming a factor limiting further growth. This is already the case over large— underdeveloped—parts of the world, from where the rich, developed, and developing countries even import part of their food. Under the conditions set by the world market, people in underdeveloped countries often cannot even buy their own produce, whereas the richer countries can. For this reason, as was recently the case in the Philippines, they sometimes sell their own rice and have to import cheaper rice from elsewhere. During the 1970s, it was thought that fertilizing and irrigating the deserts would be sufficient to improve agricultural production. But enormous problems arose in transporting fertilizers and water, and this in the face of more expensive fertilizers and more desiccation under warming climate conditions. Efforts in Libya and Saudi Arabia to pump freshwater to the desert inland, paid by large oil revenues, ended in failure. The consequence: even higher evapotranspiration from the crop plants, and salination of the soils. It seems that these schemes have now fallen out of favor.

In order to maintain our numbers or even to let them grow further, we are bound to the present areas of agricultural production, hoping that these will remain stable in the face of further climatic change, urbanization, and increasing pollution pressure. But for how long?

4 POPULATION GROWTH AND INDUSTRIAL PRODUCTION

By about the middle of the eighteenth century, standards of living had improved so much that all over Europe mortality rates had declined, whereas fertility rates had remained roughly the same for almost two centuries. The inevitable results were that population increased, requiring economies to change, broaden, and intensify, and that emigration took place. Millions left Europe, spreading out over the world. Within Europe, the old subsistence economies of local communities with only limited trade and top-down regulatory governments broke down, with the result that there were revolts, often connected with food shortages. Agricultural production became reorganized through various land reforms, leaving more freedom for rural settlement and production. Also, as we saw in the previous chapter, production was improved in various ways, such as by rotation, fertilization, and mechanization. Farmers colonized previously unoccupied areas like forests and mountains, which they denuded from their ancient trees; large-scale drainage, road building, and canalization programs extended the potential area of arable land.

However, instead of alleviating the problems of undernourishment, these measures aggravated them. Farmers had to find other sources of revenue in order to survive. One of the first was that they began to weave cloth at home. Thus, a home industry was initiated that complemented or replaced a closed guild system of manufacturing that was still based on local subsistence to meet personal needs. The new system was based on impersonal mass production and trading the surpluses. It was from this multitude of home industries that the Industrial Revolution sprang.

The next step was that both production and trade were streamlined by specialization and increasing the volumes of output and, hence, revenue. Raw material was bought in bulk as cheaply as possible, and the products were similarly sold in bulk at the highest profit possible, better than the individual home worker could have achieved. This was possible in larger units— factories—where the work of many individuals was combined. In turn, this aggravated rural depopulation and stimulated urbanization. The Industrial Revolution thus detached from its initiator, the agricultural revolution, as an answer to growing overpopulation. Overproduction combined with the expansion of the hinterland was the answer to overpopulation.

Now, however, the bulk of the earnings went to the factory owners and traders instead of to the aristocracy and the church, as had previously been the case. And the new class of workers was also denied their full share. Society changed radically. As the bulk of the population remained poor and undernourished, social reforms were instigated, and ultimately the masses were given their democratic voice. Rules of mass behavior were discovered and formulated, giving rise to both a new judicial system and population sciences, such as sociology, economics, demography, historiography, ethnology, and anthropology. Spatial knowledge increased through better knowledge of geology, land surveying, and geography. Nations were recognized as social and economic units, and nationalism arose. A new world took shape.

In terms of personal well being, new fields of science emerged to alleviate living conditions. Among these were the medical sciences and microbiology, and eventually psychology, as well as life insurance organizations based on historical demography and statistical theory. Production processes also benefited from advances in science and technology: the developments in chemistry, physics, and engineering were used for communication, illumination, combustion, and rail transport. All these developments required a modified educational system that also catered to the masses. The worldview became less local and concrete and changed to one based more on abstract understanding, which was also reflected in a new literature and in new philosophies. Quantification penetrated the new society, and mathematics flourished. The world changed radically, often within decades: first hesitantly, in the Low Countries, then more extensively in Britain, and finally in France, the United States, and beyond.

Economically, the world was changing as well, not only during the nineteenth century, but also well into the twentieth. After the concentration of labor and people from the countryside in factories and cities, the markets extended from the local to the national level, and soon to the international level as well. Thus, resource areas were initiated and built up, based on the cheap labor of African slaves in the cotton fields of the American South. And at the other end of the spectrum of producing and selling the cotton products, local and national industries and markets were subjugated and destroyed, as happened with the cotton industries of India and Egypt. National and international infrastructures were formed to cope with ordering, transport, and payments. As a next step, the internal national markets were extended by improving workers' pay and then by introducing a flow industry by giving the products a maximum lifetime. After this, the consumption per head of the growing population increased, and, finally, the growth of the number of people in the population was encouraged by stimulating reproduction.

So, population growth resulted in agricultural growth and then in industrial growth and a growth in economic hinterlands, which in turn resulted eventually in enhanced population growth. The growth of the various national populations was accompanied by an increase in the complexity of their organization, which in itself required more and more people to enable the interactions necessary for forming the infrastructures needed. As mentioned, the number of these interactions was actually growing faster than the number of people in the populations concerned. In turn, this eventually stimulated communication and computer science, so that increasing parts of the work were taken over by machines, to ensure the running of industry and for communication, trade, and banking.

Of course, the response to large numbers in the population was not a deliberate choice made by entrepreneurs to improve the efficiency of the processes concerned; what drove the entrepreneurs was a desire to make profit more efficiently. By introducing markets beyond the reach of the local ruler, then beyond the reach of the guild or the town, and finally during present-day globalization beyond the control of the nation or federation, entrepreneurs can operate outside price, labor, social regulations and laws, and increase their profit. Outsourcing and globalization partly resulted in the deregulation movement of the 1980s and after. During the Industrial Revolution, society itself did nothing to help the resulting redundant farmers and laborers who in some way still had to find work and food. Overall, therefore, more energy and materials were used, not just the traditional ones, but also new ones to support the additional activities. This is a common occurrence: the invention of new machines, whether in agriculture or elsewhere in society, increases and changes rather than reduces resource use and waste production.

It was not only more workers and machines that were required: more energy was needed too. More food had to be produced, and for this, more energy was initially put into mechanized food production, transportation, and road building to enhance the conservation, distribution, and selling of food. And more energy went into the production of fertilizers, herbicides, and pesticides, and into the production of yet other chemicals for the preservation and preparation of food. In terms of energy, however, urbanization and industrialization took an even heavier toll when obtaining the minerals needed to build houses, roads, factories, harbors, and storage and distribution depots, and to construct a huge variety of machines for mining, transport, and the production process. Soon, after wood had been depleted, coal was adopted and then partly supplemented by oil, which is only one-sixteenth as energy intensive but could be mined and transported more economically. Currently, oil is gradually being replaced by natural gas, and this may be replaced by shale oil. These fuels can either be used directly as an energy source or be transformed into electricity as another, derived source of easily transportable energy. Apart from the possibility of easy transport, oil and gas have the advantage over coal that they can more easily be purified of undesirable compounds, such as those containing mercury and sulfur, the latter of which causes acid rain, poisonous to many plants and animals. Still, worldwide, coal remains the major energy source.

We must realize that our personal lives, and all social life from the local scale to the global, depend on these external sources of energy. We live a life derived almost entirely from fossil fuels. As organisms, we are an organized flow of energy, and this has become a flow of fossil energy; we exist by the grace of fossil energy. We have gone a long way from the energy spent individually, first as hunter-gatherers and then in the subsistence economies of local farming, to all the energy spent in our grown-out, extremely complex global societies. And our grown-out societies cannot be ungrown; the margins for change are too narrow. It is the enormous fossil fuel energy supply that structures and maintains our society, a society that, in the end, keeps us living as individuals. We built this society out of necessity. We cannot do without it. Our numbers are too large.

One of the answers to food shortage was thought to be found in the mechanization and industrialization of agriculture and society. The mechanization entailed machines taking over various aspects of working the soil: plowing, sowing, weeding, and then harvesting the crop. In dairy farming, milking machines were invented and barns improved. In the case of both crops and livestock, breeding schemes were designed to improve the physiological performance of the organisms. Apart from improvements to the individual organisms, the conditions under which they lived, and how they were harvested, dramatic improvements were also made to the land itself. In Britain, where the landowners were more powerful than the farmers, ownership shifted to the magnates of the time, who expelled the small farmers in two successive enclosure movements that occurred during the sixteenth and eighteenth centuries. As a result, from Scotland alone some 120,000 people emigrated to North America, in particular to New York and Ontario; all those left behind had to find work in the factories in the small towns. The lords—in Scotland, the lairds—ran the farms as they did the factories, maximizing the yield of the land at minimal cost and selling the produce at maximal profit. In fact, they concentrated on the production of wool rather than on that of food, which aggravated feeding conditions. In order to promote emigration, even as late as 1827, Malthus recommended that Parliament refuse parish relief to farmers and children born two years after the institution of the law, and pull down cottages as soon as they became vacant. The farmers were replaced by laborers with no great experience of the land. In continental Europe, this also occurred in Prussia, but in other parts, such as in France, the small farmers retained the ownership of their farms, personally keeping a keen eye on the quality and development of the land, improving it where needed. Typically, therefore, during the French Revolution, it was the aristocracy instead of the farming community that was chased off the land. The British system of yield maximization and cost minimization was later taken over in the United States, where it was considerably elaborated, soon becoming connected with the financial and industrial worlds. Here, profit maximization has become the basis of food production. And here, in fact, lie the roots of the two types of capitalism, the Anglo-Saxon and the Continental types, which are also called the liberal and coordinated market economies. The Anglo-Saxon type based on profit maximization inevitably leads to overuse and abuse of the land and, hence, a loss of its intrinsic qualities. With the rise of a land proletariat replacing the farmers, much knowledge on local conditions was lost, a trend that continued when large, foreign food industries demanded land use reforms and enlargement, though often with the result that production fell. Now, the Chinese, for example, are renting or buying land in Africa on a large scale which is typically worked by Chinese laborers.

A next step was the integration of agricultural business with petrochemical industries producing agricultural chemicals, such as fertilizers, herbicides, fungicides, and pesticides. First the axe and then the plow exposed the humus in the soil to the air so that it decomposed more quickly, which, in turn, lowered the water retention capacity of the soil. More recently, chemicals have lowered this capacity even more. The result is soil that has to be irrigated and fertilized more heavily, salination and desertification, and erosion. Ultimately, affected areas have to be abandoned, lost for the production of food. In Ecuador and India, after a production boom thanks to new fertilizers, the soils are now eroding, and the farmers are having to leave. Large parts of China are following suit; fertile soil is coloring and choking the Yellow River. Worldwide, more phosphate, a mineral essential for our food, is being eroding away than is being mined.

The foundation of big seed factories is part of this trend to industrialize agriculture. In turn, this has led to great losses of genetic variation, particularly when such factories spread their interests to other, often underdeveloped countries as well. Farmers worldwide are being encouraged to use "improved" seed, but though disease resistance is greater and yields are higher, there is a cost. Thousands of local races, developed over hundreds of years and containing nonreproducible genetic information, are being lost forever. Fortunately, many national seed banks have recently been brought together in an underground store in Spitsbergen, Norway, to counteract further loss of this aspect of biodiversity through relentless profit-making. With the risk of increasing fuel shortages and the rising prices of energy, interest is now shifting to the more profitable biofuel crops, and this has been leading to food shortages worldwide. In Europe, this integration of the domain of agriculture with the domains of high finance and industry has not yet developed that far, but it has begun. However, given the escalating problems with food production, partly due to climate warming, we cannot afford great losses in soil deterioration, diminished genetic diversity, or from growing nonfood crops.

The industrial developments described above concern the production of food, but the food itself is prepared as fast food, and it also has to be preserved, distributed, and sold, and the resources and waste also have to be moved. So, a sector in society concerned with mobility has developed for the production and maintenance of cars, trucks, aircraft, ships, roads, and traffic. It has culminated in the tourist industry, which can account for as much as 10 percent of a national budget. The construction and service industries have also grown, as have the industrial sectors manufacturing mechanical and chemical household products. The medical industry developed and flourished, and so did the judicial sector. And the chemical industry became essential for the industrial society, as did the electronic industry. The military industry became essential for the defense of our interests and needs. Financing and regulating became industries in their own right, and so on. Together, these industries form the societal superstructure safeguarding our personal well-being: our food, clothing, and shelter. Humans have entered the industrial society, an abstract notion nobody can comprehend or master anymore, an organizational structure living its own life, following its own rules and its own dynamic that is beyond the capacity of anybody to control.

First the kings and aristocracy as well as the religious organizations, then the parliaments and governmental organizations at all levels, the big industries and the high finance have to let go; their influence doesn't reach further than the ever-narrowing margins of society.

In exceedingly broad terms, this seems to be the situation into which we have maneuvered ourselves during the last few hundred years, which have been centuries of unparalleled success and improvement for humanity as a whole—success and improvement in that we have completely reshaped our world to increase our personal convenience and happiness. We should not forget that opting for a path of ever-increasing growth and wealth has brought us benefits. However, these are put at risk unless we realize what is happening to the world and what the future holds if the most urgent remedial measures are not taken. In fact, in a book on threatening, negative trends, at least part of a chapter should be devoted to a brief account of what—at least in the developed part of the world—we have built up and what we need to defend. Too much is at stake: our numbers, our enormous achievements, and our future. Our remedial measures should be such that many, if not all, of the good things we have achieved can be safeguarded, if not expanded on under other, still unknown conditions.

The primary benefit we have achieved so far is that in many—although certainly not all—parts of the world, food has been greatly improved, in terms of its quantity as well as its safety, quality, and variety. In fact, its quality is now usually permanently controlled, both governmentally and by the food industry itself. Our housing and clothing has also improved greatly, which in turn has improved the general level of health of many populations. Many of us also live in cleaner and healthier environments, not only in our immediate surroundings, but also because of the extensive sewage systems in urbanized areas. Along with these cleaner, healthier conditions, in many countries elaborate medical systems have been put in place to prevent diseases from developing and to cure and tend the sick. Parallel with this development, among various types of national health systems and organizations, systems of health and life insurance have been organized, each requiring large funds but allowing for sick leave and retirement from work at old age. Life and working conditions themselves have become physically less demanding, because machines have taken over most of our heavy work, transport has improved, and because of greatly improved infrastructures and communication systems. Working conditions have improved, and our work has become shorter, more varied, and interspersed with leisure and vacation time. During the day and, more particularly, the evenings, streets have become safer because of police control and better illumination. Underlying many of these improvements has been an education system for all (or at least the majority of people); at the personal level this has led to a higher cultural level in a broad sense, and, in terms of society, to a higher working level and efficiency. Compared with earlier, subsistence societies, many of our present societies are enlightened in terms of freedom of thought and the equality of people with regard to gender, race, and religious background. It is also important that the social control of bygone times has been replaced by legal control.

All these changes and improvements came at a cost, not only in terms of resource use and waste production, but also in financial terms. Production costs have risen to the extent that other countries have taken over because of their lower prices. Thus, gradually, outsourcing procedures have been introduced, improving the financial and therefore the socioeconomic conditions of those countries too. Their resource use and waste production then develop in the same way as in the developed parts of the world. Inexorably, the world is globalizing, resulting in the growth of the populations in the new societies and in the growth of their per capita resource use and waste production. Entirely new economic sectors are added to these societies: the service, communication, health, and tourist industries, for example. All these developments are exacerbating global resource use and waste production. These trends have happened relatively recently in India and China in the Far East, for example, and in Brazil and Argentina in South America. India, in turn, has begun its own round of secondary outsourcing because of its own price increases.

Despite the huge improvements that have been made during the past two centuries, one basic aspect of production has remained the same: the linear production system going from resource use via processing activities to the production of waste. In fact, this very same production system has merely expanded, intensified, and sped up. With the growth of our numbers, our society has grown in size and complexity, becoming more and more demanding on the one hand and generating increasingly more waste on the other. Population growth generated societal growth at increasing rates, and the demands and waste of this growth increased accordingly. Our societal growth was meant to be an answer to population growth—to alleviate shortages of food and other resources and to reduce the amount of waste. But it generated further and further growth at increasing rates. And it demanded more of the remaining resources of Earth, and it gave them back in the form of more waste. As a specific answer to population growth, economic growth cannot simply be stopped or be replaced by another model not based on growth. In order to do that, we also need to stop population growth and the growth of consumption in the first place. A reduction of the highly demanding societal infrastructure will follow suit automatically.

Unfortunately, recycling as the alternative to the linear flow of material cannot be the answer, because, as an activity, it requires materials and energy itself. One can argue that the materials needed as the next resources can be found in the waste we make, yet this cycle is far from perfect. For example, we can make oil from various forms of plastic, but this results in a regain of only 5 percent of the original oil. As we have seen, the natural system has not managed very well in this either, given the ever-growing size of the continents on which we live, the literal mountains of biological waste on these continents, mountains of organic sediment we like to hike in during our holidays, and all the oxygen certain bacteria—cyanobacteria—and plants have expelled for the last 2.5 billion years as their waste into the atmosphere, oxygen which we breathe in for our survival and which our industries, trucks, cars, and airplanes use in order to operate. As a waste gas, oxygen constitutes at present 21 percent of the atmosphere. Bringing our own waste products back into circulation will cost an excessive amount of energy, an amount to be added to the still increasing amount we already need to live. Finally, only part of the waste can be taken up into the recycling program again; the rest still has to be obtained from fresh resources (below). Recycled glass, for example, contains at most 80 percent used glass, the rest comes from sand. Similarly, the fibers in paper wear out in each recycling round, so that there too, new fibers from freshly cut wood must be added. In the recycling of concrete, 20 percent former concrete at most can be used, the remaining 80 percent has to be made from new, nonrecycled material.

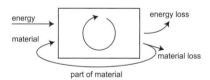

energy   energy loss
material   material loss
part of material

There are many positive effects the two revolutions—the agricultural and the industrial—have indisputably had, both as a response to and as a cause of unbridled population growth. If anything were to go wrong in our present society, ultimately due to the growth of our global population, all these achievements would be put at risk. This will happen if we fail to take the right remedial measures against further growth. Rather than taking measures against the effects of population growth—for example, against carbon dioxide accumulation in the atmosphere, against water and soil pollution, or to achieve a more economic use of energy and materials—it will be more effective to try and do something about the ultimate cause of these risks. This means that any further population growth has to be stopped as soon as possible. The next step is to put in place measures to reduce these numbers.

This does not mean that we can forget about taking measures against dumping carbon dioxide in the atmosphere to avert or alleviate a change in climate. We have to continue to fight all these problems, of course. Apart from taking measures against this prime cause of irresponsible resource use and waste production, it remains imperative to eke out our resources in a systematic and fair way, reduce the production of noxious waste, recycle whatever can be recycled, and begin a serious, large-scale search for alternative sources of energy. What is at stake, apart from the life of so many billions, is the whole system that has evolved over time to keep them alive in a humane way.

All these inevitable historical developments define the margins of the conditions to be changed when we talk about reducing climate warming due to carbon dioxide production or reducing our energy use, and so on. As already mentioned, the margins are narrow and over time will only get narrower. It will be ineffective to do something about the *results* of population growth without also addressing the root of the problem—the population growth itself. We have to tackle both fronts simultaneously.

5 AGRIBUSINESS AND CORPORATE STATES

Have you ever wondered how archaeologists know where to dig for an old grave, ancient village, road system, or battlefield? They use old maps, local field names, the location of certain plants or animals, or, occasionally, they analyze legends and age-old stories. For example, Homer's ancient Greek epic, the *Iliad*, contains clues about the location of Troy, and in and around Sicily and other Mediterranean islands you can still find actual objects that may have inspired many of the tales in his *Odyssey*. The story of the Cyclops, for example, might be based on the fossil skulls of miniature island elephants, in which the nasal opening could have been mistaken for the eye socket. Archaeologists also scrutinize aerial photographs, looking for clues such as discolorations in natural vegetation or in crops that reveal abandoned villages or villas, or even two thousand-year-old cart tracks from Roman times. Some years ago, I saw this for myself high up in the Alps near the Italian frontier, where completely different vegetation clearly marked the site of a French military encampment from the Second World War. The area covered by the camp was quite apparent, even after half a century. History casts a long shadow.

This makes you think about the effect the Green Revolution has had in countries like India, where for many generations thousands of farmers intimately knew the local growing conditions for their crops. They knew all the local races of rice that over thousands of years had been adapted to these often very specific conditions. Huge numbers of these farmers have moved to the cities in search of a better life; of those who stayed, the losses were so high that, over the last sixteen years, a quarter of a million committed suicide, or one farmer in every thirty minutes. Their local agricultural knowledge has been lost and so have thousands of crop varieties representing thousands of genetic adaptations to a great variety of environmental conditions. Instead, modern seed farms do not produce newly adapted genomes, but genetic hybrids that last for only one or two successive crops. Such hybrids are genetic combinations of two slightly different genetic types within the same species. After that, the combination of the hybrid falls apart into the two earlier, less productive types that yield less or are more vulnerable to drought, so the farmer has to buy a new hybrid instead, or an improved irrigation system. Worse, the country as a whole now depends on the political goodwill of the seed-supplying country, which may change. If the experiment of the Green Revolution eventually proves to be a failure, how can we then turn back the clock? Indeed, history may cast a permanent shadow.

With all these agricultural improvements, all the regional, subcontinental, or, occasionally, worldwide economic and political rearrangements and changing living conditions, you wonder what is actually happening to our food and our food-producing conditions. What kinds of benefits do we get, not just in the short term but over several generations? Will we ultimately win out or lose out? And if we do lose out, we will still have to meet increasing demands because of our increasing numbers and our improving standards of living. We always think that we will be able to meet our demands, and that this will last until far into the future, indefinitely perhaps, but what if we are wrong? How much can we then still reduce our demands if necessary, or reduce our numbers, and tidy up the waste we have left?

For most of our human history, peasants have worked the soil primarily to supply the needs of themselves, their neighbors, and the owners of the land they worked. The little they may still have produced in excess was difficult to transport because of the poor state of roads into town (if there were any roads). And certainly not all of them could afford a cart or an ox or horse to pull it. Also, for most of history, there were no canals or railroads or other modern ways of connecting the farmers with nearby markets where they could sell their produce. They were bound to produce for an extremely local community, and they depended on mutual help rather than favorable outcomes of competition on a market. In the French Alps, for example, bread was baked only once a year, in the oven of the local village. Food production, therefore, was scant and patchy, and supplied only a small area. Similarly, the supply of manure and human feces was also very local; nutrient recycling was a local affair. Food prices were also local, fluctuating greatly depending on the weather conditions; there was no world price for staples like wheat. Of course, there was some long-distance food transportation, especially from the fifteenth century onward. During the fifteenth and the sixteenth centuries, the Dutch imported some of their grain from Spain and the Baltic countries, sometimes exporting it in turn for profit to other Western European countries. In times of famine, however, they imported only enough grain from their enemy Spain to meet their demands, which were based on feeding the population in the growing towns. The farmers further afield in the Dutch countryside mostly kept producing for their own local needs.

This way of producing food changed radically during the nineteenth century when roads, canals, and harbors were built, and when, at a later date, railroads were incorporated into the spatial exchange system of food and other materials. Countries as a whole became the units of production. At the same time, chemical fertilizers were developed to improve crop yields. The first such fertilizers added phosphorus, nitrogen, and potassium to the soil; later, the practice of liming the soil released nutrients that until then had always remained bound to the organic matter (humus). However, the decomposition of humus reduced the water retention of the soil. At the same time, all over Europe, extensive swampy and marshy areas were being drained and brought into cultivation, greatly increasing the area of crop production. Also, the first agricultural machines were developed. Yet, their mass production and use by most farmers had to wait until after the Second World War, when the new energy supply from fossil fuels got going on a sufficiently large scale.

Of course, this exceedingly broad trend in agricultural production in Europe varied geographically. In England, rich landlords and industrialists bought the land from the farmers during two successive enclosure movements and began to make personal profit from their land. More often, however, the peasants were simply chased off the land and their dwellings demolished before their eyes. Usually, this went against the demand of the population for food; when wool made a better profit, the fields were used for sheep instead of food crops. The wool was used in the textile industry that was also owned by the new capitalists, and the textiles were either exported or sold on the domestic market; the cheap cottons went to the poor, and the expensive, warmer woolens went to the rich.

Profit-making became the underlying rationale for agricultural production. The agricultural products were used in industry, with the factories operating in a network of international markets. Former farmers formed the urban, industrial proletariat, and their land was worked or looked after by hired workers, who formed the new agricultural proletariat. However, these workers often lacked detailed knowledge of the land; they only had to work the land to help make the largest profit for the owner. In short, the seeds were laid for a new system of land use, agro-industrial capitalism, and these seeds soon germinated.

Some of these seeds were exported to the newly formed United States, which was still growing internally. Here, vast areas of prairie still existed, uncultivated and roamed by millions of buffalos. It is well known that these magnificent animals were hunted to almost total extinction. The prairies were divided up and sold to a new community of immigrant farmers who hoped to earn their living from the produce of their newly acquired fields. Soon, however, capital-rich speculators from the eastern cities became interested in investing their money in this land as well in order to make profit the way the English landlords had done—that is, not farming for their own and local needs, but instead for the state market or, eventually, for the even larger national market. Step by step, a new concept took shape, that of the world market, that would determine the price of the local produce. Transport systems for freight grew in scale from national to global, at first using intercontinental ships, and later, mainly after the Second World War, airplanes. During the nineteenth century, the postal system was established, which also facilitated commerce. It soon developed into an international telegraph system, which, in turn, was followed by the telephone and fax. And after 1980, the burgeoning electronic computation, computer administration, and communication systems meant that much of the payment, banking, and capital exchange between banks could be done electronically. At present, staggering amounts of money are exchanged daily by this means. Outsourcing parts of the production lines to countries with lower labor costs began after the Second World War too, and has continued in recent decades in the form of the electronic outsourcing of parts of the trading, insurance, and other financial transaction processes.

In these and in many other cases, intricate networks of local, national, and international interrelationships, collaborations, and production systems originated—systems without any physical basis or boundary. These networks exist entirely as computer networks or in our brains. They can extend or contract, shift, and change completely in composition from one moment to the next. As networks, they are highly dynamic, virtual systems that can originate and grow or vanish within a short space of time.

This is what happened from the nineteenth century onward, and the end is not really in sight. The local scale of tangible land use and food production was replaced by production on an ever-increasing virtual scale, the scale of the "market" with its own prices, which often went together with more concrete extensions of our world: roads and harbors were built, new branches of companies were set up abroad. Colonies were founded, which in turn were replaced by abstract business contacts during a decolonization phase. New ways of making and keeping contact were made across the world, by post, telegraph, phone, fax, and eventually by e-mail: nowadays, you may never meet a person with whom you're in regular contact and never see the product you're trading in, the conditions under which it has been grown or mined, the way it has been transported, or where it will go. Often, the food you trade in has not yet been grown, as this is irrelevant and unnecessary for your firm's production process or for the trade. You give the instructions and make or receive the payments by post or e-mail. It's just pulses along a line—occasionally still via several satellites circling Earth. Other people you don't know will look at the results, and you will receive a profit you can put into your account at the local bank. If you want to, you can then withdraw some cash and use it to buy your bread. At last, something concrete, something you can touch and eat!

Similar to what had happened in England, in the United States in the nineteenth century, the land was tilled by paid workers rather than by farmers as part of the general trend toward depersonalization and abstraction. The aim was to maximize the short-term profit. To do so, large investments in new technologies could be made, often facilitated by borrowing large amounts of money—loans that were impossible for local farmers. The industrial farmers could produce larger amounts and sell at lower prices than the small farmers. This gave them an advantage so that it became easier for them to increase the area they farmed by buying up more land. In this way, a new type of farming originated—the entrepreneurial type—directed not so much to local consumption, but to making profit on the domestic or international market. This economic organization became the core of the modern agribusiness.

As such, however, the enterprise itself remained an agricultural one. Yet, running an agricultural enterprise at long range, with unknown, hired laborers on unknown land often meant that the large areas were worked with less knowledge and care so that, over time, their reserves of essential minerals became exhausted and the yields gradually declined or the soil became saline or eroded. Moreover, it has been shown that on farms owned by entrepreneurs, the efficiency in terms of nitrogen use, irrigation, and energy is lower and the cattle do not live as long. Eventually, and as soon as the profit becomes too low, the entrepreneur will sell his land, putting his money into other, more profitable land somewhere else or in companies or industries. Overall, therefore, the greater the drive to make profit, the more the amount and quality of food produced decreases, whereas the amounts of material and energy needed, as well as the amount of waste made all increase. This pattern worsened in the next stage of development in food production and supply.

A logical consequence of this development toward agricultural entrepreneurship is that at some point, the network of these kinds of entrepreneurs merges with other networks in society, be it only in order to spread the risk of investment. All networks like these have in common that their aim is to make maximum profit. Fusing networks to create ever bigger ones could only improve on this. Indeed, it has often happened that an agricultural network concerned with making profit from the agricultural part of food production merges with networks that are above and below it in the food production and supply chain. For example, it merges with the present seed-producing industries, and with the part of the petrochemical industry involved with the production of agrochemicals, fertilizers, herbicides, and pesticides, and so forth, and with food transportation, food preservation, and the fast food industry, or with industries and chains for the packaging and selling of food, and finally with banking.

Understandably, merging on such a large scale is only possible if there are elaborate and fast electronic communication systems for coordinating all the necessary activities across the world, including the regular supply to individual stores or restaurants. These communication systems also enable rapid instructions, agreements, and payments to be made instantly in various currencies worldwide. This is how the large, complex, often multinational retail corporations like Ahold, Aldi, Woolworth, or Wal-Mart or food and drink producing corporations like Unilever or Nestlé have arisen. You may only know of them as your local large food retailer, or from the small letters on the package, but their existence has major ramifications throughout society and on a global scale. Your food store is only the tip of the iceberg. Production and supply of food is now part of the big, international business enterprise. About forty years ago, this development became alarming in Switzerland, when Migros, a very large, popular, financially powerful food corporation with socially responsible interests throughout society, proposed to form its own political party. This was prevented by the Swiss federation, because it would theoretically have been possible for the state as such to be taken over and ruled by a private enterprise. But the tens of thousands of lobbyers circling the governmental institutions of the European Union or the United States still form a force to be reckoned with.

Of course, all the varied activities of one single corporation are often not

built up from scratch; instead, the central organization incorporates existing, specialized companies one after another through mergers and takeovers.
And these large corporations need not have developed from the farming
community itself and penetrated deeper into the food industry; their roots
may be in the food handling and trading industries themselves or in the
petrochemical industry, and from there they penetrate the agricultural and
other food-producing sectors of society. Thus, several independent chains
of production develop together into conglomerate networks with branches
throughout society. As large, economically and occasionally politically powerful conglomerates, they can also have ties with national governments, col-
laborating with them in various ways. For example, local governments may
grant them special permission and facilities or adapt the rules or laws to
meet their wishes and needs—perhaps with respect to water supply or drainage, or to the felling of trees or the clearing of forests, for example—which
other farmers or conglomerates do not get. Or they can obtain prime sites
for their stores in or around towns.

What happens is that these conglomerates organize the operation and
output of existing companies in various sectors of society and connect them
with banks and the government so that their profit as the superstructure
over the entire organization is maximal and most stable. Thus, most of the
profit made goes to this organizational superstructure rather than to the actual companies they connect. Because they are large, multinational corporations, these conglomerates can connect countries with the lowest production
costs to those with the highest prices for selling their products—or with the
most favorable legislation. But, unlike the processing factories in nineteenth-
century England, they often do not add anything to the processing itself, or
to its improvement. They merely organize what is available and make a lot
of money doing this. Typically, in such an organization the units themselves
remain more or less independent and can be sold off as soon as they do not
generate enough profit, or when a similar company would add more to the
profit, or when another corporation offers a good price.

What has happened—and what is still happening ever faster and at larger
global scales—is that on the one hand there is the idea of profit maximization, and on the other there is the idea of limiting competition with other
corporations by eliminating or merging with them, both elements driving
the process of economic growth. Thus, retailers will pay the food-processing
industries as little as possible to gain maximum profit themselves, but at the
same time, because of competition at their lower level within the corporation, they have to sell at the lowest price.

For their part, the food industries have to increase their output as much as possible for the same reason and therefore will economize on quality and maximize the number of product lines. And the same holds for the retailers: they too have to minimize cost and maximize output, paying out as little as possible to their personnel and cutting out their various types of insurance and other social conditions, and they too will maximize the number of product lines in their stores. These stores, in turn, aim to have the most floor space that is economically viable. At the same time, they all have to pay their taxes, loans, rent and personnel, apart from the other costs of running their business. Yet, being multinational, they can avoid national taxes and regulations, thereby minimizing their costs. The overall effect is that they all run on minimal profit margins and exhaust the soil and their personnel's social conditions, while at the same time they overproduce low-quality products at ever-lower prices. In order to maintain their market position, they outsource to regions with lower prices and wages, and they grow and diversify to gain enough profit. They often flood the market with superfluous "new" products that have a short lifetime. The result is that an enormous amount of waste is generated and customers become overweight. The large-scale occurrence of obesity has dire medical and demographic consequences. The economic result of this chain of events is oversaturation of an ever-growing market: first the domestic market and then the international one. As already mentioned, other results are soil and water exhaustion, waste production and pollution, and a rapid erosion of social conditions and employment, and therefore an increasing social instability, both nationally and internationally.

Another consequence, apart from that of the erosion of social and health regulations, is that since the 1980s, research has been cut from the expenditure of large companies and has shifted to the universities. This is happening even in the pharmaceutical industry. This means that in the universities, long-term, fundamental, and blue-sky research has been swapped for short-term applied research, which will slow down processes of innovation. As a society, we are therefore less well prepared to face the big challenges the future will hold.

With the emergence of new markets in large, enormously populous countries like China and India, the whole process that the developed world (the United States, and to some extent Europe and Australia) has gone through is now being repeated among nations on a global scale. And when it comes to competition in bidding for surpluses, if China and India grow richer, they might outbid the United States and Europe. In this scenario, the poor developing countries (among them ones with the highest population growth), which are already, even in the face of famine, renting or selling their agricultural land to richer countries, might be squeezed out. This leads to instability at an international scale, and increases the risk of war.

Perhaps the next phase after various parts of society have consolidated into very large multinational units operating in different continents is that of state capitalism, in which production and commerce are based not on private capital but on funding by entire states. Apart from a number of smaller countries, such as Dubai or Singapore, very large, financially exceedingly powerful ones, like China, Russia, and India have entered the scene as well. They can buy up big companies and banking corporations all over the world, thus getting extra profit from their earlier earnings. The rush is also on with regard to metal ores: China has contracted or bought mines in countries such as South Africa, Congo, and Australia. At present, China, India, and Saudi Arabia, too, are renting or even buying up valuable agricultural land abroad to obtain food, in effect putting the existing shortages in the African production of food under an even greater strain. Meanwhile, as has happened in South Australia, wheat production in China is at risk of being reduced by some 40 percent because of extreme drought and groundwater lowering. South Korea, another Asian country active in Africa, has already bought two-thirds of the crops grown commercially in Madagascar. (It is not to be hoped that history will repeat itself when in Victorian times fifty million Indians died due to severe El Niño droughts, while enormous amounts of food were exported to England, necessitating the use of the railway lines being built by the starving people for this reason.) And all this is taking place at a time when in North America, Europe, and Brazil, large areas—and large amounts of water and fertilizers—have been reserved for the production of biofuel rather than food, putting global food production at risk and increasing environmental strain. All this is happening in the face of the need to increase food production to meet population growth and a projected 70 percent increase in the per capita demand for food by 2050. It is likely that in order to compensate for this loss of production, more of the remaining natural resources and the environment will be swept up in this apparently unstoppable process.

Growth and decline are the inevitable consequences of the system we have built, beginning with the Agro-Industrial Revolution as an answer to stresses ultimately originating in earlier centuries, or ultimately even the earliest millennia of humanity. It is a system based on meeting the ever-growing demands of ever-growing numbers of humans, at present whipped up by the relentless pressure of competition. It is a system that will ultimately exhaust itself, along with its resources: a system that will degrade the soil that ultimately supplies us with food. This will inevitably lead to poverty and hunger among hundreds of millions. Surely, we must be able to find other ways of resolving the problems arising from the growth of our numbers and demands. And again, can still we turn the clock back, and, if so, how? How can we alleviate the problems brought about by the relentless growth of our numbers?

A parallel can be drawn with the way a cell works. To continue operating, the cell depends on some form of disequilibrium between the contents of two separate compartments, the inside and outside of the cell, from which it obtains its energy. A similar disequilibrium generates financial profit within and among nations of our global community: it is essential to profit-making. Profit and growth are maximal as long as this disequilibrium is maximal. Our present society runs on processes that are exhausting the soil and natural resources and on marginal profit in the production, processing, and retail chain on the one hand, and the maximization of consumption on the other. But the disequilibrium decreases as soon as the soil degrades too much and the resources are depleted, the price of energy and minerals increase, the processing system itself becomes less efficient, and the environment degrades and becomes polluted—trends that all affect the resources of our food.

Similarly, corporations exist because of the existence of differences in payment and standards of living; finding and maintaining these differences is essential for making profit, and it is disadvantageous to these economic systems to smooth out these differences. Even though the overall standard of living increased between 1961 and 1997, during that period the disparity in per capita income between the richest and the poorest nations increased from about twelvefold to thirtyfold. Globalization does not seem to be smoothing out differences—it is intensifying them. Instead of a leveling off, there is an increasingly rugged global economic landscape with deep clefts. The world isn't interested in equality; it runs on maximal inequality, rapid resource exhaustion, and the cheapest way of dumping waste.

The situation is being exacerbated by formerly Communist countries such as China and Russia; though they had a system of internal power distribution, they are now using this system to internally plan new private companies and at the same time are using the total system for competitive external growth. Both industry and banking are organized in such a way that the state as such has become an international competitor among other states with a less developed internal system. This also has happened in non-Communist countries, such as Singapore. Thus, in principle, such states are rich enough to easily take over industries, big banks, or entire corporations abroad; all that prevents them from doing so is the protective legislation put in place by these vulnerable states. Yet, this kind of organization is allowing unprecedented economic expansion to take place; in China, annual growth rates have consistently exceeded 10 percent for decades. The national income, in turn, allows China to invest heavily in making this system more efficient and organize its infrastructure with respect to housing, living, education, health and working conditions, and transportation and industrial planning. China is well underway to achieving the next step—having its currency replace the U.S. dollar as an international currency of payment. This will further enhance China's development and weaken the West. The state itself has become a corporation, larger and excessively more powerful than any other corporation. The West itself is deindustrializing on a large scale by outsourcing large sections of its economic activity. The disequilibrium maintains; it only shifts its global position. What it earns from its abstract organization is at risk of collapsing when conditions change.

The consequences for aspects central to this book are evident: the building up of such enormous state corporations requires an unheard-of increase in resource use. Excessive amounts of minerals and energy are needed to build new towns for tens of millions of people, as well as houses, industries, thousands of kilometers of roads and flyovers, all other means of personal and freight transport, and for the generation of energy, as well as for the industrial products with which to flood the foreign markets. They need money to pay for their expansion, as well as for the accumulation of their new wealth. And more wealth stimulates personal demands even more, which increases the pressure on resources. Yet it has been estimated that the present generation in the developed world has already used up roughly half of the energy resources on the planet, so that only the other half remains to enable this additional economic expansion and support all the generations that follow us. The established economies, however, will not be prepared to relinquish their own high—and growing—demands, which means that we will easily use up this remaining half within the next one or two generations. But this trend is impossible to stop because of the huge economic power exerted by new, rapidly expanding state corporations like that of China and, to a lesser extent, Russia and Brazil. These emerging economic superpowers can currently take their pick of the world's mines, mineral resources, banks, industries, and food-producing areas. And note that "half of the world's energy resources" refers to all the resources together and takes no account of their relative abundance or rarity. We may already have used up more than half of those resources that are rare: rare metals, fresh water, or particular types of forest, for instance.

Yet the minerals and energy used today are tomorrow's waste and pollution. The use of fossil fuels to supply the energy for enabling this social development and expansion will unavoidably result in a further warming of Earth's climate. The environment will be exhausted, ruined, and spoiled in no time. Mines and aquifers will be depleted, agricultural land will be impoverished, and the natural and agricultural biodiversity will be eroded. Similar to the way national and international corporations of the developed world operate, the impersonal recklessness of massive societal organizations comprising and serving hundreds of millions—if not billions—of people will hasten the process.

This new economic organization of the world is the result of the development so far, the intensifying competition between entire continents for the remaining half of our resources on Earth. It is the outcome so far of the self-enhancing growth in our numbers and demands, which began in Europe during the late eighteenth century. How on earth can we still turn this tide, especially since growth is the basis of mankind's survival? When will we be willing to do this?

The race seems to be on between corporations or states with the largest amount of money to reach their goal, largely independent or irrespective of local populations and knowledge of the local conditions—unfortunately to the detriment of the long-term existence of humankind.

This is the state into which our human society has developed so far from its initial stages some ten or twenty thousand years or longer ago: a society based on abstractions of large numbers of people interacting electronically without knowing one another, alienating themselves from their environment and thrown together by the millions into vast, treeless megacities, piled up together in huge buildings of concrete, and surrounded by slums. Cities and countries draw much of their resources from vast areas surrounding them, to which they also return their polluting waste. Ours is a society based on trust—mutual trust, trust in the financial system, trust in the organization, trust in the law and the medical system, trust in the government, in the international governmental organizations, and the global market, trust in the future of mineral resources, the condition of food, and the environment. We trust—we don't think and question. People don't know each other anymore, don't know the system they form; individuals have hardly any grip on the developments themselves. We are ruled by the rules, driven and kept quiet by trust—concrete people ruled by abstractions, abstractions that live their own lives, that live our lives.

The socioeconomic and political margins to play with are narrow; we are driven forward by an unstoppable urge for growth. Ever further, ever faster. Growth because of growth, growth because of greater local efficiency and, more recently, also of competition, despite its effects. Growth itself following its own abstract rules. We are caught between the population growth, lifestyle demands, and the recklessness depleting our resources on the one hand, and the growth of our rules, institutions, organizations, and pollution on the other. Can we still turn back on this road when we see the shortages of almost everything growing, when we realize that polluting and wasting of our environment, air, soil, groundwater, and even the vast waters of the oceans is inevitable, unavoidable, inherent to using the resources we so badly need? Can we still go back when components of our abstract system will collapse, evaporate into nothing, into a vague memory at best? Do we know how to go back? Or are we too focused on growth to even ask how to reduce the rate of growth? Which parts of societal organization should be the first to go? Which can we miss with the least harm? Or are all the parts too tightly interconnected—either because of direct interdependence, or indirectly via our numbers—to allow us to make the cut? On what terms, and precisely how could we change our numbers? Back to what level? A level also dependent on the length of the future into which we would like humankind to prolong its existence on Earth? For how long yet? Do we have plans, ideas, expectations, or hopes?

The growing problem of humankind is that our speculations have gone too far: we speculated that Earth would always have enough resources to sustain an ever increasing number of humans; if one resource begins to run out, we speculate that we will certainly find another one in time. And we speculate that Earth can always absorb more of the waste we make. We can pollute the environment with whatever we want to expel, and nature will deal with it. We can reduce the carbon dioxide in the air, we can stop temperatures from rising. But what if Earth cannot meet all our speculations any longer? So far, we have actually been fighting the results of the growth processes by speculating that we can reduce our demands by reducing the amount of carbon dioxide being expelled into the atmosphere or by introducing recycling schemes. But for how long will these measures be effective while the ultimate driving forces, the unbridled growth of our already too-great numbers and demands are left untouched?

We are caught in social, economic, political, moral, and religious nets: networks of growth, depletion, and pollution; networks of organization and planning; networks of institutions and organizations large and small, local and global; networks of abstraction and alienation; networks of speculation and hope. We put our trust in fighting the symptoms rather than fighting the root cause, our too large numbers and growing demands.

Past Choices, Future Decisions

Once, thousands of years ago, we chose (though not consciously or deliberately) a path to escape from the problems of too many people. But the path proved worse than circular; it spiraled downward bringing us into ever greater difficulties, and we now need to solve this as quickly as possible. During the 1970s, we were warned explicitly at a time when we still had the chance to find reasonably easy solutions, but since then the world population has almost tripled, our demands have increased and our supporting infrastructures have grown. This means that the size of the problems now facing us has more than tripled since then. We need to act quickly, not only because of the ever increasing size of our problem, but also because, after having lost so much valuable time, we are getting ever closer to the point of resource exhaustion and of smothering in our own waste.

B

Exhausting and Wasting Our Resources

All people and their organizations require resources to keep going. Energy is a pivotal resource: no organization can run without energy, not even for a short time. However, the sources of our energy are limited and threaten to run out soon. They have done so for as long as mankind existed on Earth, but every time, people were able to change from one source to another, even to a more abundant and more energy-rich one. Unless we decide to change from our last source, fossil fuels, to nuclear energy soon, people are in serious trouble, not only because of running out of these fuels, but also because of their eventual effect, climate warming. Recycling of most mineral resources can be done, be it to a limited extent only, whereas recycling of energy is impossible in principle.

6   PEAK OIL AND BEYOND

The 1960s and 1970s were a fascinating period in recent history and are still widely remembered. The student revolts that occurred all over the world during and after 1968 were just the tip of the iceberg. What was happening underneath was a large-scale social reform of society on an almost global scale in which hierarchically imposed relationships between people were being replaced by individual responsibility. The hierarchical relationships had become rigid, so that each individual had his or her fixed role in society. The roles concerned the behavior of members of the various social strata, members of different religious or political groups, people of different ages, and men and women. It was time for a less hierarchical and less strict, more equal and dynamic set-up of society, through and across all social layers and structures, families, education systems, and religious, governmental, and industrial organizations. Another socially deeper movement was women's liberation, which changed family structure radically, as well as the working conditions, responsibilities and careers (equal pay!) of both men and women. Moreover, the responsibilities concerning the upbringing of children were revised.

Everything changed. There were significant changes in cultural, scientific, and technological initiatives—including new approaches to performing music from different periods. Think of The Beatles and all that followed. Social and cultural changes affected science as well, both in universities and industrial research programs, as well as science's outgrowth into a pivotal societal factor. Yet, freethinking, more principled people also objected to how science was being handled in government and industrial policies. They protested against the atom bomb, for example, and against the unmitigated use of chemical pesticides and herbicides. At that time, developments within science itself changed the way we view the world, and many established and cherished theories were toppled. For example, by widening our horizon to the whole world and the universe beyond, the space program changed our appreciation of life and its origin on Earth. The first photos of Earth and the moon as seen from space date from that time. Geology began to look at the world as a whole, describing its origin, structure and dynamics through the eons of time, rather than as local or regional features. Climatology and human, economic, and physical geography were redefined; in mathematics, statistics and spatial analysis were founded and applied in a range of disciplines from business logistics to ecology. Increasingly powerful computers were developed, operating at increasingly higher speeds and with greater efficiency, thus allowing large numbers of complex calculations to be made within amazingly short times. Concomitantly, new mathematical tools—if not entirely new subdisciplines—were developed to support the construction, along with (re)constructions of social developments through simulation techniques. This, in turn, was stimulated by space exploration, which had great military, industrial, economic, and political interests during the Cold War years. Finally, computer applications allowed images of global economic developments to be drawn, representing the condition the world might be in a few decades ahead.

The idea of all people living together in a global village was born out of this broadened perspective, which underlies more than just the process of economic globalization. Other ideas spawned by the global village idea concerned ongoing global processes, such as our use of resources and their finite supply on Earth. All this was made possible by the greatly improved means of electronic communication and computation. Developments in the petrochemical industry were revolutionary, changing and improving all aspects of the way we live. This happened, of course, with the unavoidable price tag of new kinds of nondegradable waste and greater killing efficiency: one aspect was the so-called war on insects. Similar processes concerned those of climate change, or of global soil and water pollution, mineral depletion, and so on.

The founding in 1968 of the Club of Rome, an international group of people concerned about the finiteness of Earth's resources, was in line with these revolutionary developments. This group included leading industrialists, academics, members of the general public, and politicians. There was a small organizing committee and a large number of people from all over the world who collected data or modeled future scenarios on computers. Its founding fathers were Aurelio Peccei, an Italian industrialist and general director of the Fiat car factory in Turin, and Alexander King, a Scottish chemist and top scientific advisor to the British government. Computer forecasts in 1972 by a group of system analysts led by Dennis Meadows gave a picture of the constraints of several of our main resources and of how their use and global limitations are interrelated. The future for humanity was bleak: one by one, our basic resources would be exhausted within the foreseeable future—that is, within the next few centuries, if not the next few decades. This forecast sent shockwaves throughout the world. In Western countries, reproductive rates dipped, and individuals restricted their resource use. Yet, the global human population of just over two billion kept growing and consuming Earth's resources at ever-higher rates.

At the same time, there were forces working counter to the recommendations of the Club of Rome. To promote their industrial and economic interests, governments stimulated resource use via their policies. In the mighty United States, some economists maintained that the population should grow, not slow down, arguing that a larger population with their growing demands would stimulate the generation and use of energy for obtaining other resources for industrial production, personal use, or for private transport. Industrial production was geared to making cars; using oil and petrochemical products; building roads, houses, and harbors; constructing pipelines and high tension lines; and expanding the scientific, communication, and military systems. World dynamics intensified and expanded.

This burgeoning of resource use, particularly of our energy sources, was especially ironic, as it went against the Club of Rome's recommendations. Moreover, during the late 1950s and throughout the 1960s, M. King Hubbert, an American oil geologist, analyzed the overall oil production in various countries. Not only was this resource finite, but it appeared that the reserves in the United States would soon run out, and so would the gas reserves a few years after the oil. Hubbert pointed out that after a new field has been taken into production, oil is initially produced at maximum rates. Often, the oil wells up with great force by itself because of the high pressures in the deep strata where it is found. Later, the remaining oil has to be pumped out to the point that the pumping itself requires more energy than the oil would supply. The curve of the amount of energy obtained divided by that of the energy expended, expressing the rate of depletion, is a hollow, downward one: a negative exponential curve. At first, depletion rates are high, but then they slow down and diminish, until complete depletion is reached after a long, drawn-out period of very little gain. In 1940, for example, the amount of energy extracted from a typical oil well was a hundred times that expended, whereas by 1970, this amount had declined to a mere twenty-three times. The same trend that occurs in a single field also occurs in an oil-producing country, and, indeed, all over the world. After an initial steep rise to a peak, oil production declines rapidly. Eventually, all oil-producing countries will give up producing altogether. Economic production costs are therefore expressed in terms of relative amounts of energy to be expended: the tipping point is that at which more energy is needed to pump up the crude oil than that oil later generates. From then on, the economic energy balance becomes negative. In 2008, it was estimated that the decrease in production of oil fields that have passed their peak would be 9 percent annually.

In principle, when net energy gain is small, the mining of crude oil could still be beneficial in financial terms, though not in terms of energy. Over time, however, the large, readily discoverable fields are exhausted, whereas the remaining ones are smaller or deeper or both, and therefore are more difficult to find. The higher detection costs mean less profit. Also, the production time will be shorter and the pumping costs higher; this again reduces the profitability. Mining then rapidly becomes financially unviable as well. However, the national depletion rate can be more complex than the international rate. This was the case in the United States, where domestic production costs overtook the production costs abroad, because the U.S. dollar was high on the international currency exchange markets and therefore salaries and equipment were more expensive. So, for financial reasons, the United States turned to other countries for cheaper oil, first to the Middle East, which has large reserves. Meanwhile, other countries that had been dependent on the rich American oil fields also looked to the Middle East to supply them with oil; the result has been half a century of increasing international strife. Though large, the Middle East reserves will take us only decades—not centuries—further.

Whether mining is beneficial depends not only on the energy to be spent to pump it out relative to that gained but also depends on what other purposes the oil or gas has. Part of the crude or gas is used for producing fertilizers and herbicides, for producing pharmaceuticals or for all the plastic we need at present, and so forth. Then, mining can remain beneficial for a longer time although it would cost energy.

During the early 1970s, when the Club of Rome was most active, Hubbert's predictions came about for the United States but were ignored there. Now his predictions are still being ignored on a global scale, as are the recommendations of the Club of Rome. However, most oil gurus agree that global oil production will peak soon, if it hasn't already. And, indeed, in recent years, the price of oil on the world market has more than trebled, reflecting the commodity's increasing scarcity. There are also indications that production is declining by several percent annually; this has triggered a desperate race for the last large fields and has reawakened interest in gas and coal as sources of energy. Ironically, gas was initially burned—flared off—in the oil fields as a waste product in meters-high, roaring flames. As our fossil fuel runs out in the next twenty to thirty years, we will be rapidly losing our main source of energy—energy we relied on totally for about a century, particularly in the last fifty years, the time of massive growth in our numbers, demands, and industries.

The effects of global energy depletion are already being felt. Mexico, once one of the world's main oil-producing countries, where oil production has been state owned since the mid-1930s, can no longer cope with the increasing costs and lower profits. Meanwhile, places that still have appreciable reserves, like Russia and western Australia as well as Caucasia, the Middle East, and Venezuela, are rising in economic and political importance. Australia is opening new wells on its northwestern continental shelf. Meanwhile, in the Northern Hemisphere, the melting of arctic ice is bringing the oil- and gas-rich continental shelves of Siberia within reach. Abandoned wells are being exploited in the Niger Delta of Nigeria, and offshore oil exploration is in progress off West Africa, Alaska, Brazil, and the southern rim of Asia. Yet, still our energy use is insatiable and increasing rapidly; this will continue as long as our numbers, demands, and social infrastructures continue to grow. It is estimated that the output from the large oil reserves in a pristine natural area in Alaska will meet only 5 percent of the United States annual demand for a period of only eighteen months. Similarly, a recently discovered, very large and rich field in the Gulf of Mexico will hardly change the picture of global supply and demand, if at all. The reserves of shale gas, a new source of natural gas found in many countries across the world, are estimated to be sufficient to supply gas for only a maximum of the next two hundred years. The North American reserves would be able to meet the demands for only forty-five years, just over half the lifetime of the average American.

Given the declining reserves and increasing development costs, you might expect we would become more economical with mining our oil and energy, but the reverse is true. The Caucasian countries, for example, hoping to get rich quick, are offering their oil at minimum competitive rates. Our economic system, based on profit rather than performance, is wasteful by definition. As the car producer Henry Ford once said, "Mini cars give mini profits," implying that for him, it was more profitable to produce larger, less efficient, more wasteful cars with shorter lifetimes. The same applies to most products we use. This idea, perhaps defensible in a clean world of plenty, underlies much of what we are doing in our present-day world, despite facing resource depletion and a stifling production of waste.

Just think of the highly disturbing effects the loss of our sources of fossil energy can have on all aspects of our immediate daily life. At present, we need large, rapidly increasing amounts of energy to heat and cool our houses, to illuminate the dark, and to power our appliances. We have come to rely on electricity to power our shavers, toothbrushes, coffee makers, electric heaters, microwaves, mixers, electric carving knives, fridges and freezers, dishwashers, washing machines and dryers, fans, cookers, ovens, vacuum cleaners, digital cameras, computers, projectors, televisions, CD and DVD players, cell phones, GPSs, iPods, electric blankets, hearing aids . . . the list goes on and on. In your shed you probably have an electric drill, circular saw, hedge-clipper, and lawn mower. Outside, you use automatic doors at store entrances, an electric golf cart or perhaps even an electric wheelchair.

Your house is full of petrochemical products manufactured using much energy and material from fossil fuels: plastic toys, shopping bags, plastic bottles, fast food packaging, pharmaceuticals, and an enormous variety of detergents and soaps, kitchen implements and utensils, door handles, lamp shades, brushes, paint, mattresses, clothes, hangers, toothpaste, shoes, windowsills, carpets . . . and what about the chemicals you use to kill insects, weeds, molds, and bacteria? And the additives and preservatives in your food? Recently, babies in China developed breasts because of some additive in their formula. Realize that all the vast areas of wheat and maize depend on large inputs of fertilizer, as the soil minerals are already largely exhausted. In a way, our food plants are revitalized fossil remains. In fact, we ourselves, as consumers of basically plant food, are largely revitalized fossil fuel.

As soon as we run out of fossil resources, we will run out of petrochemicals as the material basis of our society. Don't forget that we need staggering amounts of energy to build our houses, offices, and roads—and for the bricks, tiles, carpet, and window panes, and for the electricity, telephone network, sewage system, and shops. To move, we use cars, buses, trains, ships, airplanes, elevators, and escalators. For industrial production, we need melting ovens, rollers, and heavy machinery; and for transportation, we need cranes, trucks, tractors, spraying machines, and pumps. Their construction from iron and other metals requires astounding amounts of energy. Moreover, we need freshwater for the production of our clothes from polyester and nylon; even the organic production and dyeing of cotton or wool requires huge volumes of water that may have to be pumped up from deep aquifers, filtered, or desalinated and pumped under high pressure through long-distance pipelines.

Our societal superstructure, whether local, national, or global, adds extra costs in terms of the material and energy we use—costs over and above the costs of what we personally need, but that allow each of us to live a personal life. Only a small percentage of the population lives in low-density areas, as farmers on the land, for example, directly producing our food. But although they live on and from their land, farmers need a fridge and a car to go to town (to buy the food they don't produce themselves or to get their pharmaceuticals or spare parts for their machines or to go to the bank). We need a socioeconomic infrastructure to communicate for weather forecasting, orientation, and navigation. We need communication satellites brought into orbit by rockets and a dense network of fiberglass cables spanning the world. In our developed society, nobody lives independently anymore; we all depend on this superstructure of societal organization. Without it, there would be no energy, no water, no food, no trade, no transport, no hospitals, no police. All would immediately come to a standstill, and we would not even know what happens or what to do without radio or television. Chaos, crime, and famine would prevail. Without the energy from fossil fuels, we are completely helpless, though, as times of war have shown, we may be able to make do with less. How do the hundreds of millions of people along the mud streets in shantytowns—roughly a third of us—live? How can they manage without electricity?

We take for granted that our energy is there: for the evening meal, for example, we microwave some frozen fish, meat, peas, or a complete meal, and we wash it down with a canned drink from the fridge. What will we do when energy prices rise? What can we do without? Privatized electricity companies won't want sales to fall as a consequence of higher rates or taxation, because it would affect their profit. Should you buy a smaller car? Or move closer to your work? Climb all thirteen floors or higher to your apartment? Use public transportation to get into town? A rise in energy prices resulted in car industries in the United States losing money by the billions, air companies reducing the number of flights and merging with other airlines, and airports closing by the hundreds. What about mushrooming societies like India, or especially China, which are building up incredibly large new infrastructures and industries—thousands of kilometers of roads, flyovers, and bridges, thousands of hydroelectric dams, hundreds of thousands of factories? Mimicking the West, they have all chosen to go down the energy-intensive road.

Understandably, the looming shortages in oil-based energy have fueled interest in alternative energy sources like natural gas or the old nineteenth century source, coal. Large amounts of coal can still be mined—some five thousand gigatonnes—enough for the next couple of centuries, though then it is all over. But project this number against the present two gigatonnes used and released as waste each year: two billion tonnes annually, which over the last century have already had such a large effect on our climate. Other disadvantages are that the amount of energy per unit is low and the costs of mountain topping, mining, processing, and transport are large. In the end, only 3 percent of the energy coal contains enters your house as electricity. Other mines have to be dug and pumped dry, which may require new technology, given the scale at which this must be done. New industries must be set up to convert the coal into oil, gas, or electricity for easy transport, and networks of pipelines or high-tension lines must be laid out from the mining areas to the cities and industries. Grinding the coal first and then pumping it, mixed with water, as a slurry through an extensive network of pipelines is another option, although this again requires large amounts of freshwater and chemicals, as well as energy for pumping and material to make and maintain the pipes.

Thus, the returns on investment will diminish and the suppliers of capital will turn to more profitable investment, thereby leaving part of the reserves unused. As part of this process, some of the remaining coal will prove to be of poor quality, as is already the case with the western coal relative to that from Virginia, or it has to be mined at great costs from greater and greater depths, possibly up to a thousand meters deep, as in China. Taking everything together, this means that Hubbert's curve, developed for describing the gradual depletion of oil, applies to coal as well. For example, it applies nicely to the production and eventual exhaustion of Pennsylvania coal, having peaked already in the early 1900s, whereas the overall energy production from coal in America peaked in the late 1990s. And, similar to other energy sources, demands for coal are increasing at high rates. Over the next decade, for example, urbanization in China alone requires at least seven hundred million tonnes of coal for building material alone, such as cement, steel, copper, and aluminum. With rapidly increasing demands for coal, it is no wonder that the reserves thought to be an abundant energy resource for the next centuries, may suffice for the next few decades only.

But, of course, all that coal is eventually transformed into gaseous waste—

more gaseous waste as there is coal, waste mainly in the form of carbon dioxide. For example, by burning a hundred kilograms of gasoline in your car, you add three hundred kilograms of carbon dioxide to the atmosphere. If it is emitted into the atmosphere it will increase the temperature (the greenhouse effect) and react with water vapor to form carbonic acid. The resulting acidic rain will greatly acidify the environment, killing off much of the aquatic and soil life. Acidic rainfall can also enhance erosion by reacting with certain rock chemicals, particularly limestone.

And there are other waste products associated with burning coal: sulfides and nitrogen oxides. Both add to the greenhouse effect, as well as to the acidification of the environment. Moreover, when sulfides enter the atmosphere, they attract water vapor, forming droplets, which, when small, scatter the incoming rays of sunlight, as thin, white clouds. Being white, these clouds counteract the greenhouse effect and cool Earth. However, when large dark clouds form, they absorb the sun's rays, and that keeps us warm. A warmer atmosphere will be able to contain more water vapor, yet another greenhouse gas—one actually accounting for 60 percent of the warming. The higher the temperature of the atmosphere, the more vapor it can contain and the warmer it gets. Thus, every degree of warming due to carbon dioxide is more than doubled to an increase of 2.2°C by water vapor. One solution would be to bury the carbon dioxide and other chemicals deep underground, but so far, the technology of burying carbon dioxide has not been developed successfully. And if carbon dioxide leaks to the surface in high concentrations, it will tend to form a layer there, because it is heavier than oxygen, with the risk of suffocating animals and humans. Moreover, all the energy spent that way reduces the amount gained by using coal. Another way to lower Earth's atmospheric temperature is by injecting burned hydrogen sulfide, sulfur dioxide, into the stratosphere to generate small, light-reflecting droplets. However, this sulfur dioxide would speed up the breakdown of the ozone layer there by chlorofluorocarbons (CFCs), which protects all forms of life against the breakdown of their DNA.

Mountain topping and mining itself generates an enormous amount of waste, as well. And processing coal leaves large amounts of slag as a waste product, which is usually piled into tips, many of which are still features in the landscape. Rainwater percolating through this waste material becomes very acidic and pollutes the groundwater underneath. As we will need vast amounts of coal to meet our energy requirements, the mountains of primary waste will also be enormous—the rubble covering large areas and wasting much usable soil and natural areas, along with the groundwater underneath and in surrounding areas.

To exploit another source of carbon-based fuel, the tar sands in Alberta, Canada, vast areas are being open-cast mined—stripped from their vegetation and soil. Mining, heating, and cleaning of the tar sands is very energy costly and requires freshwater from extensive areas, which is then turned into heavily polluted waste water, pumped into large, open-air, poisonous basins. Rivers there now contain the stable polycyclic aromatic hydrocarbons (PAHs), many of which are carcinogenic; when taken up by aquatic animals, the PAHs are metabolized within the body and form toxic compounds. Thus, the costs in terms of the use of energy and groundwater, and of the pollution and destruction of the environment, are vast.

In principle, methane, an abundant and cheap natural gas, could be a new, abundantly available carbon-based energy source. (In fact, natural gas is fossil methane, also known as mine gas from coal mines, or, when it is recent, marsh gas.) Methane is a general waste product of the decomposition of plant material by certain bacteria in the absence of oxygen. As such, it is released in considerable quantities into the atmosphere by cattle and from grassland, rice fields, dumping sites and landfills, inland lakes, marshes, and the vast subarctic melting permafrost and peatlands. Ninety percent of the soil in Alaska, 50 percent in Canada, and 60 percent in Siberia consist of permafrost, frozen layers of organic matter up to fifty meters deep. From there, humus drains into creeks and rivers and eventually into the oceans, and so is found in a fringe around all continents, where it decomposes, forming methane. There and also in some of the deeper parts of the oceans, it occurs in the form of methane ice, a crystalline combination of methane and water, which is stable at moderately low temperatures and a certain pressure. Therefore, as soon as the water temperature rises, the methane ice melts and turns into methane gas, which escapes from the water into the atmosphere. The ocean floor is pockmarked by the continual submarine explosions that occur when floes of methane ice detach from the sediment and melt under reduced pressures. Its instability, however, means that its exploitation is difficult; a large volume of gas rising to the surface would be a serious hazard to shipping. And at the surface, the methane can easily ignite, producing sheets of flames on the sea. This difficulty of exploiting it is greatly enhanced by the broad, uncontained dispersion of methane ice as a vast underwater stratum surrounding the continents, rather than in high, local concentrations from where it can be mined in quantity. Moreover, being twenty times as effective as carbon dioxide in withholding infrared radiation, the heat radiation causing climate warming, it could greatly add to the greenhouse effect of the atmosphere if large amounts accidentally escaped into the air while it was being mined. Finally, once in the atmosphere, it reacts with the oxygen and begins to burn. This reaction results in the formation of carbon dioxide and water, both of which, once again, are greenhouse gases, albeit that carbon dioxide is more benign than methane itself. Despite its abundance, the difficulty in mining recent methane from the oceans or permafrost, given current technology, precludes it as an alternative energy source for the time being.

As mentioned, yet another newly discovered, widely occurring source of gas is shale gas, which is released by pumping water at high pressure into the deep underground strata in which it occurs. Some of these reserves, like those in China, are expected to last for up to 250 years. This is on top of the time the currently exploited reserves are predicted to last. Not only does it cost much energy to press gas out of shale, but it also requires the addition of chemicals, such as benzene and xylene, to free it. These toxic chemicals can pollute the groundwater, which is a serious drawback to mining this gas, apart from depleting the freshwater of large regions, particularly in dry regions like South Africa.

One alternative, currently being promoted nationally and internationally, is the production of biofuel obtained from a range of crop plants or their residues. This energy source could be carbon-neutral if the plants used take up the same amount of carbon dioxide as they subsequently release when they are burned as a biofuel. Yet, here again, the total amount of energy needed globally is far more than the energy yielded; even growing the plants does not reach a break-even point with regard to energy, particularly because the plants require fertilizers derived from petrochemicals, as well as large amounts of freshwater, to be pumped from aquifers or from rivers. Worse, farmers have to shift from food production to the growing of plants for biofuel, which can—and does—jeopardize national and global food supply. At present, the United States spends some 30 percent of its grain harvest on the production of biofuel, whereas Europe aims for 10 percent of its fuel for transport to come from biofuel. Importing food from elsewhere to meet the national shortfall can now prove to be too expensive, as witnessed by the food revolts in Brazil and the Philippines.

At present, therefore, carbon-neutral techniques are being developed as alternative energy sources, that is, sources other than those based on carbohydrates formed from plant remains of fossil or recent origin. For example, the most modern and efficient incinerator that generates energy from garbage (located in Amsterdam) covers only 1 percent of the energy requirements of the city. During the 1970s, many people still considered energy sources like wind, water, and tidal and wave movements as potential alternatives for producing energy. And faced with the impending energy shortage, many are reconsidering them as feasible alternatives. However, forty years ago, with a third of the present number of people around and, accordingly, a less industrialized world with less transport, Ehrlich had already shown that those sources would fall dismally short. Since then, the situation has gone from bad to worse, because the number of people on Earth, the energy usage per head, and societal demands have grown excessively. Combined, these alternative sources of energy can at best account for 20 percent of the total energy usage, leaving another 80 percent for fossil fuels to fill. Therefore, on their own, they are no longer considered sufficient alternatives for energy generation on a global scale, though they often have a significant local impact.

A more promising carbon-neutral source of alternative energy is hydrogen gas, which, when burned, generates a large amount of energy and an innocuous waste product: water. The most straightforward way to obtain hydrogen is to split water. This results in hydrogen and oxygen—the very same oxygen that is later used for burning hydrogen, a perfect recycling system. When salt water is used, the salt stays behind, whereas freshwater is generated by burning the hydrogen gas, which may therefore alleviate problems of freshwater shortage. Yet, water must first be split into its two components, which costs energy—more than the amount gained.

The problem, therefore, is how to minimize the energy needed to split water, how to do this in a technically easy way, and above all, how to do it on a potentially industrial scale. Furthermore, hydrogen, as the smallest chemical element, easily leaks out of pipelines through which it has to be transported, as well as out of tanks. One promising option is to find out how plants split water by using solar energy and to imitate or even improve on this process. It appears that in plants the mechanism of interest has an efficiency of 1 percent, so that without altering the process, there is scope for improvement in energy gain. Recently, such an artificial device has already achieved an efficiency of 12 percent. There is great interest in this way of producing energy from abundantly available and cheaply obtainable water and free sunlight.

Yet another alternative source of carbon-neutral energy is nuclear energy, but here too, the amounts needed are rising so fast every year that even this option will fall short. Merely to compensate for the annual increases in energy demand would mean building six hundred extra nuclear plants every year, in addition to the four hundred and fifty already present. Also, we need to remember that it takes between ten and fifteen years to build just one plant and that a nuclear plant's lifetime of forty to fifty years is short relative to the hoped-for persistence time of humankind. Still, China is building nuclear plants at a high rate to meet part of its needs for energy, and is also increasing its number of coal mines with one large plant every week, and has begun installing a large wind farm in the Gobi Desert. Two problems of general concern with nuclear energy are that they leave radioactive material as waste and their risk of exploding, whereby radioactive material is released directly into the environment.

So far, our main concern in society pertains to climate warming. But given our almost complete dependence on carbon-based fuels and our increasing use of them because of our growing numbers and societal demands, it will be difficult to reduce the output of carbon dioxide sufficiently to slow down this warming. For its part, the development of alternative energy technologies still depends mainly on research done by small research groups at universities, and by some oil companies. Broad-scale funding and global coordination of this research is not yet apparent, despite its urgency and pivotal significance for our global society in the future.

Overall, the problem concerning our future energy supply is that fossil fuels are not only running out before too long but their waste gases, like carbon dioxide and nitrogen oxides, are greenhouse gases responsible for climate warming. Their thoughtless use should therefore be terminated as soon as possible. However, the production of hydrogen gas as an alternative, carbon-neutral energy source and its large-scale transportation and use are still underdeveloped. This potential alternative source of energy can therefore not take over the present large-scale application of fossil fuels. At least for the near future, the only energy source left and able to meet our demands, therefore, is nuclear power. As mentioned, there are two kinds of fears against its use: the risk of power plants exploding and the long lifetime of its radioactive waste. Yet, during the decades nuclear plants have been in use, fatal incidences in mining for coal, oil, gas, and the like ran in the tens of thousands, whereas they ran in the tens only with regard to nuclear plants, including those due to the explosions of Harrisburg, Chernobyl, and Fukushima. Against the long-term genetic effects of radioactivity, we can put the effects of the direct, irreparable, large-scale destruction and pollution of the terrestrial and aquatic environment by mining, and the indirect ones by the release of minerals like mercury and sulfur and by the petrochemical wastes, as well as by the wastes of burning fossil fuels leading to atmospheric pollution, and this to climate warming of at least the coming two thousand years. And, as we shall see below, climate warming puts the existence of humanity and even of life on Earth in general at risk.

The risks incurred by radioactive waste can be reduced considerably by changing from slow nuclear reactors to fast ones. In slow reactors, the neutrons—parts of the atomic nucleus of a chemical element—released in the fission of uranium are slowed down. This does not happen in fast reactors. The slow reactors generate so-called transuranic actinides with lifetimes of tens of thousands of years, whereas these elements are used along with the uranium in the reactions in the fast reactors, leaving elements with lifetimes of a few hundred years. Moreover, this nuclear waste is unsuitable for making explosive weapons. Furthermore, the slow ones release only 1 percent of the energy uranium contains, while the fast ones release 99 percent, making them not only safer but exceedingly more efficient. The waste of the slow reactors produced during the past decades, the transuranic actinides, which represents an enormous store of energy, can still be used in the fast reactors.

One could say that before we have solved and installed a sustainable hydrogen economy, we are left with a choice between two bad solutions. But in such cases, it is always better to solve the problem underlying those two solutions, which in this case is ultimately the growth of our numbers which already are much too large to be sustainable. The real solution toward sustainability is therefore a drastic reduction of these numbers rather than remaining complacent about their further growth toward some stabilization level that we know is unsustainable.

7 LIMITED RESOURCES

The most basic requirement for living is, of course, food. Whether rich or poor, we all need it. Without food, we cannot survive for long. In the richer part of the world, we may not always realize this: we need food, and therefore it's there. We just go and get it—from the fridge, the cupboard, or the store. We all know that we can get hungry, which indicates that we should eat again. However, for many, eating has also become a matter of good manners at social occasions, where the nutritive value is of secondary importance. Many of us don't know the persistent feeling of acute hunger anymore, although it is a daily experience for hundreds of millions throughout the world, one that has dominated human history until recently, even in the Western world.

Yet, food can easily become limiting on a large, global scale. Human populations have collapsed because of one, two, or more years of severe food shortage due to cold or humid weather or a disease that destroyed their staple, as in nineteenth-century Ireland. There, potato blight struck, and over a million Irish died from famine; almost as many fled the country—many ended up in the United States. At the end of the thirteenth century, food shortage and adverse living conditions crippled a large proportion of the last Norwegians living in Greenland, most of whom had stunted growth and shortened lives. Often, famine was a permanent curse, responsible for life spans being only twenty to thirty years on average and for high mortality, especially among the poor, thus keeping the number of people in the population as a whole in check. And when weaving was mechanized, thousands of English and Irish home workers were thrown out of business, causing many of them to starve.

Similar disasters still happen in the world, in response to changing weather and socioeconomic conditions.

The social conditions in one part of the world can also affect those elsewhere, for example, when war breaks out suddenly. Populations grow under certain favorable environmental, social, or political conditions, but these conditions can change, triggering mass migrations. In towns in the richer part of the world, food supply and other living conditions are artificially stabilized, whereas in the natural world on which we depend, nothing is constant, nothing remains the same. As a result, countries can be underpopulated and live in wealth for some time, but as soon as conditions change, they become overpopulated. The same can happen when society itself changes. Monetary inflations, slumps, depressions, and wars come and go; countries or entire continents rise and fall. Up to recent times, many societies had social customs and rules to prevent overpopulation, ranging from delayed marriage to euthanasia (particularly with regard to the elderly), or to infanticide (especially with respect to girls, as it still happens on quite a large scale in India, for example). In that way, famine was averted or fought, even under changing conditions.

In fact, we are still living in a highly changeable world. Think of the great depression of 1873–1896 in Great Britain, the depression in Austria immediately following the First World War, and then the crash of 1929, which started in Germany and then spread across Europe and then to the rest of the world. There was another general crash during the 1980s, one in the 1990s, and then the latest, which began in 2007. Meanwhile, there have been many national crashes. Or think of one of the worst battles ever, in which millions died, fought during First World War in Champagne and Lorraine in northeastern France, because of its local resources and industries. But then, within a couple of decades, resources and industries elsewhere in the world took over, leaving the once-flourishing area silent, its remaining resources unused. Houses, banks, and the Maginot Line became derelict, and once-important canals and railway connections were grown over.

Now, we are entering an era in which many of our present resources will run out one after the other, shortages potentially bringing great international unrest, destabilization, and international rearrangements of interest and importance. For example, in 2009, an estimated one billion of the six and a half billion people on Earth were living below the hunger line. And although some assure us that food prices will decrease, others estimate that they will increase with the rising production costs and as fossil fuels dwindle and become more expensive, and are therefore even partly replaced by growing corn and maize for biofuel. Then, many of the poor in the world won't be able to buy food anymore, even if it is still available. For them, food will become limiting. Ultimately, they will run out of energy.

Our highly developed society runs on energy and petrochemical products, which, as I've already pointed out, are both running out. If we reckon that humanity in its present abundance wants to prolong its existence on Earth by, say, another one thousand years, then most of the world's minerals will have been totally exhausted well within that period. The few that remain will be of no use. For example, if energy is generated in sufficient amounts but phosphorus has run out, human life is still impossible to maintain.

Since time immemorial, people have noticed that they exhaust the soil by cultivating it; yields decline, eventually forcing farmers to change to other crops before stopping altogether. This happened in many parts of the Mediterranean and in Central and South America. In other cases, such as in Mesopotamia, Egypt, China, and northern India, rivers brought in new, nutrient-laden silt with their annual inundations. But every year in Mesopotamia, groundwater and irrigation water brought some salt from deeper underground to the surface, which crystallized into a crust when the water evaporated. This happens when the groundwater level is near the surface (alongside rivers, for example) or when it is elevated by irrigation, as it still happens in Australia, China, or the American Southwest, for example. Many crops can't grow in such hard, salty soils. After early farmers noticed that using animal manure and night soil improved yields, they began to fertilize their fields. In other communities, farmers burned stubble to release its nutrients to facilitate their absorption by the next crop. This returned some of the nutrients taken away from the soil, and so was an ancient way of nutrient recycling. Of course, this system of nutrient recycling only works in closed subsistence systems; the cycle breaks down as soon as corn, flax, or cattle are removed and traded between communities or even—as occurs today—across continents.

Another way of counteracting nutrient depletion, albeit temporarily, is to plow. This mixes the exhausted top layers of soil deeper into the soil. Ancient subsistence systems used this method. When combined with fertilizing the soil with nitrogen-rich manure, plowing becomes a link in the recycling scheme. In earlier communities, after harvesting the soil lay fallow for one or more years and wild plants invaded the plot; when these were plowed in, the nitrogen content of the soil was raised. Later, crop rotation was introduced. Clover was one of the rotation crops, as it increased the amount of nitrogen in the soil. As communities grew, it became essential to fertilize with manure and with nitrogen-fixing crops like clover.

The expanding communities generated more trade, which meant that the nutrients were exhausted more rapidly and that other means of fertilization had to be introduced. In Friesland and Groningen, provinces in the north of the Netherlands, people began to fertilize their fields with soil taken from "terps"—the artificial mounds or knolls on top of which farmsteads and churches were built as a protection against tides and rising seawater before dykes were built around the area. As they consisted partly of rich clay and partly of humus, garbage, and detritus, they were ideal as a cheap fertilizer. Where the soil was peaty or humus-rich, farmers could release the nutrients by liming the soil, an indirect way of fertilization. Later, European farmers fertilized their fields with nitrogen-rich guano transported all the way from Chile. Later still, guano was replaced by artificial fertilizers. In the mid-nineteenth century, the German chemist Justus von Liebig had discovered that the elements potassium and phosphorus were also crucial for plant growth, and today's chemical fertilizers are still based on compounds containing these three essential elements. However, all three will run out when both the petrochemical products and the phosphate deposits in China, Morocco, and to some extent those in Florida and the Kola Peninsula in Russia, are exhausted. The first fights for Canadian potassium mines are already on. The exhaustion of nitrogen depends on the exhaustion of natural gas, fossil methane, supplying considerable amounts of energy needed. This is done in the so-called Haber-Bosch high-pressure catalytic reactor in which the chemically very stable atmospheric nitrogen gas binds with hydrogen, forming ammonium and carbon dioxide. The ammonium can then be used to fertilize the soil for crop growth.

Thus, earlier fertilizers were discarded one by one and replaced by others when rural communities grew into towns, and as trade in agricultural products between these communities intensified and broadened. At each stage, the fertilizer used reached some limit. Overall, we inevitably exhaust the soil, removing its nutrients and expelling or dumping them elsewhere as waste. Most of our essential nutrients end up in the oceans via the rivers or in environments we cannot cultivate. Given the limited availability of the soil nutrients and the finiteness of our mineral resources, only global planning and recycling can soften—but not eliminate—the limits to food production set by our linear, exhaustive resource use.

The same story can be told for many of our other essentials, the fibers of our clothing, for example. At first, people used animal skins and leather for clothing. Grass and bark were other early fibers, but wool was more important. During the Industrial Revolution, wool was largely replaced by cotton and exported abroad in great quantity. During the mid-1960s and again after 1990, wool fell out of fashion as Australian production collapsed in the face of competition from cotton and synthetic fibers. Linen is another ancient fiber; because its production is labor-intensive it was replaced by cotton, which is easier to mass produce and can be combined with synthetic fibers. But cotton needs high temperatures and plenty of water to grow. This unfortunate combination of factors can become limiting under the presently shrinking availability of water, as has occurred around the Aral Sea in southern Russia, the American Southwest, and western China. Two other plant fibers, jute and hemp, are most commonly used for making bags and rope and have also been replaced by synthetic fibers. Finally, natural silk has always had a restricted use because its production is labor-intensive, which is why it is imitated by synthetic fibers.

Thus, growing global demands led to a shift away from the natural fibers toward synthetic ones derived from fossil fuels, or, recently, also from waste plastic. Except for cotton, most of the natural fibers are difficult to grow on a mass scale and they compete with food crops, whereas the synthetic ones derived from still easily available petrochemicals have no such shortcomings. Moreover, the growing of increasing amounts of cotton depends on how climate will change in the future. Similar to other materials, a return to natural fibers seems unlikely, given the increasing demand of our growing numbers; the continued production of their synthetic substitutes, however, depends on the continued easy availability of fossil resources.

At present, iron, probably the most widely used metal, is also the commonest and most easily available. At one time, it was dissolved in high concentrations in the reducing water of the primeval oceans in which the earliest forms of life appeared, but after one to one and a half billion years, the water became more oxidized and the iron precipitated out in thick layers on the ocean bottom. The resulting banded iron formations can be found all over the world and are virtually inexhaustible. The problem with the use of iron is not its own availability, but the availability of other metals with which it forms many alloys to give it a specifically required quality.

Finding the right alloy has always been very difficult. For a long time, iron could not be used on a large scale because it melts at high temperatures, rusts readily, and is brittle. Yet, it was already widely used by the Etruscans, who produced it for the Romans. Bronze, the alloy of copper and tin, did not have iron's disadvantages, although tin has the disadvantages that it is not abundantly available and had to be mined in faraway Cornwall in the west of England. In fact, bronze can be produced at the relatively low melting temperatures that wood burning gives, so the real mass production of iron had to wait until coal, which has sufficient heat capacity, was used in the foundries. Its mass production in the form of steel began in England during the second half of the nineteenth century, thereby giving impetus to the Industrial Revolution. One of the components of steel that makes it hard and nonbrittle is carbon, which is added in different proportions to make different types of steel. Metals can also be added, giving the steel yet other properties with respect to melting point or hardness. However, as these metals become more difficult to find or to mine, such alloys have to be replaced by other materials.

The supply of copper, once available in such quantities that an island in the eastern Mediterranean, Cyprus, has been named after it, is already becoming limited. For pipes and communication it has widely been replaced by PVC and fiberglass, which are also cheaper because of their wide availability. Fiberglass also transfers information better than copper, and at less expense. Still, copper survives in great quantities in the electric wiring of houses and power lines. For short-distance electric transfer within your computer, cell phone, or smart phone, however, gold is used because it does not corrode, allowing it to be packaged in very high densities, which enables the connections to be miniaturized, and, hence ensures the high speed of the apparatus. However, gold is expected to run out within twenty to thirty years and is therefore recycled on a large, industrial scale. In 1970, South Africa produced two-thirds of the amount of gold used in the world, but its present production is only one-tenth of what it was then. Understandably, China is investing billions of dollars in Congo alone in the form of infrastructural projects in order to obtain copper, gold, tin, radium, uranium, diamonds, and cobalt. And it is doing the same in Zambia, South Africa, and Australia.

Genetic resources, though only recently of industrial interest, are also limited and can therefore run short. This may sound surprising, since people have always manipulated their animals and plants genetically by interbreeding, usually with the result of an overall increase in genetic material. Many deviating genetic properties were kept and used at some point. Thus, in agriculture and husbandry, as well as in pigeons, pets, and indoor plants, many, often rather weird-looking, local races of animals and plants originated that were bred for a particular adaption to their environment or function or for their appearance. An enormous array of breeds of dogs originated in this way—for going down rabbit burrows or tracking down prey or criminals by their scent or herding sheep, and so on. In plants, literally thousands, if not tens of thousands, of local types of rice (just to name one example) have been selected over the centuries across Southeastern Asia, each adapted to often very local conditions. In dogs, performance was enhanced; in crops, yield was improved. The mechanism of improvement was selection: any new genetic deviation cropping up and exhibiting a certain trait or quality was kept and combined with existing ones. By keeping and multiplying all sorts of novel deviations, the total variation in genetic resources increased bit by bit. Often, this took decades or centuries (as in the case of dogs and tulips), but in other cases—in many crop plants, for instance—it took many hundreds, if not thousands, of years. When we lose those local breeds, they are in practice lost forever.

In the past, one had to wait until a new variant appeared by chance and was recognized as useful. Recently, this time-consuming procedure has changed radically. Local races of plants are collected from all over the world and hybridized in foreign seed factories for maximum yield under certain, man-made environmental conditions. Thus, new hybrids are now grown over vast areas. This procedure contrasts with that followed traditionally, where hybridization was prevented from occurring because hybrids lose their favored trait when this particular gene combination falls apart, which usually happens in the first generation after its application. Hybrids then return to their nonhybrid states, less adapted to the new, man-made conditions, and become obsolete. Moreover, apart from selecting a small number of genetic types from the thousands of existing races, the breeding criterion is to produce a high-yielding genetic combination, giving the highest profit within a short time, that is in the one or at most two years that the hybrid combination is stable. So, new hybrid races and new pesticides have to be produced continually and from a rapidly narrowing genetic base. Because of the rigorous choices made during the process, the result is greatly reduced genetic variation. The rich genetic resources, until recently available to humankind, are therefore being rapidly exhausted, which enhances the risk of disease epidemics, maladaptation to changing conditions, and thus, large-scale crop failure on the longer term.

Another laboratory procedure is to deliberately introduce individual genes or gene complexes that originate from different species of plants or animals. Thus, one can introduce genes for improved milk production or for immunity against certain diseases or pests. This reduces the genetic diversity. Some of the manipulated individuals can develop disadvantageous side effects, because the transferred gene finds itself within an alien genetic and biochemical context with which it interacts in one or a few developmental stages, often producing undesirable traits. Individuals with the wrong biochemical context are then discarded; this reduces the overall diversity.

Finally, the new strain with the wanted yield or resistance against a disease or pest is immune for only a certain period: the disease or pest may adapt itself during the next few years, finally developing outbreak proportions. Then a new resistant type has to be developed to replace the previous ones, which are destroyed.

In these cases of hybridization and genetic manipulation or engineering, existing genetic material is used; nothing new is added. Instead, with each step, part of the existing variation drops out, often in large amounts. Thus, the new procedures of improving the quality and quantity of our food reduce our limited biological resources. Worse, the initial genetic variants used or reserved in advance are often patented by the seed factory and therefore cannot be used anymore in their country of origin.

Resources can reach their limit of usability in various ways. As well as exploring new and more efficient alternatives of processing, broad-scale recycling schemes have to be evaluated in terms of how they reach their limits. Or we may restrict ourselves to one application only, leaving other applications to alternative or artificial sources. The remaining reserves of fossil fuels, for example, could perhaps better be reserved for petrochemical applications like food production, and we should use alternative ways of generating the energy now obtained from their combustion. We will inevitably have to turn to large-scale recycling fed by artificially generated energy. Sooner or later, however, all the resources and recyclables will run out.

8 MAN-MADE WASTE

As children, we were always told to wash our hands, to tidy up after playing. And the same happened at school, at work, or after a meal or working in the garden, the shed or the garage; you always had to tidy up the mess you had made. Later, our domestic chores included doing the washing, vacuuming, dishes, polishing, tidying up toys, emptying wastepaper baskets, getting the children's bikes repaired. Never ending. Outside too, plant remains, leaves, and old branches have to be tidied up in the garden or nearby woodlot—to the detriment of all the animals, fungi, and beautifully specialized saprophytic plants that live on them. They have vanished from most parts of our towns and villages—tidied up. And with them have gone all those other animals, birds, beetles, and spiders that lived on them. In our village, most gardens are tidied up so much that they have become sterile, leaving much lifeless, bare ground. Our own garden, instead, with all the birds singing, with all its hidden life of creepy crawly insects and slugs, with its screeching owls and its hedgehogs shuffling in the night, is now known as "the wilderness." There has never been any real, positive interest in dirt, untidiness, waste, wild animals, and plants. We are not interested in weeds and waste. Birds are good because they eat insects, and spiders are beneficial because they remove flies, which have certainly wallowed in refuse with their filthy little feet. Yet we want those spiders to stay outside. They are dirty because they ate something dirty. Dirty inside. We don't like to know about dirt, about waste. We'd rather wash it away in the river or a lake, or expel it into the air. Gone. Clean again.

Yet, waste is the counterpart of all the things around us that are useful or delicious, useful when we are playing or, later, when we are working, and waste is also the natural counterpart of eating. Whatever we do, we are constantly turning resources into waste; there is just as much waste as there are resources. No more, no less—just the same. Resources are potential waste. People worry about the world running out of resources, but not about it suffocating in an equal amount of waste. As any physicist can tell you, we cannot make new matter, and we cannot destroy it either. The same applies to energy. And this is as true for the natural world as it is true for our own, man-made world. There is no difference. The only difference is that part, if not most, of the waste we ourselves are presently generating cannot be recycled as happens in nature. In nature, only a small part of the waste remains as it was—nondecomposed waste—but most decomposes. Sooner or later, it is broken down by thousands of other organisms, large and small, from vultures to fungi and bacteria. It is taken up again in the great biospheric cycles of life as new nutritious food, new resources for other forms of life to live on. But most of our own waste remains waste, piling up around towns, being transported across the world to be deposited in unknown, far-away places—some in underdeveloped countries. There, as in our own environment, it acts as pollutants, blocking the natural nutrient cycles. Eventually, therefore, we leave these pollutants behind in those places, and this to our own detriment, as we, like any other organism, are still completely dependent on the persistence and proper functioning of those biological nutrient cycles. Without these cycles, there will soon be no resources left. Unfortunately, whatever we feel, whatever our upbringing, we have to develop some interest in the other half of the dynamic of the world, the dirty side of the making and processing of waste.

With the growth of resource use per head, the amount of nondecomposable waste follows exactly the same trend of our growth toward higher and higher numbers. Each of us has increasing demands—a trend, too, in a society of increasingly greater complexity with its own dynamic of growth, and thus, with its own increase in demands and waste products. This has, of course, always been the case, and it has been said before. But in the past a large part of the waste left by people decomposed naturally. For, actually, how much has been left by the millions of people before us, apart from the rubble and ruins of their houses and temples, and some of their rusty swords? Hardly anything. Look at an old photo of thousands of people together on a square to protest, or to commemorate somebody. They have all eaten, they are all dressed in shoes, trousers and skirts, coats, caps, or hats. Thousands of shoes. Year after year, century after century. Where have they all gone? Where are all those people themselves? But in the last century and especially in the last few decades, the kind of waste we leave has changed. It can no longer decompose in a natural way. Most of the recent waste is nondecomposable. Moreover, the staggeringly large world population, having grown threefold since the Second World War, is recent. Because the increase in the population and production of this waste, and the rapidly increasing amount of space it requires are so recent, our experience in dealing with it is limited. But so far, the overuse of our limited resources has been of greater concern than their consequent turning into waste.

During the last 3.8 billion years or so (roughly the time life has existed on Earth), organisms have always produced organic waste, most of which could have served as food for other organisms. Each of those organisms contained enzymes to decompose the waste left by other organisms, and yet other enzymes to build up their own body from the components thus obtained. From early on, all the relevant minerals were recycled in this way in a perpetual merry-go-round. There was also waste that could not be decomposed; this precipitated into the shallows of some lakes and on the bottoms of the oceans, over time forming thick layers of sediment. Over the millions of years that passed, this sediment slowly turned into the rock of past and present-day mountains, or into the fossil fuels found deep in the geological strata. Thus, apart from the digestible part which passed through those endless cycles, indigestible waste has been produced for as long as there has been life on Earth. There has always been an overconsumption of the resources of Earth, leaving behind some minerals and chemical compounds that did not recycle.

When, much more recently but still long ago, humans arrived on the scene, therefore, nothing new or exceptional seemed to happen. Initially, humans too fitted in the grand scheme of the natural, biospheric nutrient recycling. Like all other animals, humans breathed out carbon dioxide and water, digested from the starch of the plants they ate. These waste products were then taken up again by plants that, once again, with Sisyphean dutifulness, produced starch from them. And the indigestible parts of humans' food formed their feces, which were digested by bacteria and other organisms, which split up some of the remaining components into carbon dioxide and water. Yet, there were still some leftovers. Around the world there are many sites with tens or hundreds of shells or bones of the animal species early humans ate: fish, shellfish, mammoths, horses, cows, goats, dogs, marmots— all thrown together as prehistoric human trash. Under suitable conditions or given more time, these bones could have eventually decomposed too, so that some of this waste would also have been recycled. But this was soon to change. At some time, arrowheads, axes, and scrapers made of flint appeared among the trash left by early humans. And soon afterward, charcoal appeared, together with the earliest remains of pottery and other "kitchenware" made for cooking or storing food. This material could not decompose and so it was not reincorporated into the grand scheme of natural recycling. It remained where it was left.

Pottery introduced an entirely new type of waste: an artificial kind of stone, ceramics, produced under high temperatures. The fuel used to obtain these high temperatures was wood. Its remains can still be found in ancient kilns as bits of charcoal, which are as nondegradable as the pottery. Gradually, other materials appeared as well, metals and glass, both of which also require high temperatures for their manufacture. Apart from the pottery, charcoal, metals and glass, there are remains of buildings like houses and temples. Usually only their foundations are left; the rest has eroded away or been reused. Over time, all this nondecomposable waste began to mount up; each new layer deposited on top of all those laid down previously. In the Middle East, people continued to live on top of these early garbage heaps, which could be tens of meters high and are known as tells. Thus, ever-more waste, and ever-more kinds of it, were produced by the ever-more sophisticated human populations.

Roughly the same kinds of waste still prevailed during the next few thousand years, except that the amount generated grew concomitantly with the population. However, particularly from the second half of the nineteenth century onward, wood was superseded by coal as the main source of energy for fulfilling our requirements. Then, particularly in some places in England, coal could still be mined at relatively shallow depths. Its introduction as a new high-energy source triggered a revolution in technology and, hence, in the way of living of humankind: the Industrial Revolution. Now, metals, mainly iron, were mined, processed into steel, used, and exported in rapidly increasing quantities. Bridges and stations were constructed of iron; iron trains ran on iron rails, forming extensive and dense networks of railroads; and canal boats and marine ships for the transportation of all the coal, metal, and goods were made of iron as well—as was the Eiffel Tower, which was built for the World's Fair in Paris in 1889. Drainage and sewage pipes were, again, made of iron. Also, vast suburbs of multistoried houses and factories were built, all made of brick and with glass panes. Such buildings had been known since the mid-seventeenth century, but now they were higher and larger. Huge greenhouses entirely made of glass and steel were built for tropical plants, as were glass-covered markets and galleries throughout Europe, like the spectacular exhibition center in London, aptly called the Crystal Palace.

Not long after this ubiquitous use of coal had begun, oil was found and began to make its way into human economy. At first, oil was mainly used to fuel vehicles and illuminate factories, but soon the large transports (trains and ocean steamers), as well as the ovens in steel plants, and the heating, cooling, or lighting in households, also ran on oil. Not long after this phase, increasingly higher temperatures were needed in industrial processes, which required even larger quantities of coal and oil. All this resulted, in increasing amounts of nondecomposable waste, such as carbon dioxide and the sulfur and nitrogen compounds expelled in great quantities into the atmosphere, which resulted in the formation of fog—and, inevitably, lung diseases. From that point on, the many nondecomposable products thrown away—such as ceramics, bricks, iron, and glass—were joined by new ones, including inorganic chemicals and early petrochemicals.

Particularly during the second half of the twentieth century, oil and its products began to permeate all parts of our society. Although, globally, coal remained our main fuel, many mines were closed down. Roads were paved with asphalt—a waste product of refining the crude oil. Soon, many household utensils and appliances, as well as industrial equipment, were being made of plastic. From then on, the new petrochemical industry, run on fossil fuels, began to make an enormous amount and variety of new products for very broad use, from fibers, clothes, furniture, building materials, pharmaceuticals, pesticides, and fertilizers to shoe polish and toothpaste. At present, there is hardly anything that does not contain petrochemical material; if it is not entirely made of plastic, it will contain components made from oil derivatives. Even the production of our biological, slow food is partly dependent on petrochemicals. An advantage is their relatively low energy cost of production, but the disadvantage is that as fossil fuels run out, we will have to revert to more energy-costly materials. And in food production, we will have to fall back on exhausted soils, depleted oceans, and human labor again.

All these changes resulting from the use of oil were revolutionary. First of all, both the energy and the greatest variety of materials imaginable were all cheap: their raw material was welling up for free from deep down in Earth, at first often even without pumping. Our personal wealth, our health and longevity, and the wealth and diversity of society is running on the pocket of the past. But, apart from oil's limited occurrence on Earth, the waste thus generated cannot be decomposed naturally. Large-scale recycling is therefore the only option. However, although this is cheap, so far making new material directly from oil is even cheaper and therefore remains preferred.

Living structures are always eking out their energy, step by step, reshuffling and reconstituting little bits of molecular material through the metabolic cycles in our cells, every step requiring only a minimum amount of energy. One of the main tricks in their energy conservation is that they keep to the same electrochemical energy, rather than converting it into another form and losing an enormous amount of energy by doing so. In this way, life was able to build up organisms of extremely complex structure, functioning, and behavior. However, when an animal eats a plant or another animal, thereby breaking it down into its digestible and indigestible components, it destroys the chemical pathways and cycles as well as most of the molecules. But from the components of those molecules—some elements, carbon dioxide, and amino acids—it builds its own molecules, some chemical elements, carbon dioxide and amino acids, and inserts them into its own chemistry. In this kind of conversion, 80 percent to 90 percent of the energy of its prey is lost so that only 10 percent to 20 percent is used. Similar percentages apply to us eating meat, independent of the energy spent on transport and cooking, a loss which is often not accounted for. Still, these biological conversions remain structural, not conversions from one type of energy into another.

Humankind does not use such intricate, energy-conservative, multistep pathways and cycles. Worse, humans use excessive amounts of energy obtained directly from burning. They do not restrict themselves to chemical energy but rather convert chemical energy to another kind, usually via the two conversion steps of heat energy and mechanical energy, to electric energy. Ultimately, therefore, it is heat energy on which all our activities are based. This conversion process causes enormous losses of material and energy at each step, all wasted, both into the atmosphere (the carbon dioxide and many poisonous fumes) and into outer space (the conversion energy). Moreover, at each conversion step, large amounts of energy are lost, easily up to 60 percent, leaving 40 percent for our use. After two such steps, only one-sixteenth of the original amount is still available to be used. The rest, 84 percent, is wasted—gone. And the conversion of electricity into light energy in a light bulb results in an energy loss of 95 percent; that is, only 5 percent is used as light and the rest is lost as heat into the environment. Even within one kind of energy, electricity, conversion losses occur; converting alternating current into direct current means a loss of some 40 percent. By using your computer, this expensive electricity then converts into heat, which dissipates from your transformer, leaving abstract "information" behind.

Apart from loss of energy due to conversion from one form of energy into another, there is also the loss due to inefficient energy use. Your air conditioner or your fridge, for example, has an efficiency of only 25 percent, the engine of your car, about 30 percent. Combined with the loss due to conversion, this means that of all the energy in fossil fuel, a fridge uses only 10 percent (40 percent × 25 percent = 10 percent) and loses 90 percent. With more conversion steps, such as a mere currency change, the loss, the pure waste of energy, is even larger.

Losing large amounts of energy due to conversion happens all the time, at each of the billions of energy conversions happening in our energy-hungry society. At night, our towns can be seen from space as glowing spots of infrared light, of heat energy. And many rivers are dead because of all the hot cooling water they contain. All energy wasting away. Gone. Forever. Enormous amounts of energy are lost in the face of a rapidly approaching energy crisis, a growing population with growing personal and societal demands.

Apart from the direct loss of energy in these ways, there is also the indirect loss by the large-scale food waste. Between the harvested crop and the final food product bought by the consumer, some 30 percent to 40 percent is lost in transport and at the supermarket; a quarter of what remains is thrown away by the consumers themselves. This chain of events already starts with the plant itself, which takes in only a small percent of the incoming sunlight and eventually uses only a fraction of this: just one percent. The fertilizers the crop needs are produced from fossil fuels that have to be mined, transported, then processed into fertilizer that, in turn, is transported and applied; at each of these stages there are large losses of both energy and material. Some of the fertilizer lost pollutes the soil and water.

Over the course of history, at each step of humankind's development, people have increased the amount of energy thus wasted by burning fuels. Moreover, apart from wasting larger and larger amounts of the energy sources of Earth, many of the products we produce cannot be recycled, because they cannot be broken down by enzymes made by living structures. There are no enzymes to do the job. There is no enzyme for breaking down pottery, no enzyme for decomposing the hubcap of the wheels of your car. As our presently used materials are completely foreign to products of life processes that typically operate at temperatures well below 100°C, they will hardly decompose or decay, if at all. Since the molecules to be broken down are obtained by applying unnaturally high temperatures, each of our man-made materials adds more to the vast amount of waste already found on Earth: enormous amounts of brick, glass, iron and other metals, chemicals, plastics of all sorts, old cars and many hundreds of thousands of used tires, household rubbish, discarded ships, concrete, and so on. All these materials together fill up deep landfills at staggering and ever-increasing rates or are piling up high as mounds in the landscape, affecting the health of the people living nearby with their unhealthy acrid stench. Water seeping through these pits or mounds pollutes the groundwater. The air, soil, and water now contain increasing amounts of our gaseous and liquid waste, poisoning any form of life. For decades now, our pesticides have even been found in penguins living on and around seemingly pristine Antarctica. Vast areas have become virtually or wholly denuded of any life. Our global numbers may stabilize or slowly decrease in future, but in order to persist on that level, we have to keep using our limited resources—and keep returning them to the environment as polluting waste. How will this work out in the end? Similar to the limitations set on the amount of resources to be found on Earth, there will also be limits to the amount of waste produced and recklessly thrown away (preferably dumped in places where nobody can see or protest against it, at places where the producer of the waste can look the other way, not liking to see useless waste and dirt, as none of us does). We have completely changed our living conditions and are unavoidably continuing to do so at ever-increasing rates. We are affecting both the chemical and the physical living conditions on Earth at local and global scales—our own living conditions and those of billions, trillions of other creatures of largely unknown species around us.

The use of fossil fuels is balanced by the amount of waste produced: we inevitably produce ever-increasing amounts of carbon dioxide and dump it into the atmosphere from where it enters into the oceans and penetrates into the rock of the mountains, acidifying all environments. We also dump petrochemical products like old slippers, chairs, mattresses, plastic bags, and pollutants into the local environment, where they smother all life. Similarly, we dump metals into the environment in high concentrations, sometimes as large pieces—entire airplanes and ocean steamers—and we allow entire factories and nuclear plants to become derelict. The metal parts of your car, for example, wear out; at some point, the pistons don't fit anymore, and the doors rattle. The metal has unnoticeably dissipated away into the environment in minute pieces, which is why you need to change the oil every so often. At some places, you can smell metal, evaporated into the air, or sticking on your fingers, where it kills off bacteria. Thus, life is also killed off around crematoria, because the metal waste of teeth and the like evaporate at high temperatures, then accumulate in plants and insects, and in the animals that eat them. Fish are affected by minute amounts of metal in their environment, as these change the electric conductivity of the water and disrupt the functioning of their lateral line, the sense organ with which they orientate and communicate. This, in turn, affects their chances of reproduction, and, hence, their local or regional survival. We still do not know all the ways that tiny, hardly measurable amounts of metal can affect many forms of life.

As well as altering the electric conductivity, metals and many other chemicals also change the acidity—or pH—of soil and water, allowing some species to multiply to outbreak proportions but causing most others to die out. Present agricultural practice is to apply hundreds of new chemicals each year: fertilizers, herbicides, fungicides, vermicides, insecticides, bactericides, and other pesticides, as well as veterinary pharmaceuticals, joining the human medicaments. These remain in substantial doses as pollutants in the environment. In the long-term, they all form part of our man-made chemical environment, as they are largely nondecomposable. Eventually, they end up in our bodies via the food we eat or the air we breathe. Surplus—and thus wasted—animal and human medicaments discharged as effluent by some Western European countries into the Rhine and other rivers and eventually reaching the North Sea meant the death sentence for a great many life-forms naturally occurring there. In the Netherlands, the increasing amounts of soil bacteria that are now resistant to bactericides because of the large amounts of pharmaceuticals in the environment may affect public health. Resistant bacteria are even found in our meat. Similarly, certain pathogenic bacterial strains have become resistant to antibiotics, rendering these worthless for further use.

Other chemicals used and produced in industry and research laboratories are daily washed away directly as waste, often coloring the sky, ditches, creeks, and big rivers. I have also seen small rivers that contained so much waste that the "water" was sluggish. The same happens with all sorts of chemicals—often bactericides—used in amazing amounts in modern households for cleaning and disinfecting. All eventually end up in the sewage system or the air. And what of the leftover paint in the tin you throw away at the end of the day, along with the brushes? And if you did clean the brushes, what about the water or the solvent you poured down the drain afterwards? From the sewage system, these pollutants eventually reach the sea too. Over the years, the Rhine has been cleaned up, but people are still forbidden from swimming or bathing in it. What about the major rivers in the tropics, which are used for bathing and for drinking water for villagers, farmers, or all squatters in shantytowns who can't afford a sewage system? What about the millions taking a ritual bath in the greatly polluted Ganges in India?

Similar to much of our industrial waste, the oily and gaseous waste of modern traffic is expelled into the groundwater, surface water, and the air. The Smoky Mountains are now really smoky, and have, as a result, largely

been denuded of their forests by this smoke, just like the forests of Central
Europe and many other parts of the world. Our gaseous environment, which
we inhale with every breath we take, is becoming one large open system of
smoky clouds of poisonous fumes; our terrestrial environment is dotted with
sewage pools, garbage pits, and mounds. Waste is deposited free of charge
just as it always has been throughout human history, but now of such proportions and chemical stabilities that it will affect our own living conditions,
along with those of all the species on which we depend as biological recyclers
and those to be appreciated in their own right. We are using ever more and,
hence, wasting more.

In fact, in terms of the processing unit of figure 1 (see page 000), we are
speeding up the processing by deliberately enlarging the unit by stimulating
the growth of our numbers and our personal demands. And we have deliberately made it run less efficiently and faster by building in short lifetimes
for our products, our cars, clothing, and furniture—by following the latest
fashion. Our economies are based on profit maximization within a competitive market, which means a diversification of often useless and short-lived
products, which are marketed through extensive advertising. Cars, like laptops, have a maximum lifetime of five years of less, cell phones last less than
a year. But our consumerism stimulates not just unnecessary inventions but
real progress as well, so that our equipment, appliances, and machines have
high turnover rates, our living conditions are healthier, more certain, and
less energy demanding. This consumerism also implies that our means of
communication, kitchenware, houses, factories, entire town districts, and
entire agricultural, financial, technological, and transportation systems have
to be replaced by new systems. We chuckle when we watch movies from
fifty years—or even ten years ago. We need to replace our computers every
few years. We send up satellites for convenient orientation, more efficient
weather forecasting, measuring climate warming, and military control. The
last marine fish, now even those from the deep sea, are hunted using computerized technology. But our overheating processing unit, thus enlarged,
has sped up, partly made to run more efficiently and partly less, and is costing ever-more resources and producing increasing amounts of waste. The
economic and military system we have built is inherently wasteful, and is
growing more wasteful by the year. Within this system, recycling has no
place, being too costly.

Until recently, waste production wasn't an issue, although locally it was certainly of some concern. This has clearly been shown by archaeological finds
from old shipyards and other early industrial sites. Even prehistoric work-

shops can be recognized from the thousands of flint chips lying around or forming great heaps. And ancient towns were often dirty and stinking; the local streets and roads between towns were muddy and full of potholes; and streets, canals, and creeks in and around towns were open sewers. When under Dutch rule, Batavia (present-day Jakarta, Indonesia) was a hellish pool of disease, therefore officially to be avoided by English sailors.

Nondecomposable chemicals have been around for some time. Building materials and metals have long been freely disposed of into the environment. However, as mentioned, most of the early human-made waste (including textiles, sails, wooden carts, houses, furniture, and thousands of ships) decomposed naturally, because it was organic so that bacteria, fungi, and wood-eating animals could digest it, and because it was produced in relatively small quantities. There were enough decomposers to keep up with the production of this kind of waste. In fact, the rapid natural decomposition of many materials is the curse of archaeologists, as it deprives them of critical clues. And of how many of the uncountable millions of people who have walked this Earth do we still find the skeletal remains? Or the remains of the ever-larger numbers of all the animals they have eaten, let alone of those of all the hundreds of millions not eaten?

As described above, all this changed radically during the last couple of centuries and especially during the last fifty years, with the production of nondecomposable metal and petrochemical materials. This has had two consequences. The first is that we are extracting more and more of our resources without replacing them through recycling, because recycling is often impossible for reasons of economy, energy expenditure, or material dissipation. So, these resources are lost forever. In that case, using them is wasting them. The same applies to all the energy we use. Recycling has its limits. The second is that our kind of waste is occupying a rapidly increasing amount of our environment from which we also have to get our food and other resources. Waste is generated not only by producing food by present-day agricultural practices, but also by transporting and processing it. Food production, moreover, can be considered as the wasting of land: sooner or later, fertile land becomes wasteland because it is exhausted, polluted, urbanized, or mined, and it erodes, desertifies, or salinizes. Wasteland is another kind of human-made waste. This environmental loss is an indirect loss on top of the direct loss resulting from the processing of resources. And we cannot recycle a region.

For several reasons, therefore, we have to take any loss of the remaining environment seriously. We can't afford to lose any part of our environment anymore, especially because the population and society of humans on Earth will keep growing for at least another half century. In that time, half as much again of the present number of humans will be added to the world's population, after which the figure is expected to stay at this very high level for the remaining part of our existence. All that time, we will keep adding nondecomposable waste and pollutants, year after year. All that time, we are wasting the land, the water, the air—that is, the environment on which we depend.

Our demands and wastes will grow much more than the number of people. A large part of the world's population is currently lagging behind, relative to the more demanding and wasteful Western world. The most populous countries, primarily India and China, will increase their resource use and hence their production of waste, and their per capita resource use and waste production will develop along the same lines as those in the West. This means that in the coming years we will need still more of the existing resources and will require more land for their production and to store the nondecomposable waste they leave. And the areas of occupied land and wasteland will grow, leaving less and less to serve as a basis for our further existence.

Our wasteful use of resources means that we are wasting our human world, wasting all we have built up in society through the past millennia, the past six million years of existence on Earth, all our opportunities for the future. We are wasting, in fact, the amazing result of billions of years of biological evolution leading up to the origin of humanity and to our unique level of civilization, the variety of all our different cultures. From this perspective, the various ways of turning resources into waste, of their production, and its deposition into the environment, are only details of this larger and deeper kind of wasting our uniquely human opportunities of living, of enjoying and understanding our world, its past and its future, the way it works, our unique existence in the universe. That is what we are wasting by our carelessness, our reckless throwing around of trash, by leaving deep trails from our heavy machinery in soft, marshy soil, by polluting the soil and the groundwater, by dumping billions of tonnes of poisonous gases into the water and the air, by warming the earthly climate. Wasting our future, wasting our world. Wasting the beautiful life on Earth. And our Earth is wasting away.

9   WHEN IT'S GONE, IT'S GONE

Take a little lump of sugar and put it in your tea (I assume that you take your tea black, as I do). Then stir the tea, and you will soon see that the sugar has all disappeared—not a grain to be seen anymore. Easy. But now try to get that sugar out of your tea again in the shape of that initial lump. That won't ever happen. Impossible. Of course, you can evaporate your tea using energy. This will leave a layer of small crystals of sugar on the bottom of your cup. You could then compress all the little crystals you scrape from the bottom into the shape of that initial lump. Again, you have to use some energy to do this. In that way, by using energy, you can get a lump of sugar back again, but a different one, of course. It will never be exactly the same lump with the crystals in the same place as in the old lump, and with all the molecules of sugar in the same place in the same crystals. And it will never be without traces of tea.

But getting all the molecules back again into the same crystals is not important for practical application anyway. It could be more interesting to know whether the new lumps are pure, containing no traces of the tea from which you extracted them: some of the tea will prove impossible to be removed. And it can also be interesting to know that you obtained the same amount of sugar, and that no sugar escaped with the evaporating water in some way, or disintegrated in the process. Such questions can be relevant for recycling, not particularly for sugar, but for recycling other materials.

The fact that you really can't get exactly the same lump of sugar back again expresses a general physical principle, applicable throughout nature, living and nonliving, in your tea cup, and in the universe as a whole. This process is called entropic dissipation ("entropic" meaning disorganized, from entropy, disorder). When your lump dissolved, it lost its spatial organization; the organization of the molecules in the various crystals, and the crystals in the lump. In this case, no material needs to have been lost, dispersed, or diffused out of the system, your cup, but in other cases this can happen after the material has disintegrated into smaller units. Those disintegrated units can drift away in all directions, out of sight, in ever smaller concentrations. They dissipate into the environment, like many pollutants do in the soil and groundwater, or into outer space, which is what happens to hydrogen and to all the energy we use. Therefore, the term "dissipation" is often used together with that of entropy; it is the next step after that of the disintegration of something—whether that "something" is material or energy.

The organization of your sugar lump is lost in your tea. The entire universe has been losing its organization since its origin some fifteen billion years ago. All order in the universe is gradually disintegrating, getting lost. And the universe will continue to disintegrate for billions and billions of years until it develops into a vast space of a completely unstructured form of energy: heat, the lowest form of energy. During these billions and billions of years, the universe will be dying a heat death, as it is called, heat of only a few degrees above the absolute minimum, if that. According to some theories, after that stage has been reached, the universe can never be restructured again. Organization will be replaced by disintegration, order by disorder, order by entropy: all the structures you see around you—mountains, buildings, statues, and chairs, all the organizations and organisms, you yourself, down to the molecules, atoms, and subatomic particles of which they consist—all the energy in fact. Absolutely everything. Nothing escapes this general, ubiquitous tendency toward disintegration; everything disintegrates into the vast, silent, unstructured heat. Some believe we are now roughly halfway between the birth of the universe and its heat death.

Because of its universality, this tendency to disintegrate is a basic law in nature. Everything that exists is subject to it. So it is relevant to our considerations about turning our waste back into resources again. Although the principle of the law is simple, its scientific description is complicated, a reason why this process has hardly ever been discussed in the context of all the problems we are dealing with in this book. The principle of the inevitable production of waste at any step we take, and the limitations this entails to recycling. Whatever we do always results in the production of waste, and this waste can only be taken into use again by adding a certain amount of energy.

The same applies to the iron in our cars, ocean steamers, heavy machinery, or kitchen mixers. At some point, they all have to be taken apart into small, manageable pieces and taken to a place where they are melted down, so the metals can be used again through some recycling process. But there is always some loss of tiny little bits and pieces. Since individual parts wear out because of friction, the movable parts in the engine of your car need oiling. This oil has to be refreshed every so often, the waste oil being heavy with metal. Other metals simply rust away in a shed, a derelict factory, or in a storehouse where eventually small wafers of rust are blown or swept away, thus getting lost in yet a different way. Bit by bit, we are losing material; all these larger and smaller bits have to be replaced continually and their loss supplemented.

Worse, when building a car, we do not use pure iron, which is brittle, but always some kind of steel: iron mixed with carbon and other metals. In different proportions, these constituents give the steel the specific properties you need in the various parts of your car. Often, however, unlike the iron, the other metals are rare, found in low concentrations in rocks in a few parts of the world only. When steel gets lost, these rare metals get lost along with it.

Take platinum, a catalyst in your silencer. To obtain it in the amount and concentration wanted, huge amounts of rock have to be destroyed at the cost of enormous amounts of energy. Then, it still has to be freed from the ore, which again takes a large amount of energy, water, and chemicals. And your silencer has to be made and transported, sold and installed—all activities needing energy. When the silencer is finally thrown away rather than kept and dismantled, the little platinum it contains gets lost in the environment, where the silencer decays and the platinum scatters in useless concentrations. From that point on, any attempt to retrieve it will be hopeless because of the amount of energy that this would cost.

Or take gold. In Nevada, for example, heavy machinery removes tonnes of gold-containing rock from open mines about five hundred meters deep. The rock is then transported by trucks, ground to small pieces, and thrown on piles some sixty meters high. As a next step, cyanide is percolated through these piles to amalgamate the gold. After this, the cyanide has to be chemically removed from the gold again. Obviously, all this costs an enormous amount of energy, the chemicals used sometimes leaking away, polluting the environment. Also, the mine has to be kept dry, "dewatered," at that great depth, which means that in a large radius the surrounding land, springs and rivers, plus part of an aquifer containing water thousands of years old, all dry up and will remain so even after the mine has been closed. But now realize that despite the permanent loss of large amounts of freshwater, and apart from the amount of energy and chemicals used and pollution too, three tonnes of ore are required for the production of a single wedding ring containing eighteen grams of gold. And usually, you need two. But then? Well, if you keep your ring on after you have died, this valuable little piece of gold gets irretrievably lost in the environment.

Where have all the tonnes of tungsten gone, used over the decades for filaments in the many millions of light bulbs throughout the world? To obtain nickel, whole mountains were topped off in New Caledonia: to obtain 2.5 million tonnes of metal, 100 million tonnes of ore had to be mined, which meant in turn that 500 million tonnes of rock had to be removed—a mere half a percent of metal from all the rock taken from those mountains. Minable molybdenum makes up just 1 percent of the metal removed from rock. Think of all the energy this has cost, all the pollution it created—the rivers dying and ancient forests cut, families in the local communities uprooted and poisoned. Yet, recycling this metal was not economically worthwhile, which makes one wonder about the future when less energy is available, pollution is more widespread, and the richest ores exhausted. And where has all the metal gone? Dissipated into the environment, the soil, the water, the air? Irretrievably gone?

To give an idea of the dissipation of some metals by evaporation, between 1850 and 1990, the estimated amount of nickel evaporated into the atmosphere was 133,140 tonnes. And during the same period, the amount of cadmium lost this way was 29,780 tonnes, that of copper was 226,700 tonnes, zinc was 1,318,000 tonnes, and lead was 1,591,000 tonnes. All poisonous heavy metals, by the way, probably were washed away by rain or blown away in fumes over thousands of kilometers. Irretrievably gone.

We often use only tiny amounts of metals or minerals to give iron the specific properties we need as an alloy. Moreover, these metals often come from different rocks from all over the world. They were there originally present in minable concentrations of one or a few percent in those rocks, but now, they occur in low concentrations in all sorts of instruments, cell phone screens, machines, engines, or airplanes. And when these artifacts are lost or discarded, a great variety of metals used in the alloys become irretrievably scattered. The following passage from the first page of Amory B. Lovins's *Openpit Mining* shows the huge variety of metals used in an ordinary old-fashioned typewriter, and where they came from.

> The typewriter I am now using probably contains Jamaican or Surinam aluminium, Swedish iron, Czech magnesium, Gabon's manganese, Rhodesian chromium, Soviet vanadium, Peruvian zinc, New Caledonian nickel, Chilean copper, Malaysian tin, Nigerian columbium, Zairian cobalt, Yugoslav lead, Canadian molybdenum, French arsenic, Brazilian tantalum, South Af-

rican antimony, Mexican silver, and traces of other well travelled metals. The enamel contains Norwegian titanium, the plastic is made of Middle Eastern oil (cracked with American rare-earth catalysts) and of chlorine (extracted with Spanish mercury); the foundry sand came from an Australian beach; the machine tools used Chinese tungsten; the coal came from the Ruhr, the end product consumes, some might say, too many Scandinavian spruces.

Once, all these metals and other minerals, though often found in low concentrations in a few locations in the world, were minable, but now they have been scattered across the world to even lower concentrations than before. This example of a typewriter does not invalidate the argument, because all those metals now found in alloys do not occur only in that kind of machine, but in all machines; all metals used are alloys of some kind. And not only that, metals are also used in glass; iron colors your wine bottles deep green, and a little lead gives your fine crystal wineglass a beautiful slightly bluish shine and its delicate tinkling sound. And metals are used in plastic; in Lego®, for example, poisonous cobalt once gave the pieces their nice red color, until its health risks linked to children licking and chewing the pieces and to fish in aquariums became known. Thus, in many products, many metals—often rare—became locked up in small quantities, gradually getting lost somewhere in the environment.

The worry is that these rare metals are running out all over the world. Many of them occur at the bottom of the periodic table of elements; they are the rare-earth metals that are relatively heavy because their atoms are so large. Also, in general, the larger atoms are, the less stable they become internally, and the easier they fall apart into one or more smaller elements, plus a certain amount of energy. Thus, there is a constant natural loss of such metals, though very slowly. Their instability also explains how nuclear plants and fission atom bombs work: they release an enormous amount of energy, thereby leaving a certain amount of nuclear waste: newly formed, unusable radioactive elements. Furthermore, such large heavy metals are more difficult to form, hence, their rarity. Because of their rarity, they are difficult to find and to mine, which means that mining usually costs a large amount of energy, water, and chemicals, and leaves a large amount of polluting waste. However, they are often obtained as contaminants of more abundant elements. Selenium, for example, is a contaminant of sulfur, tellurium is a contaminant of copper, and uranium is a contaminant of phosphorus in rock phosphate, and therefore in fertilizers and on the DNA of a consumer.

Also because of their rarity, their occurrence in the world is usually patchy and very local; tellurium comes mainly from Peru, which produces over 50 percent of this element used in the world. This very heterogeneous occurrence means that only one or a few countries dominate the world market, which inevitably leads to political tensions and instability. China, for example, produces 83 percent of the gallium, 79 percent of the germanium, and 95 percent of the rare-earth metals used worldwide; Australia and Brazil produce 60 percent and 18 percent of the tantalum, respectively; South Africa produces 79 percent of the rhodium and 77 percent of the platinum; and Brazil produces 90 percent of the niobium. Overall, the main players in the world for these rare heavy metals we need for alloys will be China and the United States. Europe has none, so part of Europe's industry will shift to China (where an added advantage is that labor costs are lower).

How can we get around the depletion of these rare heavy metals? A nonphysicist once suggested we make copper ourselves when it runs short. Though possible in principle, this is unpractical. Like the other metals, copper is a chemical element, the production of which takes time and costs large amounts of energy and money. Its production half-life, starting from another element, nickel, is 100 years, which implies that to meet present-day annual demands would cost many thousands of years of production. And those demands would take 1.5^9—1,500,000,000—tonnes of coal. All in all, in 1990, it would have cost 2.5^8 times its current market price. So, for practical reasons, it's out of the question to make it ourselves. The fission products in nuclear plants could be used in other instances, but their radioactivity is disadvantageous. When an element is gone, it's gone.

But we can't do without these elements. Take tellurium: we use it in the manufacture of solar panels, which we increasingly need for our energy production. We also need indium, which is running out; in the last twenty years our use of this element has tripled. It is a contaminant of lead and zinc, both of which are *also* running out. Not surprisingly, in six years the price of indium has increased almost tenfold: from $100 per kilogram in 2003 to $980 in 2006. Or take selenium: unlike other rare metals, we use it in pharmaceuticals or in the flat screens of our cell phones and computers. Moreover, for one application, we need two or more metals to make it work. Hybrid cars in particular gobble up a good number of rare metals—cerium, dysprosium, europium, lanthanum, neodymium, platinum, praseodymium, terbium, yttrium, and zirconium, of which China delivers 95 percent. Half a kilogram of neodymium is needed for the optimal working of the magnet in its engine—as those in wind turbines and other electric engines. This amounts to a million kilograms for the two million hybrid Toyotas Japan produces annually. It can be replaced by dysprosium, but this is less efficient and much more expensive. Terbium is used in the increasingly used energy-saving compact fluorescent lightbulbs. Erbium, ytterbium, and neodymium amplify infrared pulses in fiberglass cables. Tantalum is a pivotal metal in many applications. As an insulator, its applications include capacitors and filaments of cell phones, digital cameras, DVD players, laptops, and computer and television screens; because of its strength it occurs in turbines of power stations and in aircraft; because it is noncorrosive it is used in medicine.

All these—cobalt, gallium, germanium, indium, lithium, molybdenum, niobium, palladium, platinum, rhodium, tantalum, tellurium, titanium, and other rare-earth metals—will run out within decades. All dissipate into the environment at various rates. They all occur in high-tech applications essential for running our future society (and often, those new applications are meant to save energy). Again, how can we get around their depletion?

Furthermore, because of their low concentrations, it is costing increasingly more energy to retrieve the last percentages from ores and alloys than to retrieve elements that occur in high concentrations. As a comparison, it is relatively easy to retrieve water from alcohol, but it costs lots of energy to retrieve the last few percent. Similarly, it is extremely difficult to purify gold for the full 100 percent. So, in practice, there will always remain some water in your alcohol, and some other minerals in gold. This is why your wedding ring, chain, or watch is said to be made of, say, eighteen-karat gold. Also, despite their low concentrations, these "impurities" affect the properties of the alloy you make next after melting it down. This process is exponential, this time not because of an increase of some sort or because of growth, but because of the decreasing amounts of minerals left at subsequent applications. Moreover, retrieving the last percentages out of an ore requires more machines, and produces exponentially more waste.

The same story applies to the phosphorus and potassium used in fertilizers, as well as to helium. Like neon and argon, helium is a nonreactive gas, a so-called noble gas, and is relatively rare. As the second lightest gas after hydrogen, it easily escapes from Earth, dissipating into outer space. It is formed in nuclear reactions such as those that occur widely on the sun, Helios, hence its name. In the 4.7 billion years of the existence of the sun, small amounts of helium have also formed on Earth. The largest reserve is in Kansas, from where it is mined and exported. Because of its very low boiling point, helium is used for cooling the magnetic resonance imaging (MRI) scanners used in hospitals and research, as well as in liquid crystal display (LCD) screens. In neither of these applications can it be replaced by any other element. Another—wasteful—use is in balloons (instead of using hot air from burners). As its price is low, it is considered cheap, although given its limited reserves, it is a gas that is expensive to use. It is estimated that it will run out in one generation, that is, in 25 to 30 years.

In the subsistence economies of the past, some of the nutrients were returned to the same field as manure, thus cycling 'round in the same environment for a long time. Yet, from early on, the soil nutrients did become exhausted, except along big rivers like the Nile, where the nutrients lost—dissipated—over the year were replenished at the end of the year during the annual inundations—replenished by nutrients from other areas that lost them, that is. In all other cases, animals had to supply new nutrients with their dung after having taken them up in neighboring areas. But as soon as trade in food began to build up, nutrients were removed from the local fields at higher rates and were replaced only partially, if at all. The nutrient contents of those fields declined slowly but surely, until the fields were completely depleted of their minerals, after which people abandoned them. In cases like these, as in those of iron, uranium, platinum, and other metals, the chemical elements and minerals concerned, our nutrients, were gradually scattered over ever larger areas, or over large volumes of water, the oceans, never to be retrieved again. Dissipated. At that point, the natural fertilizers had to be replaced by artificial, chemically produced ones, some made from fossil fuels, and some, like phosphorus, from rock. Thus, similar to our energy, we are living on the pocket of the past. But, of course, these chemically produced nutrients too are spilled into lakes and rivers, the sea, and eventually into the oceans via the sewage systems of our cities. There, they dissolve, diffuse, and disappear forever from sight. Nonretrievable. Nonrecyclable.

In many cases, however, metals or their alloys can be replaced by other materials: copper has been replaced by fiberglass in regional and submarine telephone cables; nylon derived from fossil fuel is used in cables and cloth-ing. Some of the parts of the keyboards of our typewriters, like the one men-tioned a few pages before, are now made of different materials. Yet, in other cases, this isn't feasible. It is just as impossible to replace some chemical ele-ments in our own physiology or in the biochemistry of organisms in general. The compounds concerned often evolved from ancient chemical structures formed in the chemical environment of the time, an environment which has changed over the billions of years since then. Iron, for example, was once very abundant in solution in seawater, which explains its presence in many chemical compounds in present-day organisms, such as in the hemoglobin of your blood, however rare it is relative to its former abundance. Its present concentration in seawater is ninety thousand times less than in our cells. As an essential element of the hemoglobin in your red blood cells for transferring oxygen from your lungs to all cells of your body, strong chemical pumps are now needed. Yet, it is still abundant enough in our food not to give problems of acute iron shortage.

Contrast phosphorus. Probably from early in the evolution of life onward, this element was present in certain enzymes, coenzymes, compounds that facilitate reactions between other compounds, speeding them up and thus making them possible within a biochemical context. From a simple, still ubiquitous little molecule, adenosine triphosphate (ATP), concerned with the transfer of energy within all cells, a great number of other, often highly complex molecules evolved, collectively known as nucleotides. Their name gives a clue about where they are found in great abundance: the cell nucleus, where they are the backbone of deoxyribonucleic acid—DNA. This large and enormously complex molecule is pivotal for the working of all cells in the world—all bacterial, fungal, plant, animal, and human cells—and in the transfer of our genetic properties. It is probably the end product of the long evolution from the primeval ATP, which itself evolved from initial phosphate strings. However, there will have been a stage in-between, before DNA evolved and in which another, very similar but less stable and less complex molecule, ribonucleic acid (RNA), had similar properties and a host of biochemical functions. In recent decades, a whole suite of different RNA has been discovered, suggesting that underneath the chemical stratum of the nitrogen-based proteins, there is another, still-active one based on all those primeval nucleotides and ancient RNA. This biochemistry based on all those RNAs still reminds us of an ancient RNA world previous to our present biochemistry which one could call the DNA world.

What is important from our perspective is that a very large number of different molecules is active in many key functions in the cell, all based on phosphate, which is a compound of phosphorus, hydrogen, and oxygen. And those key functions are irreplaceable, deeply embedded in the chemistry of each of our cells. They form the basis of our chemistry, indeed of the chemistry of all known organisms on Earth, primitive viruses included. We cannot do without them. But we have depleted most soils from their phosphorus content and now have to add this element to those soils with our fertilizers. This makes us, in turn, completely dependent on those fertilizers and, ultimately, mainly on the mines in China and the western Sahara and Morocco from which we obtain most of our phosphorus, that is, 27 percent and 43 percent, respectively. And these mines are expected to be depleted before long, that is in the next seventy to a hundred years. In principle, though, 80 percent of the nitrogen and 90 percent of the phosphate can also be retrieved from wastewater by means of the bacterial anammox—*an*aerobic *amm*onium *ox*idation—process, intended to convert ammonium directly into nitrogen gas, but this technique has not yet been widely adopted.

What is happening is that here too we are following the linear way of phosphorus processing rather than the cyclic way farmers and townspeople obviously followed in the past when they returned manure and night soil to the fields. At present, we waste our sewage by disposing of it into the ocean from where we cannot retrieve it. Not so long ago, we even polluted the environment with large amounts of detergents containing phosphorus. Furthermore, much of the topsoil containing some phosphorus is eroding away, also ending up in the oceans. Overall, at present, more biochemically essential and irreplaceable phosphorus dissipates out of our environment than we are mining. We don't retain anything; our budget is negative. All the phosphorus be gone by the end of the present century. Given foreseeable food shortages, the race is also on between big mining and agricultural firms for potassium, another irreplaceable element in our food and, hence, in fertilizers. It is found in only a few main mines in Saskatchewan, Canada; Brazil; and Kazakhstan. And still no alarm bells ring when we talk about reaching some stable population level of nine to ten billion during the next few decades, or—according to some—possibly one billion more than that by the end of the century, or when we talk about having to produce 70 percent more food in 2050. How can we reconcile these facts?

Similar to our lump of sugar, the initial concentration of our metals and nutrients is lost during "digestion" and dissipates into the ocean. To restore their low concentrations—even if this were feasible—would need vast amounts of energy. Yet energy itself is subject to the same process when it degenerates into heat. Energy is not something uniform; rather, it occurs as tiny packages, energy quanta, which are not identical. To describe them, physicists use an equation in which a constant is multiplied by a variable. The constant, of course, is always the same, but the variable is not: its value varies, as the word indicates. Since such quanta can be considered as a wave with a variable, yet specific frequency of undulation over time, the variable component varies with the wave frequency. There are long wavelengths with low frequencies over time, such as heat radiation or those we use for receiving radio messages, and short wavelengths, such as those in harmful ultraviolet radiation, or in x-radiation as used for making radiographs—x-rays—in hospitals. X-radiation passes readily through most of our body tissues, except for our bones, and got its name because at the time of its discovery its effect and origin were mysterious. Heat and radio waves contain only a little energy in the resulting package, and ultraviolet and x-radiation much.

Heat radiation is therefore energy of long wavelengths close to radio waves, representing only small, energy-poor quanta. When energy is used in a reaction, the large, energy-rich quanta are "chopped up" or degraded, resulting in a greater number of quanta, each of which represent less energy than the original quanta. This means that energy loses some of its earlier, higher organization and concentration.

Some people think that we could restore energy-poor radiation by adding other energy. However, if we could do this, the other energy quanta being used would lose their own organization in the process, so that in the end we would be none the better. So, in practice, we cannot make any new, organized material to replace what we have lost by dispersing it, and neither can we make any new energy-rich quanta. We are always losing high degrees of organization and concentration, and moving to disorder. Thus, materials can to some extent be recycled, but energy cannot; it always follows a linear route of processing.

On the other hand, the fact that materials cannot be recycled for the full 100 percent but often for much less than that means that we have to allow for time: given enough time, all recyclable materials also follow the linear processing route. And this applies to all renewable materials. Every year, we can harvest a new crop; every year, we can slaughter a new cohort of animals. But we can do this only for so long because bit by bit, we are reducing the amount of soil nutrients on which the next yields depend, eventually depleting them. This has happened to all large civilizations, and it is happening to ours. The only thing is that usually we don't see it happening, because artificial fertilizers are compensating for the loss. A simple and rough calculation shows our dependence on fertilizers. In their absence, concerning nitrogen and phosphorus, each of us needs on average 0.6 hectares of land to feed, which amounts to 6 billion hectares for 10 billion people. However, of the 15 billion hectares of land on Earth, only 10 percent is cultivated, that is, 1.5 billion hectares, which might be increased by another 0.8 billion hectares using modern technology. This amounts to 2.3 billion hectares, leaving a shortage of 3.7 billion hectares, which has to be covered by fertilizers obtained from fossil fuels. Over time, therefore, all renewable resources prove to be nonrenewable. The distinction often being made between renewable and nonrenewable, or recyclable and nonrecyclable materials is not absolute but depends on the time period chosen, as well as on the rate at which we use the material concerned. This rate, in turn, often increases, depending on the number of people and the consumption per head, which are both increasing. Therefore, the question is how long do we envisage humanity to exist on Earth, in what numbers and with what demands?

So, order is continually replaced by disorder, but we also know that life, as it develops over the billions of years in one individual or between individuals, has become increasingly more complex—much more highly organized and therefore well ordered. And, together with social bacteria and animals, we humans are building up and maintaining highly intricate, well-organized social systems, apart from highly intricate instruments, computers, or high buildings. This costs energy. Somewhere else in the process, therefore, energy degrades and increases in entropy and thus loses its order. According to the general law of physics, the net result is a loss of order, an increase in entropy. Whatever we do, how much we recycle or build up, on the whole we are breaking down. We continually need to add energy to keep up the organization of our own body, to keep its order, and we need more to make it grow and adapt to the situation. And the same applies to the organization of our society. As well-ordered organizations, as organisms and as a society of organisms, therefore, we represent a certain amount of energy used up over many years of building, maintaining, and adapting: a certain amount of virtual energy. We are a dynamic flow of energy with a history—a complex, ever-changing form of channeling entropic decay of energy and material dissipation. To replenish this continual loss of energy, this flow has to be kept going, which happens by plants capturing solar radiation, and by ourselves by eating those plants and by burning their fossil remains.

When life originated, the environment was energy rich, but over billions of years with hydrogen bound chemically and with poor electron-pulling, electron-rich elements being substituted by strong electron-pulling elements, the environment became energy-poor. During this long geochemical process, the constituent compounds of organisms became more stable, consisting of ever-stronger electron pullers, a trend which may have been a major, long-term driving force in the evolution toward ever-more complex living systems. As the chemical energy in the environment ran out, to keep going the living systems also had to accept more and more, new, rich energy from the sun's radiation.

With the help of solar energy, first bacteria, then algae, and later, plants on land produced oxygen by splitting it off from the hydrogen in water molecules day after day, eon after eon. Most of this hydrogen, being the lightest element, escaped from Earth into space, never to form part of the earthly set of molecules again, leaving oxygen behind in the atmosphere. That is how, first, the atmosphere and then the oceans began to contain oxygen—becoming aerobic—followed by the soils of our terrestrial environment. The soils became the realm of aerobic—oxygen-loving—bacteria, fungi, and earthworms. All of them brought dead plant and animal material back into circulation by breaking it down, forming a large part of the living sphere on Earth, the biosphere, organizing, reorganizing, thereby gradually degrading energy. Just as in the case of attempting to reconstitute the sugar lump, this biological recycling needs much chemical energy, and would not have been possible if the sun had not have supplied life with its daily light.

By breaking down plant material, fungi and animals release the solar energy stored in the plant's molecules then apply it to their own use. When we burn gasoline to make our car move, we use the same energy but this time from fossil algae and plants, and, similarly, the reaction products of carbon dioxide and water are released as hot gases, mix with the air and warm it up: waste gas and waste energy. If we needed to use these gases again—to make a plastic bag, for example—it would cost much energy to retrieve them and then we'd need more energy to be able to recombine them into the kind of plastic we want. Our human way of using energy is extremely wasteful, even when it is used for recycling material.

Many people feel that we ought to replace our linear resource use by a circular one in which waste is permanently recycled. But to rely exclusively on recycling of material to save the world raises false hopes. Recycling efficiency can never reach 100 percent, and it costs a large amount of energy when we do it ourselves. And most of our waste cannot be decomposed by bacterial action; bacteria neither have the chemicals nor the energy to break it down. And if they do have the chemicals, they take a long time to do it—longer than we can wait before we can apply their products as new resources again. Our numbers and habits require high turnover rates that cannot be attained by bacteria and fungi, however numerous they are. On the whole, however, in terms of energy, our waste consists of end products exhausted of chemical energy; they are extremely stable. So, one of our problems is that we are rapidly running out of energy. Moreover, given the limited amount of material to be recycled, we also need a very high turnover rate, and therefore a large amount of extra energy, which we simply don't have—at least, not soon enough.

Apart from the need to allow for the energy recycling costs, we also need to look at the material waste we are still making: recycling is only one phase in the flow of materials used. Broadly, every tonne of garbage to be recycled has cost twenty tonnes of waste to its mining, pumping, logging, or its transport and another five tonnes at its manufacturing stage. As recycling cannot be perfect, we are still making those material costs, plus have to account for the pollution the process as a whole incurs. Other people feel that, because of the limited amount of material left, remanufacturing old material instead of recycling it might solve the energy problem inherent to recycling. This is true but old material, fridges, computers, heating systems, et cetera, use more energy than the improved, more efficient apparatuses do. This, therefore, can't be a way out either.

By isolating the problem of resource depletion from all the others we will face, proponents of the cradle-to-cradle principle feel that our use of materials can continue to grow. Simply put: just speed up the rate of recycling to keep pace with our growing numbers and demands. However, our overall consumption isn't a cyclic one and our numbers and demands are growing linearly. But apart from this, the law of physical degradation means that purely cyclic systems can't exist but always contain a linear component of material loss. Even the artificial generation of energy from solar radiation requires materials to be used, and these will wear out or decay over time. We therefore always need to search for new resources and for new places to dump our waste responsibly.

As the hunter-gatherer populations grew, it became necessary to settle permanently, then to organize the settlement, and so on, requiring ever-more people and material use, and reducing the number of places still unaffected by waste. Population growth required further population growth; it required growth in material and energy use, which in turn introduced a self-accelerating growth of social and economic organization. At present, our vast and complex global socioeconomic superstructure is growing faster than the human population itself, requiring more and more people; more, faster, stronger, and more automated machines; more energy, and more chemicals. Thus, both population and socioeconomic growth are self-accelerating, and, on top of this, are enhancing each other's rates. Socioeconomic growth cannot be stopped or brought back to more amenable proportions by us being more economical or by introducing recycling of materials yet leaving our numbers to grow or even to stabilize at the current level. This all follows from the basically linear way of processing of resources. The linearity of the system makes growth unavoidable and, hence, unstoppable—unless we manage to control its basic driving force: the growth of our numbers and personal demands.

Proponents of the cradle-to-cradle principle isolate our problems from their central driving force: our growing numbers accompanied by our growing demands per head. They negate the basic physical law of entropy. What is speeding up our wasting of our world is a combination of several processes that lead to energy depletion, to the depletion of plant nutrients and water, to deforestation and its consequent problems of erosion and inundation, and to problems of weeds and diseases. All issues that are discussed in this book.

The way we use energy and material resources is wasteful, so in the end, we will have to pay the price. We are degrading the physical organization of those resources and are dissipating them faster and faster, without thinking about a longer-term future. But the price of our profligacy will have to be paid by future generations, without them having gained anything from it. We will leave them with very little; soon energy and resources will be irretrievably gone, degraded and dissipated.

Limitations Beyond Our Capacity

When large numbers of individuals of any species—plant or animal—live close together, their standard of living, and, ultimately, their eventual local survival are affected. This applies to us too—we are no exception. And unless we take measures ourselves to reduce our reproduction and consumption, we will inevitably feel the consequences. These consequences are beyond the capacity of each of us as individuals to deal with. And although self-limitation in reproduction will often be difficult, if not painful, the consequences of not curtailing reproduction will be far worse.

Obviously, our growing numbers are closely intertwined with the depletion of our resources of food, energy, and materials; generating waste and pollution; and the problems of large-scale recycling. Although recycling can only be a marginal and temporary measure, it must become a central activity, just as it has become central to all chemistry within and between organisms. So far, we have used and exhausted our resources without significant recycling. But with the exhaustion of our resources in sight, humankind will be forced to eke out its existence by recycling, even though this will use up more and more of our dwindling energy supply.

C
Exhausting and Wasting Our Environment

Our environment can be exhausted in various ways: by direct use, such as in the case of freshwater; by destroying natural vegetation; and so forth. Vegetation destruction leads to erosion, for example, as well as to warming and drying up of the soil of large expanses of land where the vegetation is missing. Also, our environment can be wasted by overexploiting the land, that is, by removing minerals from the soil on which we depend, or by polluting it with our waste. Apart from wasting the land, we also waste the air we breathe by expelling poisonous gases into the atmosphere, or we leave large amounts of plastic to drift into the waters of the rivers and oceans.

10 OUR FRESHWATER IS RUNNING OUT!

Most of us know the tale of Robinson Crusoe being shipwrecked on an un-inhabited island. The first thing he did was to find some freshwater. Fortu-nately, he soon found a creek from which he could drink, and surely it must have been difficult for him to stop! You know yourself how thirsty you can get after a single hot day with hardly anything to drink. In many climates, it's easier to go for one or two days without eating than without drinking. It has been found that even fish have to drink.

It would not take long for Robinson Crusoe to discover the place where his creek originated, its wellspring. From time immemorial, springs have been known as places where the most delicious, cool water mysteriously emerges from the soil or hard rock. Understandably, many myths and leg-ends have found their origin in this mysterious welling up of water from underground, the underworld. According to one version of a classical Greek myth, Europe may have been named after the Phoenician princess Europa. She was chased by the supreme god, Zeus, went underground, and finally surfaced from a spring on the island of Sicily at the spot where later the town of Syracuse arose. You can still visit the little pool there where this is sup-posed to have happened, long, long ago.

Because of water's crucial importance for our survival, from the earliest times of the most ancient civilizations—and actually from millions of years before then—human settlements have always been found close to a river, and often near its mouth into the sea. Rivers, lakes, and seas supply us with drinking water and fish that are easy to catch; freshwater still irrigates the land on which we plant our crops. The greatest diversity of life still occurs in the oceans and seas where life originated. Unfortunately, rivers and seas are also where we have always dumped our sewage, pollutants, cooling water, plastic, and garbage.

Some four hundred million years ago, after life had been evolving about three billion years in the sea, the first plants began to invade the land. This was a completely new environment for them; they were no longer immersed in water but were surrounded by dry air and exposed to direct, hot sunlight that could dry them out. Lying there in the drying mud of the last little pools, they soon lost all the water from their cells, and many shriveled up and died. So, their first problem was how to develop thick, supporting cell walls, with the thickest walls on the outside, to protect them from desiccation. These proto-plants were weak, still unable to stand upright as most present-day plants do. After some time, however, they also developed mechanically strong tissue that could support their weight, although this made them even more susceptible to desiccation, as they were then much more exposed to the drying air, which now surrounded them on all sides, instead of only from above.

Yet, it would not have been possible to stop evaporation completely, so they needed tissue to transport water from the moist soil to their exposed parts standing high in the drying air. And, as had been the case in the sea, they also needed water that they could break down in order to obtain the hydrogen they needed to create their carbohydrates from carbon dioxide. Then they could store their solar energy in starch and build up their supportive tissues from other carbohydrates, cellulose, and lignin. Those compounds store most of the energy they need in each of the numerous chemical bonds in each of their large molecules. To release the energy, we as animals burn this hydrogen with the oxygen we breathe in. Energy we use for walking around, for talking, for everything we want to do. But the carbon dioxide had to come from the air, and any inlet for this gas would allow some water to escape. Obtaining and retaining water was a major problem for those early plants, for sure.

Water is not only of concern to plants and animals and humans because they have to take in enough of it not to dry out, but it is of pivotal importance to their chemistry. In their chemistry, water is continually split up and fused together: plants split up water molecules, and animals put them together again. Plants use solar energy for splitting up water, and animals use it by reconstituting the water. The bond in the water molecule is very strong and releases much energy when it is made, which is why the mixture of oxygen and hydrogen—oxyhydrogen—is so explosive. The energy "stored" in the chemical bond of the plant's carbohydrates is mainly the energy that will be released in a future reaction between hydrogen and oxygen; it is potential

energy carbohydrates preserve. So, strictly speaking, we use the energy released by making the bond between hydrogen and oxygen; for their energy
economy, it is the hydrogen of the water molecule that plants and animals are
particularly interested in. Thus, all of biology, all our life, runs on an ancient
hydrogen economy, no more, no less.

Soon after the plants invaded the terrestrial parts of Earth, the first animals also arrived on land. These animals, similar to the plants, were at first
still living in pools, and then more and more outside them. Typically, the first
vertebrates on land after the fish, which were still restricted to water, were the
amphibians, living in wet terrestrial environments but still reproducing in
water where you can find their first developmental stage, tadpoles. Similar to
plants, their problem was that on land, they unavoidably began to evaporate
the water from their body tissues and dried out quickly. Thus, they became
thirsty and needed to drink, which also tied them to the pools. Yet, as life
evolved over the next millions of years, their skins became more and more
watertight, protecting them from desiccation. In the end, they evolved into
reptiles, which you can even find in deserts. As tadpoles, amphibians evaporated part of the water through their gills, and later in adult life through their
lungs, which we as humans also do. The main way we still lose water, in fact,
is by evaporating it from our lungs. Many of you know this from walking in
the freezing cold of the winter: then, you can see you breathe out those little
clouds of water vapor. Also, think of those film advertisements with horses
fiercely rearing up, breathing out steaming water vapor. And when it's really
cold, the tiny drops of that damp from your breath freeze immediately, forming beautiful, shiny crystals of ice, exquisite tiny little stars of all shapes.

Actually, we lose a much larger part of our water by breathing out than by
urinating. Depending on the climatic conditions under which we live, tropical or boreal, overall, we lose two or more liters of water each day and therefore need to drink the same amount to compensate for that loss. We take in
water not just by drinking but, without realizing it, also by eating. All our
plant and animal food mostly consists of water, often amounting to around
90 percent of its weight. And when we want to eat hard, bone-dry peas, rice,
or pasta, we first need to boil it in water, which means that we add roughly
the same weight in water to it. Then, by digesting those boiled peas, rice,
and pasta, or our potatoes and onions, we break down the chemical bonds
of the carbohydrates this plant food contains, whereby, unintentionally, we
also produce some water. As said, this water comes from releasing the hydrogen from the carbohydrates, and then it reacts with the oxygen we inhale,
producing water. Some desert animals, such as the big oryx in Saudi Arabia,
never drink water but use the so-called physiological water they obtain by

digesting dry grass. You may also recall elephants in the zoo, munching away their daily ration of a tasty bunch of crunchy branches. Desert animals also have special water-retaining tissues in their noses, kidneys, and intestines so that they hardly lose any moisture. Not in their breath, not in their viscous urine, and not in their paper-dry feces. And they don't sweat.

Therefore, either by drinking water directly, or indirectly by taking it in by eating, each day we obtain the water we need. However, when we eat a plant or its seed, it has itself taken up water during all the weeks or months it has been growing, just by its uptake of water and carbon dioxide for making its carbohydrates. All the time, it needs to grow its spinach or cabbage leaves, its carrot or sweet potato tuber, its wheat or maize kernels, its fruits, apples, cherries, and plums. And this is the same for oranges, bananas, and pineapples, and for all the other tropical plant products you eat. All the time the apple, the cherry, and the orange trees are also forming the wood of their stems, their roots, and year after year, their new leaves and juicy fruits. In that way, plants not only have stored the—potential—energy obtained from the sun into their chemical bonds, but in order to do this, they have also stored the hydrogen split off from the water taken in through their roots. They stored it chemically. Moreover, in order to be able for us to use the real and the physiological water, we also have to allow for all the water that has been evaporating during that time. Surely, we can impossibly directly see or weigh this latter part of the water it had taken up to survive, just because it has lost it already partly by evaporating it back into the atmosphere. It's gone!

All this unseen physiological and evaporated water has been called "virtual water"; you cannot see or weigh it, but you know that you've given it to the plants. In contrast, the water you can squeeze out of a plant or obtain as a juice from a fruit like a cherry or an orange you might consider "real water." But although you can't see or feel it anymore, farmers have to allow for this virtual water because it's they who have to water the plants all the time or irrigate the land on which they grow, and so forth, before anything can be harvested. To the farmer, this water is not virtual at all—it is damn real. They have to work and pay for it. And the land has to supply it. In fact, the amount of virtual water can often be many times the amount of real water of which the plant or its fruit you are eating or digesting actually consists. Eating a plant therefore means that you are drinking directly—the real water in the juices and tissues—and indirectly, the physiological water stored chemically, and even more indirectly as evaporated virtual water. Because of the farmer's efforts, and because, ultimately, the environment has to bring it all up, the recognition of virtual from real and physiological water is more than an academic point to be made only.

It has been calculated, for example, that by drinking one single cup of coffee, and accounting not only for the little real water in your cup, but also for the virtual water, accounted for by the annual and life history of the individual coffee bush, you are in fact drinking no less than 140 liters of water. And the production of one tonne of wheat requires a thousand tonnes of freshwater. That is all for the farmer—and the soil—to supply. You can imagine how much water—real plus virtual—you and I, and all the other animals, are actually taking in by eating plants or by consuming their products in general. But now imagine, as a next step, eating the meat of an animal like a sheep or a cow. What applies to plants also applies to the animals having eaten the plants: they had to grow one or more years before they were slaughtered, and during that time their tissues formed and had to be maintained for many weeks or months or years before they could be eaten by us. All that time, they drank and ate their plants (and therefore drank their large amounts of virtual water). And all the time, just like the plants they ate, they evaporated water, breathing it out and compensating for its loss by drinking and eating. Now think of a lion, eating a wildebeest or its calf. And by eating plant food and meat all our life ourselves, we too are worth a really large amount of virtual water as well. In fact, this chain from plant to wildebeest to lion can only be so long, which is usually not more than five subsequent links.

So, animals too, represent a large amount of virtual water, apart from the relatively little real water they contain. Throughout their life, plants accumulate virtual water, and sheep and cows accumulate that accumulated virtual water. One glass of milk, for example, represents about 200 liters of water, and when we make cheese of this milk, we lose a large part of the real water that the milk contained. This means that the amount of virtual water we are actually using by eating cheese is even greater. Furthermore, one egg represents 135 liters, whereas one single hamburger, including the bread, represents 2,400 liters. Did you realize? And your comfortably dry leather shoes are equal to 8,000 liters! This is staggering. It is this amount of water that the land has to bring up, not at one single moment, but, depending on the type of food, over one, two, or three years. Independent of the temperature. Increasing temperatures and their consequent desiccation of the soil only make it more difficult to obtain enough water all the time. Not all parts of the world are therefore equally suitable to produce any kind of crop, of fruit, meat, or cheese.

The amount of water we are thus using unwittingly by eating plants or fruits is hundreds of times more than the amount of real water we directly take in from them, and by eating meat or cheese, it is thousands of times more. In fact, for producing one kilogram of wheat flour, you need a thousand liters of water, and for one kilogram of beef, you need thirteen thousand. All this continual use of water puts a heavy burden on freshwater—and limits the amount available on Earth. Yet, one can say that a large part of the human population on Earth has always used fish as a protein source and hardly any meat from cattle, if at all. They mostly confine themselves to eating plants, their roots, or their little fruits, the kernels of rice, maize, millet, sorghum, or wheat. Yet, during the last few decades, in those parts of the world like China, India, and South America, as people earn more, they are eating meat and other animal products, just as the rich in the West have done for centuries. And most Africans will do the same, although perhaps at a later date. Overall, during the last fifty years, the world population has trebled, but meat consumption has increased fivefold. The burden put on our freshwater resources will therefore greatly increase in the near future, and the increase will be hugely magnified if the world population does indeed increase by another 50 percent in the next couple of decades.

What applies to our food also applies to our clothes. Formerly, most clothes were made of the wool we obtained by shearing sheep. And sheep and their wool also represent, apart from real water, a large amount of virtual water. Therefore, like the leather of your shoes representing those 8,000 liters of water, the wool of our clothes also represents a large amount of virtual water, however dry those clothes may feel. Therefore, your woolen sweater is real, virtual water. The last couple of centuries, our clothes also have been made of cotton, a fiber obtained from the fruits of a cotton plant. This plant actually needs a large amount of water to grow, so its dry fibers consist almost entirely of virtual water, similar to wool.

Together with the globalization of industry, its products, and its way of life; trade; and agricultural production, one important aspect is now the globalization of water—mainly of virtual water, that is. This is obvious in the case of cotton, for example, being used for clothes all over the world but which can be grown under Mediterranean conditions only. Thus, Europeans consume very large amounts of virtual water from India in the form of the cotton of their clothes, and North Americans from both India and China. Being Mediterranean, however, these parts of the world are relatively hot and dry and don't have large rivers from which they can draw water for irrigating the land. Yet, these often intercontinental flows of virtual water are measured in millions of cubic meters, often amounting to hundreds, or tens of thousands of millions of cubic meters. In fact, a large, inland lake, the Aral Sea, disappeared along with its thriving fishing industry when the rivers feeding it were pumped dry in order to irrigate cotton fields. The United States still spends a large amount of their exhaustible fossil water from deep aquifers on growing cotton.

Because of cotton's great thirst for water, large amounts of virtual water are now in fact transported over large distances, often from water-poor parts of the world to those rich in water, such as Europe or the United States. This applies equally to other plant products, and even more so to those originating from animals. Thus, some parts of the world save water, whereas others gain it. On average, India and China are the greatest water importers, and water poor Australia is the greatest water exporter. Yet, the cotton-growing farmer has little to do with those averages; he has to irrigate his fields from the very scarce water around, water that his plants evaporate in no time. Consequently, large multinational food corporations, for example, grow their crops only for so long as water conditions permit, leaving the ravaged, desiccated land as soon as production becomes uneconomical.

The use of and trade in virtual water in particular can only increase greatly in the coming decades, whereas the amount of fresh surface and groundwater and that from aquifers will decrease. Worse, one may expect global shifts to occur in water-saving and water-gaining parts of the world, particularly as a result of warming conditions because of climate change. For example, the already dry parts of southwestern United States, northeastern Africa, India, and Australia will suffer from warmer, dryer conditions from both a generally warmer climate, as well as during droughts from more intense El Niño effects.

In principle, the same reasoning applies to minerals and energy: plants use many minerals and much energy, which, similar to part of the water they took in, we export from their place of origin to elsewhere. Before we harvest and export them, the plants accumulate a certain amount of virtual minerals and energy. We saw this happening already with phosphorus and potassium which are taken up by the plants over their lifetime and of which a part is taken away to the market elsewhere. This means that each year the soil needs to be fertilized with mined phosphorus and potassium again to compensate for this loss. Thus, plants and animals represent more than virtual water alone, but so do minerals and energy. In coal, crude oil, and natural gas, moreover, these virtual components are fossilized; the nitrogen in our fertilizers comes from splitting atmospheric nitrogen by burning natural gas.

As to real water, we use it not only for drinking but in many other ways, and this too in staggering amounts. Thus, we are also using water directly for cleaning ourselves, for cleaning our food, and for cleaning our clothes, or our house and car. And we are using it also in large amounts in our bath tubs or in the swimming pool some of you may have in the backyard, or we use it for watering the lawn. Think of those town conglomerates and megacities, often surrounded by extensive shantytowns, think of how much water they need from one day to the next. And most of those large cities lie in the dry subtropics. Similar to ancient Rome with its extensive and dense network of aqueducts, but even more so, they have vast catchment areas for getting in freshwater, as well as large areas for depositing their sewage waste. When you drive through them or fly across such vast areas where millions of people live, you wonder where all the water comes from, where all the waste water is taken.

Moreover, as society becomes more complex, office buildings, industries, hospitals, and research laboratories arise and grow bigger and bigger, up to the endless industrial and science parks we see today, stretching from one horizon to the other. And from their start, they rapidly use increasing amounts of water. Industrial plants use it, for example, for mining and washing metal ores or for the cooling of metals in subsequent phases of working them. Your car represents many thousands of liters of virtual water.

Open mines for metals, minerals, and coal, moreover, lower the groundwater level considerably, and this over large distances around the mine. They have to be "dewatered" continually, and the water is expelled as useless waste. And the mining of tar sands requires enormous amounts of freshwater as well, not only locally, but regionally around the area affected. Other industries and most laboratories use water in some phase in the development and production of the multitude of chemicals. And ever-increasing amounts of water are used directly in the building of our megacities, in all the towering apartment buildings made of concrete and in building the roads connecting them, in the sewage systems, and so on. Last but not least, farming uses much freshwater for irrigation, which then evaporates, not only from the animals and plants but also from bare soil.

All these forms of using real freshwater, apart from drinking it or taking it in indirectly in vast amounts from the plant and animal food we eat, is typically restricted to human society. No other animal in nature has a car to be washed, no one has a shower or a bath in the morning (except for some birds, blackbirds, for example, splashing around in a little watering place). We, however, are unique in rapidly using up all the freshwater resources found on Earth—rainwater, river water, surface water, groundwater, and fossil water pumped up from hundreds of meters deep.

Human habitation is therefore mostly confined to rivers, as well as to the coasts of all continents. And, of course, from his earliest days, mankind also needed the rivers and the coasts for trading and for traveling by ship. Trading and traveling has been happening since the first civilizations, such as Mesopotamia, Egypt, China, and India; and, of course, the eastward expansion across the Pacific could only be done by canoes. But still, people mostly needed the creeks and the rivers for their fresh drinking water, and formerly also for obtaining fish, game, and later for growing vegetables and crops. In those ancient civilizations, surpluses of food and water, obtained through irrigation mostly by water from the rivers, allowed towns to grow, and industries to be founded.

Yet, even in those days, towns and industries were given priority for using freshwater over the irrigation of the fields for the growing of food. Surprisingly, this is what is still happening today all over the world. More recently, priority is usually also given to the generation of energy in the form of hydroelectricity by constructing vast water reservoirs behind often incredibly huge dams. Reservoirs fill up extensive mountain valleys, and dams reduce wildly rushing mountain torrents to mere trickles, occasionally being endangered by useless floods. All but one of the rivers in Sweden have been harnessed for hydroelectricity. (One river was spared, as a reminder of the past.) No fish can live there, no farmer can tap water for watering his animals or plants. And thus many fishing grounds and much of the best agricultural and cattle breeding area in those valleys are lost—as are areas where human habitation had been concentrated for centuries. Thus, the eternally fertile Nile lost its millennia-old annual irrigation and fertilization capacity, as well as part of its historical background. In that case, whole mountains with temples and sacred statues built inside were occasionally sawed to pieces, removed, and reconstructed elsewhere. Some temples were donated as a reward for help and thus scattered across the world. And similar to most other reservoirs, the one behind the once-famous, but meanwhile infamous Aswan Dam has rapidly been filling up with the badly needed, fertile silt, thus reducing the utility of the socially, economically, and financially expensive dam and reservoir for the generation of electricity. However, the silting up of the reservoir led to a reduced production of electricity and to the rapid erosion of the Nile Delta and thus the loss of much agricultural land and fishing grounds. All reservoirs all over the world have a restricted lifetime, because they gradually silt up. One fears in this respect for the fate of the gigantic Three Gorges Dam in China as one of the many hundreds of hydroelectric works built there, which is ruining the livelihoods and lives of many millions, as it stands in the place where former residents' productive land and their food, as well as the biotope of many species of wild plants and animals, had been for millennia.

Because of the need for water for irrigating the land and also for supplying towns and industries with water, groundwater levels all over the world are dropping. In northern China, for example, the groundwater level drops by one meter each year because of overdrafting for wheat production. In many parts of Europe, this came after we first drained the wet and marshy land on a large scale as a way of land reclamation in order to increase crop land. Thus, the Dutch became widely known in England and Prussia for their land reclamation and drainage skills, then for canal building in northern France, and finally for the regulation (or "taming" or "correction") of big rivers, such as the upper reaches of the Rhine. Also, traveling through Italy, you typically recognize the ancient habitations, its pretty, old cities perching high on top of isolated hills, and the modern ones built from the nineteenth century onward in the drained, fertile valleys in-between. The townspeople there are genetically ancient and diverse; apparently, it was hard to go from one town to the next. Up to that time, many valleys contained marshes, so people were almost driven up the mountains to live (the marshes being the home of malaria for centuries, whereas the isolated hill towns also kept relatively free from the disease). All those marshes have been drained, and now need to be irrigated. At present, the former frog-rich, marshy country of the Netherlands too, which literally means the Low Countries, has dry ditches and needs to be irrigated. The same happened to Ferrara in north Italy, once famous for its frog dishes. This is similar in many other countries in northwestern Europe, whereas formerly, they were all dominated by marshes, fenlands, and meandering rivers. Large areas across Europe, once rich in freshwater for drinking, agriculture, and fishing, are now facing occasional water shortages during the dry season. The Low Countries now import more than 80 percent of their water as virtual water.

Moreover, all over the world, many of the large rivers are at present seen to run dry. At first, the smaller creeks dried up, but this gradually shifted to the larger ones as well, and it now applies to the main rivers draining vast parts of the continents. In this way, for example, a major river draining the western part of the North American continent, the Colorado River, does not reach the ocean anymore, and the Mississippi, which drains the eastern half, appears to be following suit. The same happens to the Murray River in eastern Australia and the Ganges in India, whereas in Brazil, the legendary Amazon, the river of all rivers, is in danger. In fact, the farm lands and towns being supplied with water by the once mighty Yellow River in China now have to get their water from elsewhere, for which the digging of canals thousands of kilometers long is being contemplated or carried out. The Yellow River does not contain enough water anymore to supply for drinking water, and it is too shallow to navigate, which now affects the surrounding industries. And as one of the largest rivers on Earth, the Yellow River, too, does not always reach the sea anymore. Ancient bridges spanning big regional rivers, such as the Po River in North Italy, as well as in many other places in the world, such as Nevada, only show their magnificence of the past; at present, you can cross those rivers on foot for most of the year, reducing the bridge to a curiosity of exaggeration.

In western China, growing thirsty cotton had actually to be given up because the area dried out. This also happens in the southwestern part of the United States and in Mexico, where the bordering states are fighting their rising water conflicts. As mentioned already, for the growing of cotton, the Aral Sea has run dry completely. In order to save the lake and, with it, the cotton growing fields, even a redirecting of the majestic Siberian rivers from north to south has once been contemplated. The water level of the famous Dead Sea, widely known from biblical times onward for its high concentration of salt, is dropping by one meter each year because of the magnesium and potassium it contains, which is mined. And so on.

At present, because of the depletion of surface and groundwater, aquifers, deeper strata containing fossil water, are now supplying the water needed in agriculture, towns, and industries in many regions the world over. However, these aquifers, although often very large, are being used up at frighteningly high rates as well. This happens, for example, in the southwestern United States, Australia, and China (in the latter case often resulting in the sinking away of the land over large areas or even of towns lying on top of it, and resulting in dangerous pits of tens of meters deep). The same happens in the San Joaquin Valley, where already between 1925 and 1977 the whole area had subsided by some eight meters. Parts of Mexico City, for example, have subsided almost twenty meters, and Jakarta is subsiding ten centimeters a year. In other cases, deep cracks have formed due to the drying out of the soil, running through the countryside. In Arizona, drivers are warned for deep cracks across the road because of land subsidence. Mexican farmers often cannot pay the pumping equipment for irrigating their land, and are giving up. In other cases, such as in the United States, marshes and large rivers run dry, their water leaking away into the drying soil above the emptying aquifers. With them, millions of people all over the world thus depend on the past for their living, and this limited resource of the past is running out. Their feeding, drinking, and cleaning situations and sanitation will be aggravated seriously with the steep rise of prices to be paid for agricultural chemicals and energy, because of rising shortages of fossil fuels. What are we doing?

One solution to solve all these rising problems of freshwater shortages has been sought in desalinating seawater, that is, by extracting the salt from it (as is done for Perth, Australia). Desalination, however, incurs heavy costs, both financial and in terms of energy and, hence, in those of fossil fuels. Much energy is used, first for extracting the salt, and then for transporting vast amounts of fresh, desalinated water over large distances. Often, this is economically unfeasible. Desalinating seawater is obviously out of reach of most poor countries, which are also unable to purify water from rivers and lakes from the many poisons they contain. In Egypt, for example, 40 percent of the drinking water of 80 million people is polluted in some way, but their number will rise to 120 million in 2050. At present, their annual water consumption per head, 860 cubic meters, well under the minimum of 1,000 cubic meters, will drop to 582 in 2017. Water is insufficient, which will be worse with one half of the present number of people more on Earth. On average, each of them, moreover, will be consuming much more, particularly in the form of the virtual water in meat, thus increasing the large amount of virtual water they are already using by consuming plant products. Increasingly, all this water will have to be transported from far away from the seas and oceans to deep into the interior of continents. There, it has to supply all the equally increasing needs of agriculture, cattle growing, and house holding, building, and industrial use, et cetera. This may prove to be hardly feasible for most inland uses. In many cases, it is already uneconomical to desalinate large amounts of water, and this will worsen when in future the price of the fossil fuels will go up when these run out.

This situation of us running out of clean freshwater is worsened by the heavy and large-scale pollution of surface and groundwater by towns, industries and mining operations, by the chemicals used in agriculture and in pharmacy, as well as of the pollution of the water of minor creeks upriver. For example, independent of household sewage mentioned, Egyptian industries annually spew some 550 million cubic meters of waste water into the Nile, whereas agriculture adds another amount of 2.5 billion cubic meters of water contaminated with pesticides and fertilizers. The farmers themselves use sewage water on a large scale instead of clean water, thus affecting the quality of their produce. At present, some streams do not undulate anymore, whereas other ones take any color imaginable. Therefore, this part of our freshwater also becomes completely unsuitable for any use, be it human, agricultural, and industrial use, or the use by animals and plants, many pollutants killing off anything. And when people desperately do use this water, they develop all sorts of diseases, developmental aberrations, malfunctioning, and physical discomforts. Again as an example from Egypt, some 17,000 children annually die from diarrhea and the dehydration resulting from it.

With the loss of freshwater of rivers, lakes, and marshes worldwide, we lose a large and significant part of the biodiversity inland and in the brackish water in the oceans at large distances from the mouth of even the largest rivers on Earth. Part of the brackish and marine fauna of the North Sea, once the fish basket for thousands of people, died out decades ago, not only because of overfishing, but also because of the chemical load from agriculture and industry, as well as the pharmaceuticals in use for humans and cattle, the rivers draining off the surrounding land through streams and canals. Many household bactericides and detergents also find their way into surface water and groundwater and eventually reach the sea, where not all are decomposed. In other cases, aquatic life suffers from the amount of silt the rivers carry from massive erosion. The Yellow River and the Yellow Sea, for example, have long been known as such because of all the silt the river transported over the centuries. Already for decades, rivers in Australia are opaque streams of mud, hardly containing any of their endemic fish on which many aborigines once lived. Thus, for many decades, and throughout all continents, freshwater fish can often not be harvested anymore—the Mekong is but one example.

The scarcer and more polluted freshwater becomes—the higher the temperatures, and the higher the cost of energy to pump it up, purify, and to transport it—the more expensive it will be. This may jeopardize the production of food and make the food that can be produced much more expensive, thus growing out of reach of millions of the poor. Therefore, not only can we expect a crisis to occur in the production of energy and some irreplaceable minerals, but also one due to water shortage.

Worldwide, the use of bottled water is already increasing rapidly, but this requires not only the use of plastic bottles, but also its transportation over long, often international distances. Saudi Arabians are now drinking water from Finland at a distance of 4,300 kilometers, whereas nationally, some 300,000 trucks are busy transporting drinking water within Italy alone. In the United States, no less than 30 billion plastic water bottles are used annually, which all have to be produced and which are later mostly discarded into the environment. As mentioned, when returned, they can be recycled, but this requires moderately high temperatures for which, again, fossil energy is needed. And for the initial production of each of these bottles, apart from the 160 milliliter of oil for the plastic, another five liters of freshwater is needed, five liters for one liter of drinking water. Another kind of virtual water being added, therefore.

Where does all this freshwater go as the aquifers are getting depleted, and marshes, lakes, and rivers or whole inland seas like the Aral Sea dry up? It has to go somewhere. In fact, part of all this water gets lost by evaporation by our crops, by the drying up of bare soil and irrigated fields, or by the loss of the wet humus of the thawing permafrost and of that from the vast primeval forests when these are cut, and so on. Eventually, all this evaporated water comes down as rain, mostly into the oceans, as these cover the largest part of Earth. And another part is lost by the run-off of water through the rivers so that this also ends up in the oceans. Thus, there is a net transfer of freshwater from the terrestrial part of the world into the oceans, a waste of valuable freshwater, that is. Another part remains as water vapor in the atmosphere and adds to its warming. It has been estimated that 97 percent of all the freshwater we extract from Earth ends up in the oceans, thus accounting for no less than one-quarter of the present rise in seawater level, the remaining three-quarters coming from the melting of frozen freshwater so far locked up in polar ice, and in the ice caps and glaciers of the world. And like retrieving minerals from waste by recycling, desalinating salt water is expensive in terms of energy, especially as this concerns vast amounts of water, some 283 cubic kilometers annually. Soon, therefore, most of our reserves will all be gone, and it will be too expensive to get it back again. Dissipated, like the many minerals and the energy we use.

Like our food, in several ways, both directly and indirectly, the availability of freshwater as an essential resource is becoming intertwined with the availability of fossil fuels. And the use of those fossil fuels connects with the warming of our climate, which itself reduces the availability of both our food and our drinking water. This forces us into a vicious circle, a downward spiral, driven by the fossil fuels, which are running out at the same time. Yet, ultimately, of course, it is the increasing number of people giving all these problems of overuse, depletion, and the production of waste and pollution. Basically, each of us needs to be fed, watered, dressed, housed, and to move about, and we all want to improve on what we already have. And we all reproduce. Demanding more. Ever more.

Not only should the problems in our present global society that are gradually coming up as a result be stated integrally, but their solutions must be tackled integrally as well, from their very basis onward: our growing, already unsustainable population.

11 POLLUTING THE AIR AND WARMING OUR CLIMATE

This morning, we heard a mistle thrush singing in the garden. It's November, and we don't expect to hear this thrush until February. Is he early because of climate warming, just like many trees develop leaves much earlier in the year and shed them later? Or have individual thrushes always sung at this time of the year, even when the winters were colder? Although on the whole our winters in Holland are warmer—skating isn't always possible—our summers aren't that much warmer than before. And weather is changeable anyway, varying from one day to the next, between seasons, and over the years. What is going on? Is climate really warming? Many people say it is, but others deny it.

After many years of study, an international group of about two thousand scientists united for the Intergovernmental Panel on Climate Change (IPCC) has reviewed the recent literature, cautiously concluding that climate is changing, and that this change is actually due to human action. But can they also predict what is going to happen? So far, their predictions have been very general. Due to changes in climate there are already food shortages, particularly in Africa. But what does that mean to me, to my own immediate environment, my way of life? Most shops in the Western world are still overflowing with food products. If climate change is indeed happening, is that bad news or could it also help feed our overcrowded world by stimulating crop growth? What can we believe, what can we predict, what can we expect to happen?

We can predict what will happen, but only in general terms, not in detail. That's what's confusing. That early singing of a mistle thrush may the result of climate warming, but not necessarily. What will happen exactly in one particular part of the world in my lifetime or yours remains unknown, though we can make statements based on statistical probability. That is, we can say it's likely we will get warmer winters more often, but the summers will often remain the same, or they may get somewhat cooler. There will be parts of the world that are expected to be warmer more often, whereas other parts will be cooler more often. We cannot predict with certainty, but only in terms of "more often" and only for large regions rather than for exactly that part of the world you or I live in. We need to think in terms of statistics, the mathematical science of "more often," and "in more places," in terms of relative frequencies. Therefore, climate may warm up considerably, even though you don't experience it every day, or each year, or in your part of the world. In Holland, despite many warm winters without severe frost during the last twenty years or so, in 2010 we could skate on the frozen canals. In that particular instance the North Atlantic became much warmer.

The dissipation of heat from Earth into space is a naturally occurring process. Without it, Earth would soon heat up to unbearable temperatures because of the continuing incoming radiation from the sun. This is what happens on our neighboring planet, Venus. It has temperatures of hundreds of degrees Celsius because of the high concentration of carbon dioxide in its atmosphere, which stops much of the incoming radiation from the sun returning into space. As a result, its rocks are very hot and dry, making it impossible for any form of life to develop on it. Once, Venus had oceans as Earth does, and most of its carbon dioxide was contained in its rocks. But overheating, a self-accelerating warming process, and runaway greenhouse gas warming occurred putting an end to all moderate temperatures, turning the planet into a hot oven instead. More water vapor and carbon dioxide in its atmosphere due to increasing temperatures meant that temperatures increased even more, causing more water vapor and carbon dioxide to dissipate into the air, and so on, putting the self-accelerating process to work. In contrast, our earthly atmosphere retains only a small part of the solar radiation so that for the last four billion years our temperatures have been neither very high nor extremely low. The temperature on Earth is relatively stable at an average of some 30°C, which is only about 300°C above that of the surrounding deep chill of absolute zero of the universe, about 270°C below the freezing point.

Still, could the same process that once happened on Venus also happen here on Earth? Calculations show that this is indeed possible when we continue using fossil fuels the way we have so far. Our oceans too will evaporate; our rocks will get bone dry and lose their carbon dioxide. That, then, would be the end of humankind, in fact, of all life on Earth. The woods and the whales, mankind and his intelligence, all the variation between people and their civilizations—no history, no future, it would all be gone. Gone due to our neglect, recklessness, and greed. A barren, lifeless, overheated Earth, not that different from Venus. And we know of no other planet in the universe with life of any form, whatever our hopes or speculations. The only thing we know for certain is that Earth is unique in this respect, and this is what we are risking to destroy forever, without a chance of it ever recurring in future again.

The temperature rise on Earth with all its great risks for the future is achieved as follows. After penetrating our atmosphere, sunlight warms up the surface of the planet—the soil and the water. However, by warming them up, the incoming radiation loses part of its energy, so that radiation with a lower energy content, infrared or heat radiation, is reflected back into space from Earth's surface. On its way back, however, the heat radiation, particularly the infrared component, is blocked by certain chemical compounds in the atmosphere. These chemicals—carbon dioxide, methane, water vapor, nitrogen oxides, sulfur compounds, and some others, such as the more complex, human-made chlorofluorocarbons—reflect the infrared radiation back to Earth, which then warms up the soil surface, the rocks, and the oceans. This also happens inside a greenhouse, where the glass blocks the escape of heat radiation; the responsible gases are therefore known as greenhouse gases. Apart from blocking solar radiation transformed by Earth itself into infrared radiation, the greenhouse gases also block heat radiation generated by all our activities which release heat. This heat forms thermal pollution and can easily amount to a temperature increase of 2°C during the day and some 12°C at night within our cities on top of the general increase due to the release of greenhouse gases.

Snow, ice, and small particles in the air, aerosols, reflect incoming radiation and bounce it back into the atmosphere. But when the air temperature warms up and the snow and ice on top of mountains and in polar regions melts, the sunlight reaching those parts of Earth is no longer reflected back into space. Instead, the sun heats up the darker surface of the soil, rocks, and water of the oceans, thus adding to the greenhouse warming. Polar ice reflects 70 percent of the solar radiation back into space, whereas the dark surface of the oceans absorbs 95 percent. Moreover, part of the warmth from the greenhouse effect is now absorbed by the melting of the glacial and polar ice, which stops when this ice has melted away, so that the atmosphere warms up even faster. Because of this self-enhancing process, polar ice and the mountain glaciers are actually melting faster than expected. This becomes worse when aerosols are cleaned from the air, whatever good reason to do this may exist. The heating of the atmosphere enhances its own further heating and so the process becomes self-accelerating. This self-accelerating process can develop into the disastrous runaway process that once happened on Venus.

Since the Industrial Revolution of the mid-nineteenth century and its large-scale use of coal and the spread and speeding up of deforestation, the temperature on Earth has risen. In the far past, some eight thousand years ago, there was also seemingly a temperature rise, thought to be partly because of methane production in the first rice fields and partly because of the burning of forests and the decomposition of timber with the first deforestations. But since the Second World War in particular, both the carbon dioxide concentrations and the temperature have increased more rapidly, despite natural temperature fluctuations. After a short spell of lower temperatures during the 1950s and 1960s, the temperature curve began to bend upwards sharply, so that in many parts of Earth warmer years now occur more often than before.

As mentioned, to make carbohydrates (their main constituents), plants use carbon dioxide from the air and hydrogen obtained from water in the soil. In the process, they exhale oxygen as a waste product into the air. Next, animals eat and digest the plants, burning the carbohydrates with the oxygen they inhale, thus reconstituting and exhaling carbon dioxide and water. Later, it rains when the air becomes saturated with water vapor. In contrast, the carbon dioxide stays longer in the air or is first taken up by the oceans before they return it to the atmosphere. These compounds form the permanent global carbon and water cycles. Similar to other nutrient cycles, such as those with nitrogen, phosphorus, and sulfur, they have been sustained for thousands of millennia. Without those cycles, life would have depleted the materials in the thin biosphere within a couple of thousands of years.

The carbon cycle is not perfect, however. Over millions of years, instead of being returned into the atmosphere, huge amounts of plant, animal, bacterial, and other remains settled on the seabed, became buried under other sediment and slowly changed into coal, oil, and gas. For example, during the Carboniferous, three hundred million years ago, too few fungi and animals existed to decompose all the plant material formed under the warm and humid conditions of the time, so eventually, coal was formed. However, after first destroying much of the forests and peatlands of the world to obtain fuel and agricultural space, humankind began to mine and burn these stored forms of carbon, releasing carbon dioxide into the atmosphere. This belated return of the "ancient" carbon dioxide raised the concentration of this gas in the atmosphere, enhancing the greenhouse effect and raising the temperature.

The gaseous waste from the burning of wood, peat, and fossil fuels now builds up as atmospheric pollution. We have broken the balance on both sides—that of the absorption of carbon dioxide by plants by deforestation and that of increasing the production of this waste gas. Consequently, carbon dioxide is now accumulating in the atmosphere at accelerating rates. And by warming up the atmosphere, the process becomes self-enhancing, self-accelerating. So far, much of the "extra" carbon dioxide has been taken up by the rocks and the oceans, but when they can absorb no more, all the carbon dioxide will stay in the atmosphere, warming it at higher rates. To the contrary, when the oceans warm up, their water can contain less carbon dioxide and will release part of it back into the atmosphere, where it can add to the warming, another self-accelerating process. The bad thing of self-accelerating processes is, first, that they determine their rates themselves beyond our capacity. As soon as they start, we lose all control. Second, they accelerate the process going on, causing it to progress at increasingly higher rates. The danger of climate warming of which many of us seem to be unaware is that as soon as it begins, several other self-accelerating processes are ignited as well, first only slowly and hardly noticeable, but then ever, ever faster.

There is another, less obvious source of carbon dioxide that is increasing rapidly: concrete production. Limestone—another ancient geological sink of carbon that accounts for a substantial part of the rocks on Earth—is used for producing concrete. But when concrete hardens, it releases carbon dioxide. About 10 percent of all the sedimentary rocks on Earth (these are rocks that originated as sediments that were laid down under water) is limestone. Most limestone is formed from the skeletal remains of marine life; it is calcium carbonate, the same material our bones are made of.

Concrete may seem a minor source of atmospheric carbon dioxide, but don't underestimate the enormous amounts of concrete used each year in various construction projects worldwide. Concrete now largely replaces wood, stone, and brick, the construction materials of previous centuries. The tall buildings and skyscrapers, the offices, apartment blocks, and factories of our towns and vast megacities where millions of people live and work all use concrete. Think of the road networks in and between those towns, of the barns, runways, harbors, mega-stores, and huge hydroelectric dams: these and many more are made of concrete. It is easy to build with, and when reinforced with steel as prestressed concrete, it has strong mechanical properties. Unfortunately, as mentioned, only 20 percent of concrete can be recycled, leaving 80 percent as nondecomposable waste, and necessitating the production of the same amount of new concrete. The emissions from setting concrete, combined with the carbon dioxide from fossil fuels, forests, and peatlands, mean that our total release of carbon dioxide into the atmosphere is considerable. There is no way that replanting part of the former forests can compensate for all this carbon dioxide emitted into the atmosphere. Even the surge in absorption of carbon dioxide resulting from the emergence of leaves and new growth of all plants each spring in the entire Northern Hemisphere makes only a small dent in the broad, ever-steepening rising curve of the concentration of carbon dioxide in the atmosphere.

Dozens of other gases in the atmosphere are also known to have warming effects, some of them stronger, and others weaker. Among them are methane, nitrous oxide, and water vapor, all of which have a stronger warming effect than carbon dioxide. As mentioned, water vapor builds up to a certain saturation point, after which it rains. Methane, which is twenty-one times more effective per kilogram than carbon dioxide in heating the atmosphere, is emitted by marshes, peatlands, the melting permafrost, rice fields, cattle, landfills, garbage dumps, and coal mines. Methane is produced when plant material decomposes in conditions where there is no oxygen (under water in swamps, for example—hence the name "marsh gas"); if oxygen is present, such as in parts of the peatlands and permafrost, other bacteria decompose the plant remains, producing carbon dioxide and water directly. In landfills, methane is produced by the bacterial decomposition of food remains and other household waste; in Britain, about one-third of a kilogram of such waste is thrown away per capita each day. This compound also occurs naturally in very large amounts in all ocean fringes due to the continuous continental run-off of humus, like from the Canadian and Siberian peatlands and when part of the permafrost melts. Being stable under certain water pressures and temperatures, it there forms methane ice—also called methane hydrate, or methane clathrate. With an increase in temperature, it begins to melt, after which methane bubbles up to the surface in very large amounts which dissipate into the atmosphere. Contrary to the stable carbon dioxide, it only stays there for up to about fifteen years, during which it burns with oxygen, with carbon dioxide, and water as final waste products. This period, however, is increasing because of a recent decrease in atmospheric ozone which facilitates its burning.

Under cool or freezing conditions, marshes, peatlands, and areas with huge layers of permafrost thus form enormously large, temporary stores of humus and peat; when temperatures rise, these stores are in danger of decomposing, either directly into carbon dioxide or indirectly via the temporary formation of methane. The methane ice, marshes and peatlands across the world, and the vast permafrost peatlands of Alaska, North Canada, and Siberia thus form a gigantic reservoir of potential carbon dioxide. Part of that reservoir is releasing its contents already, for example, when marshes are drained, and when the permafrost peat of the high north thaws. Methane will warm up the atmosphere significantly before it adds those large amounts of carbon dioxide after reacting with oxygen. The carbon dioxide produced in this way is added to the carbon dioxide generated by burning fossil fuels, deforestation, and using of concrete. In the mid-1980s, the concentration of methane in the atmosphere had already doubled since preindustrial times around 1800. The problem is that the release of methane adds to this warming because of warmer conditions, which results in a self-accelerating process beyond our capacity. It is important to realize that with a temperature increase of 1°C at the equator, temperatures at high latitudes with vast expanses of peatland and permafrost increase 5 to 6°C. Because this self-accelerating part of the process is natural, methane release is known to have caused the abrupt ending of geological periods.

After carbon dioxide, nitrous oxide, a predominantly biogenic gas, is the second-most important greenhouse gas. It is nearly three hundred times stronger than carbon dioxide and originates from the nitrification and denitrification of ammonium and nitrate released in natural and industrial processes. At present, synthetic nitrogen fertilizers used in agriculture and ammonia released from livestock manure, as well as released in traffic, along with the industrial production of nitric acid, are the main sources of nitrous oxide; they will all increase in importance in future. For example, the surface area needed for crop growing, meat production, and biofuel production (processes all requiring a greater use of fertilizer) will increase.

Water vapor is the most important greenhouse gas, at present contributing 21°C to global warming, whereas carbon dioxide adds only 7°C. It follows the rises and falls of temperature; more water turns into water vapor when the temperature rises, and less when the temperature falls. Therefore, when the concentration of carbon dioxide increases, the atmosphere's temperature—and hence its capacity to contain water vapor—also increase. Thus, water vapor enhances the effect carbon dioxide has on the temperature of the atmosphere and can even double its effect, depending on the atmospheric stratum in which it occurs. Furthermore, because it enhances the increase in temperature, similar to what happened on Venus, the process is self-accelerating: higher temperatures result in more water vapor in the air, which further increases the temperature, which results in more water vapor, and so on. Apart from this, when the atmosphere contains more water vapor, it can rain more; this means that globally, under warming conditions, precipitation will increase. Moreover, the water vapor contains energy that is released during rain and causes great turbulances in the atmosphere and, thus, a higher frequency of occurrence of storms and tornados, drought, and heavy rains, with all the consequences of disastrous floods and greater erosion. Therefore, higher temperatures not only result in a direct increase in the likelihood of more severe droughts and an enhanced frequency of heavy precipitation, but also indirect ones. However, which parts of the world will be affected by drought or by floods can be predicted only in very general, statistical terms, because rain and wind patterns are capricious. Moreover, different combinations of individual hot or cold years in a row or of two or three successive hot years produce different geographic patterns in terms of crop harvests—depending on whether the crops are annual or perennial. Geographic patterns also differ greatly over the seasons, so crops with a different seasonality are affected differently.

This unpredictability cannot be overcome by a greater understanding of climatological processes, as the problem is not so much the number and intricacy of the processes involved, but more that chance plays a major role in them. It is possible, therefore, to reconstruct and understand a particular outcome, a continental or global pattern of droughts and precipitation regions, but it is impossible to predict it. In fact, because chance plays such a large role, every pattern is unique. For example, we can understand the pattern of droughts and floods across Eurasia during the summer of 2010, but we cannot predict when the same events will recur in the future.

The extreme events in Eurasia in 2010 are related to the vicissitudes of a current of air flowing high in the atmosphere: the jet stream. At a latitude roughly between 50° and 70°N and an altitude of 9–14 kilometers, this air current usually blows strongly at a speed of 160–240 kilometers an hour. Depending on its speed, it makes large latitudinal waves or undulations with a succession of either shallow or deep troughs and ridges, which adopt different positions relative to the ocean and continents in the Northern Hemisphere. Usually, too, the pattern of troughs and ridges moves, but occasionally it remains stable, it seizes up. This is what happened during the summer of 2010 with the result that the weather remained very stable for several weeks. Furthermore, south of a ridge in the jet stream, the weather is warm and dry, in contrast to the north side (inside the trough, so to speak), where the weather is humid with much precipitation. Because of the pattern of undulations, the weather in Western Europe became hot as long as the blockage

lasted, whereas in Eastern Europe it rained. Then, further to the east, a region of very hot and dry conditions developed around Moscow, and further
east still, there was extremely heavy rainfall, causing the floods in Pakistan.
This pattern continued further into China. Around Moscow temperatures
rose to above 40°C for several weeks, resulting in a severe drought in which
dry peat in drained areas ignited spontaneously and thousands of hectares
of peatland, forest and agricultural land burned. The same happened in the
remote, northeastern part of Siberia, north of Vladivostok. Because dry
peat underground continues to burn after the surface fires are put out, peat
fires are almost impossible to extinguish and can go on for years. After a
month without fire, for example, new spontaneous ignitions occurred again
south of Moscow at the beginning of September. The flooding in Pakistan
was also aggravated by the drainage of peatlands—this time, on the Tibetan
plateau—and from the deforestation of the Himalayas and the floodplain of
the mighty Indus River.

It is unlikely, however, that this location and pattern of geographic undulations of the jet stream, fires, and floods will recur in the same way and
intensity, because the pattern depends on the location of regions with high
and low air pressures, and this depends in turn on patterns of water temperature in the Atlantic Ocean, as well as on local conditions due to human activity (such as deforestation and drainage of bogs). Furthermore, the latitude
and intensity of the jet stream depends on the energy exchange between the
warm equatorial and the cold arctic parts of Earth, that is, on the temperature gradient of the Northern Hemisphere, which varies with climatic and
atmospheric conditions.

Apart from these factors and processes, all of which vary according to
chance, there are others; it is the combination of all these that makes the
end result unpredictable in particular cases and years. All we can say is that
something similar may happen again sometime in future. Maybe it will happen more often. Yet, uncertainty within the framework of certainty of an
overall warming of climate.

One thing that is clear is that one way or another, large amounts of new carbon dioxide are constantly dumped as gaseous waste into the atmosphere,
thereby altering its composition and, hence, its temperature. We know how
much fossil fuel and material is used, and how much peatland is decomposing. We therefore know how much carbon dioxide is being released. From
this, we can calculate the rate of increase of carbon dioxide in the air, and
from laboratory measurements we know how much the earthly temperature
will rise. There is no discussion about this. And since the mid-1950s, from

measurements on a Hawaiian island, we have had a long series of direct observations running on the increase of the atmospheric concentration of carbon dioxide about which there is no discussion either. And from Antarctic ice we know that the concentration of methane in the atmosphere has more than doubled. And so on. Therefore, there can be no discussion about whether or not we are altering climate. In fact, in 1996 already, it was proven definitively that climate warming is human induced. When the sun warms up the atmosphere, namely, it warms up all strata, which is different for the greenhouse warming. With greenhouse warming, only the lower strata are expected to warm up whereas the upper ones cool down. Measurements showed that the latter, greenhouse warming happens. Unfortunately, serious discussion has become difficult, because some industries are paying out hundreds of millions of dollars annually to bribable scientists to throw unjustified doubt about the reliability of the data or analyses (similar to often the same scientists who introduced doubt into the research around the dire health effects of smoking). The only thing being discussed seriously is how much we are changing the climate, what kinds of things can happen, and where. Of these points of real discussion, it now seems certain that the human-induced greenhouse warming has a distinct, measurable effect, and we have to take measures soon before self-accelerating processes take over. Taking measures is necessary, as any further change in our climate endangers our food supply directly.

Apart from affecting our food supply and risking the destruction of all civilization and life on Earth, climate warming has many immediate effects as well, some of which can be counteracted, whereas other indirect, interactive, or branching effects cannot. These effects can already cost the lives of hundreds of millions whereas other millions have to emigrate to other parts of their country or to other parts of the world. Soil salination because of modern agricultural practices—another phenomenon that will be intensified by global warming—will also vary geographically. So far, salinated regions are found in parts of the world with a Mediterranean climate, and in areas where at times (at intervals from three to five years) the climate is affected by the El Niño current. One hopes that salt-resistant races can be produced in time, but this is not feasible for all crops. Another direct effect of climate warming could be more widespread outbreaks of disease in crops, livestock, or humans. In temperate Europe, for example, crop pests that are killed off by winter temperatures that dip below freezing will survive and could become endemic. Another direct effect of climate warming is forest dieback in certain parts of the world, trees not being able to withstand changing conditions; a spinoff from this is that when forest cover is removed, the soil is more likely to be eroded. Higher temperatures of the water in the oceans, inland seas, lakes, and marshes will similarly cause aquatic or marine animals to die out, such as coral reefs all over the world. There are fears that the Gulf Stream in the northern Atlantic may stop flowing; if this happens, the climatic conditions of the neighboring land masses will change drastically. Since this warm current now keeps large parts of Europe at moderate temperatures, these regions would then suffer from harsh agricultural and general living conditions with the possible northward expansion of malaria into Europe.

Indirectly, climate warming can affect cloud formation—and thus rainfall and regional or continental hydrology, in turn enhancing soil erosion by runoff (if rainfall increases) or reduction in agricultural production (if rainfall decreases). Rising water temperatures can affect the living conditions of fish directly, but also indirectly by lowering the oxygen concentration in the water. When carbon dioxide dissolves in water, it makes the water more acidic, which affects fish directly by weakening their bones and indirectly through food loss when the calcareous skeletons of, for example, microorganisms on which they live either do not form anymore or dissolve in the acid—effects that have already been observed. Fish dependent on corals are also affected by higher water temperatures, because the algae living inside the corals cannot survive in warm water, and without them the corals lose part or all of their color (a process known as coral bleaching). Worse, by losing the algae supplying them with oxygen, they lose their oxygen source, causing both the corals and fish to suffocate and die. The resulting reduced fish stocks around tropical islands impacts the thousands of islanders who get their protein from fish.

A more complex set of indirect effects of global warming is triggered by the rapid melting of the polar ice and the ice cap of Greenland, which, according to some calculations, is expected to be ice-free in summer by about 2020. It is already possible to sail in summer north of Canada from the Atlantic to the Pacific Ocean, and in the coming years the same will be possible north of Siberia. Ice reflects much incoming solar radiation back into space, so without the ice, more of this radiation will be absorbed by the land and Arctic Ocean, thus increasing their temperature and the temperature of the air above them. Moreover, the warming of arctic water will affect the deep ocean currents: the thermohaline current, or conveyor belt, that transports water of different composition and temperature from the Pacific to the Atlantic Ocean and vice versa may reverse its course. This will disrupt the temperature and humidity, and therefore the rainfall patterns, of the bordering continents, drastically altering the living conditions for humans, plants, and animals. The melting of the arctic ice, together with the thermal expansion of the ocean water in general, raises the sea level, causing coastal areas, such as in Bangladesh, and low islands, such as the Andaman Islands, around the world to flood. The last few years, the Greenland ice cap has become instable because melt water at its surface goes down to the rock, where it reduces the friction between the ice and the rock so that it slides off at an increased rate. Moreover, so far, the cap is so thick (about three kilometers) that its top layer is permanently high enough that precipitation always occurs in the form of snow. There are fears that this ice cap will get thinner, which means that it will no longer be maintained by snow but will instead melt during rains. Together, these two effects result in a global rise in sea level. Taking them, plus the increased melting of the Antarctic ice into account, a 2011 report estimates a rise of between 0.90 and 1.60 meters by about the year 2100. Fertile agricultural coastlands and coastal cities may be submerged, along with their infrastructure of roads, harbors, factories, sewage systems, metros, electricity and telephone cables, and so on. Remember that some 80 percent of the human population lives in coastal areas or further inland along tidal rivers.

Furthermore, some indirect effects also interact, thereby enhancing the process. As mentioned, the loss of polar ice enhances climate warming, because less radiation is reflected back into space and none of the heat will melt the ice any longer but will further enhance the warming of the atmosphere. But the warmer ocean water will melt the methane ice around the continents, and this methane will add to rising temperatures through its own greenhouse effect. Later, as permanent carbon dioxide after its reaction with oxygen, it enhances temperature rise that way—a rise that melts more methane ice and decomposes more carbon dioxide–producing permafrost peat. The process is self-accelerating; the temperature rise accelerates.

Many people wonder why we are concerned about a temperature rise of only one or a few degrees Celsius when the differences between summer and winter, or even between night and day are much larger. Why be concerned about a rise of even a fraction of a degree above the annual mean temperature over the next decade? What in the world are we talking about? The answer is that increases in temperature are cumulative rather than instantaneous; small changes during successive moments have to be added together over all those moments. Think of the origin of the Ice Ages, where a drop of only five or six degrees in the average global temperature resulted in an extension of polar ice sheets of hundreds of meters thick and covering large parts of North America and Europe. In that case, the winter snow did not melt in summer because of those lower temperatures, so the surface area of ice gradually increased. Thus, more solar radiation was reflected back into space, which made it even colder, a typical case of a self-accelerating process (one toward cooling). The opposite process happens at the moment with climate warming where small temperature increases accumulate in the form of continued melt, resulting in further global warming, in increasing ice melt, and thus in sea level rise. Also, soil desiccation is cumulative; the next day the soil dries out further, starting from the humidity deficit of the previous day, and so on. And not just soil dries out: plants dry out too because of the cumulative effect of soil desiccation, and so forth. Similarly, plant growth due to a rise in temperature is cumulative. Each day, plants receive a certain amount of light energy, which is used to build up their tissues. Since they don't shrink during the night, this increase in plant body size accumulates over time. Higher temperatures speed up chemical reactions day after day, enhancing the storage and construction processes. Agronomists therefore reason that a certain minimum amount of energy plants accumulate over the year is needed for a plant seed to make another seed. As this energy is received over a certain number of days, they call this minimum amount the minimum number of degree days. In cooler parts of the world or high up a mountain, this minimum number cannot always be reached, so that the crop concerned cannot be grown there. A slight increase in daily temperature of only a fraction of a degree Celsius can make a large difference, because over a number of days it can add up to that amount of energy still missing to reach the required minimum number of degree days. Thus, our food production greatly depends on these cumulative effects of slight changes in temperature, humidity, and rainfall, rather than on some average fraction of a degree more or less at midday. To us, a fraction of a degree higher is itself not important; we can't even feel it. But that's not the point. It's what the plants and the environment do with it.

Small temperature differences thus accumulate, resulting in large environmental effects. Similarly, sea level change leading to the inundation of many low-lying islands and of coastal regions and towns is mainly the result of the gradual warming of the water of the oceans, which are then expanding, just like an iron rod expands—getting longer and broader—when we heat it. Pollution, erosion, the loss of the humus layer and forest, or the loss of genetic diversity and biodiversity, as well as the depletion of many of our mineral and energy resources are other examples of cumulative effects we have on the environment. Those effects cannot be turned back; turning them back is beyond our capacity.

Salination or desiccation of soil can also happen on a large, geographical scale. Desiccation makes the soil vulnerable to erosion by rainfall and wind and can result in topsoil being lost from large areas. This is already happening in parts of Earth. A rise in temperature or in carbon dioxide concentration can be favorable to crop growth, but only if other factors also resulting from climate warming do not interfere. For example, temperatures can become too high or too low for the normal biochemical reactions to happen (this is why tropical species cannot grow in temperate regions and vice versa). Or evaporation due to high temperatures can make plants wilt, undoing the positive effect of an increased carbon dioxide concentration. Similarly, temperature rise may mean that certain crops could be grown in other areas, which is positive, but these areas may have unsuitable soil conditions, such as a thin layer of peat on bedrock, or may have insufficient water supply. The best option, therefore, is to stop climate change from progressing.

But climate change is difficult to handle. Of course, people are concerned about its developments, speed, and how quickly the environmental impact is felt. Many measures have been proposed, from reforestation to recapturing carbon dioxide at its source in power stations. This recaptured carbon dioxide could be pumped into the ocean, aquifers, salt mines, or depleted oil and gas fields. However, recapture is energy costly, as is the transport of the carbon dioxide through long pipelines, which reduces the efficiency of energy use, making it necessary to produce more energy. Other proposals suggest cutting down on car travel or making smaller and more fuel-efficient cars or more efficient machinery. Yet another proposal is to use electric or hybrid cars, but the electricity they use is generated from fossil fuel, which, together with conversion loss, generates even more carbon dioxide. Which of these measures would help, and how much would they help relative to the enormous amount of carbon dioxide produced each day? Given our insatiable need for energy, they probably won't help that much, however good our intentions are.

With regard to energy used, our model in figure 1 (p. 000) now reads: fossil fuels are the input for the processes happening in the central processing unit, and carbon dioxide, water, and heat are its output, its waste. Reducing the carbon dioxide output, that is, tackling part of its output, seems unfortunate as a preventive measure, because a process cannot be steered or stopped by regulating its output, its waste. Waste production of any form is a symptom of the activity of some processing unit, and reducing the amount of waste is treatment of symptoms. I cannot navigate my car by capturing its exhaust. Steering or stopping through output regulation results only in an uncontrolled suffocating of the processes happening within the processing unit, the engine. No wonder that so far our measures for carbon dioxide reduction have been ineffective, since they left the processing unit—the growing number of humans, their increasing demands and productivity—untouched; in fact, this unit was left to run faster and to grow even more. Input reduction is equally ineffective: when fuel reduction is not planned well, or operates via price rises dictated by the oil producing countries or by the depletion of oil or gas itself, other processes happening in our global society can go astray. Under the stress of higher energy prices, this has already happened in the transport sector and the car industry in the United States. Also, it affected the food supply in poor countries directly or indirectly, for example. All society can run into disarray if no choices are made with regard to the processing unit itself, that is, as long as the processing unit (our numbers and personal demands) itself is left alone.

The only effective thing is to use the steering wheel, the accelerator, and the brake—to manipulate the processing unit itself. This can be done by reducing our demands and, especially, our numbers. Reducing our demands can be achieved in the short term by eating more economically, by eating less meat, and so forth. Or we can reduce the turnover rate of our clothes, furniture, private cars, apparatuses, computers, cell phones, and the like. And we can drive less, or use better insulation or more economical lighting. Trucks, buses, cars, ships, airplanes, and heavy machinery can be made to run more economically as well or be used less. Or we can use less packing material, take public transport, or have shorter commuting and traveling distances. And so on.

Of course, we can also develop an alternative energy source for running the global human processing unit, such as the burning of hydrogen. This leaves only clean water as its waste product, which is climate neutral. In that case, as said, we have to develop techniques for splitting water into hydrogen and oxygen by solar energy at an industrial scale as well as for a transporting system. As discussed above, the problems still to be solved to reconstruct our fossil-based economy to a hydrogen-based one will take time. On a shorter term, it will be necessary to know how to generate nuclear power and to apply it, especially the combination of third- and fourth-generation plants. It cannot be denied that they have safety problems, but these are negligible relative to risks incurred by further use of fossil fuels. In this connection, we have to realize that carbon dioxide pollution of the atmosphere is cumulative, since the residence time of carbon dioxide in the atmosphere is about 2,000 years. All carbon dioxide added to the amount already present therefore increases its concentration all these years, and this concentration already approaches the calculated levels of risk rapidly. The maximum concentration, namely, is set on 450 parts per million (ppm), whereas a concentration of 400 ppm will be reached in 2012. Without us doing anything about it, the limit of 450 ppm will be reached in 2020, after which it will be unavoidable that we reach the 500-ppm level before we can stop it.

However, the level of 450 ppm may already have been set too high, and a concentration of 350 ppm seems to be safer. Moreover, the present calculations do not allow for self-accelerating processes, as these are too complex to be taken up. Not taking them up keeps us at the safe side of the calculations, which is good, but it also leads to serious underestimations of our risks in an already dangerous situation. We are losing valuable time for taking adequate measures. After 4 years, we already had to double our estimations for the rise in seawater level in 2100, from 0.25–0.76 meters in 2007 to 0.90–1.60 meters in 2011.

Particularly those self-accelerating processes are the most dangerous because they can speed up the warming processes considerably and because they are typically out of our control. The idea of avoiding the 450-ppm limit is that we have to avoid a temperature increase of 1 to 2°C by the end of the century, whereas a rise of about 6°C might bring us close to the zone of a running greenhouse process. However, when we do account for self-accelerating processes, we will end up with an increase of 4 to 6°C by 2100. Therefore, the margin left before we enter the high risk zone is getting narrower, especially when more coal is going to be used. Remember not only that our use of energy increases exponentially whereas the efficiency of western coal in the United States is only 3 percent, leaving 97 percent to be used for its mining, transport, conversion, and so forth.

Through the unmitigated growth of our population and its supporting infrastructure, we are maneuvering ourselves into a situation, the seriousness of which cannot easily be overestimated. It is serious enough for at least the American military to already be taking measures against instability and strife that will possibly arise from changes in climate. If, however, this reasoning and model estimations are correct, our situation is not only serious, it is getting critical. The problems facing us are now primarily those of whether or not we allow humanity and even life on Earth to continue, leaving all other problems behind, however serious they are by themselves—problems such as those concerning nuclear waste are minor relative to those of a possible running greenhouse process when everything gets lost. Of all things, in the time left to us, we need to solve the problems of energy generation and climate warming in the first place because their solution can at least be done on a short term, contrary to the general driving force of our numbers.

Ultimately, however, whatever else we also have to do, we must reduce our numbers. The longer we allow our numbers to grow and the longer we would exist at a roughly constant or increasing level of overuse, the more and the longer climate will warm up. Our use of energy and material expressed by its inevitable climate change endangers our existence in all its forms. Climate change is both a miner's canary and a threat.

If we don't control our burgeoning numbers and temper our demands and effects, we will be accountable for all that follows. No murder can ever be justified with the argument that for one reason or another life has always ended in death. We are responsible, and we must take our responsibility by reducing our numbers and demands.

12 DEFORESTATION AND
ITS CONSEQUENCES

Last summer, we vacationed in southeastern France, in Drôme, in the Pre-Alps. Here, at the western end of the mountain belt, the mountains gradually become lower. They consist of crumbly limestone, with abrupt, perpendicular cliffs hundreds of meters high. The stone walls of the local farmsteads and houses there are typically eight-tenths of a meter to a meter thick to keep out the cold in winter, and the entrance doors are on the first floor, accessible via a balcony above the highest snow level. Although the mountains are lower than the Central Alps of the Savoy region nearby, it can still be pretty cold during winter, although during previous centuries when the winter temperatures in Europe were on average much lower than they are at present, it must have been even colder. Surely, over the centuries, the farmers must have decimated the surrounding forests to keep themselves warm. So we thought.

Yet, when we read up on the natural history of the region, we found this was not so. Up to the French Revolution, around 1800, there seems to have been only relatively little tree cutting, first because the numbers of people in the area remained rather low, as the region was agriculturally marginal. Second, deforestation remained relatively restricted, because much of the land had always been owned by a few local aristocrats or the Church. In the centuries before 1800, though, the French population had grown to an extent that ultimately led to severe food shortages and therefore to the revolution. During and after the revolution, matters changed radically. With the aristocracy and the Church in power no longer, the farmers asserted their rights to the region, chopping down trees to get some arable land, from the valleys to high up in the mountains. The few photos of the area taken during the late nineteenth century show bare, eroded mountainsides, entirely cleared of the

ancient forests; one photo shows the remains of a solitary, branchless tree—
only the trunk was left—with a couple of poor people at its foot and eroded
slopes all around. Since then, new forest has grown again, but different from
around the ruins of an old monastery, where there are different old trees and
different rich meadows.

Old descriptions mention the torrents into which the former rivulets
and creeks had turned, carrying down stones and big boulders, which can
still be seen lying scattered about. And, of course, the torrents also brought
down much valuable soil that had accumulated on the forest floor over the
centuries. We now understood why the soil under the present forests, when
there is any, is very thin; it takes thousands of years to form. It had either
decomposed when the sun scorched the arable fields, or had been washed
away during heavy showers. The present forests were young, with no old
trees except for a few exotics or the occasional old apple tree that had clearly
been planted. The area must either have been reforested, or there has been
large-scale regrowth, mainly by pines. This forest recovery happened after
most of the people had abandoned the area, because they had been unable
to make a living under the marginal conditions and the erosion that followed
deforestation.

The late eighteenth century and all of the nineteenth century were in
many ways turbulent times, when revolutions broke out all over the world,
not only in France, but also in North and South America and throughout
Asia. And everywhere, populations were exploding. Of course, they had always grown at roughly the same low, slightly increasing rates. But in absolute
terms, a doubling of the population now meant that within decades tens of
millions of people, rather than only a few million, were added to some region
or country. Such increases in absolute numbers of people obviously had a
greater impact on the environment than the same doubling rates did before,
an impact that worsened rapidly. Inevitably, populations were spreading into
formerly uninhabitable marshy or mountainous land, as well as concentrating in the towns.

Until well into the nineteenth century, no alternatives had been found for
wood as a fuel or as a basic construction material for houses, factories, ships,
and carts. Wood was also used in road building and initially in the production of incredible numbers of ties in railroad construction (at first, not only
the train carriages, but even their wheels and rails were made of wood).
And wood was still used for all kinds of applications in society: for making
furniture and tools and even to the production of tar and varnish. Wood
remained a fuel in heavy long-distance transport; initially, locomotives and
steamships used huge amounts of wood and early agricultural machines still

ran on steam generated by burning wood. Similarly, steam power was used for lifting heavy loads, as in mines; the steam was obtained from heating water, again, by burning wood. Before coal was introduced as an alternative fuel during the first half of the nineteenth century, the melting of iron still required large amounts of charcoal; using oil and gas for these purposes was still far beyond the horizon. Initially, though known and used by the poor, coal was thought to be dirty, and it was too heavy to be transported along the poor, muddy roads.

Ever-greater numbers of people thus meant an unprecedented, increasing rate of deforestation throughout the world. All these people had to be fed and dressed, and they needed houses and transport, all requiring the enlargement of arable fields at the cost of ever-larger areas of forest. Increasingly, too, people began to trade in wood, even on an intercontinental scale. In the sixteenth century, wood was already being shipped from America to England to build houses in London. For wood, it was worth waging war to found and enlarge colonies or (as happened in North America) to break colonial ties. The colonizing, rich nations also needed land for growing cotton or luxury products such as tea, coffee, pepper, cinnamon, and rubber, to name only a few—in fact there are too many new products to mention. Each of these products required vast plantations to be laid out and so more forest had to be cleared. These activities, in turn, required a workforce that had to be fed and housed, so again forests had to be cut. And the products had to be transported from the plantations to the harbors; for this, after using wooden rails, iron railroads were built from the mid-nineteenth century onward, such as the extensive networks in South Africa, India, and North America. In turn, the railroads required tens of thousands of ties to connect the rails. There was a huge demand for iron, which had to be mined and melted from the ore. Bridges were made of iron and so were the frames of stations, down to their drain pipes. The lines themselves had to be fenced off, which required large numbers of wooden poles, and many, much longer, straight poles were needed to carry the telegraph (and later, telephone) lines for communication between stations and towns. Telegraph lines could be hundreds—even thousands—of kilometers long. All this required incredible amounts of wood.

Although for very long times, people throughout the world had eaten small amounts of meat as a luxury, it soon became more widely available for greater numbers of people. The interiors of North America and Argentina were used for producing meat, which was exported as canned meat even during the nineteenth century to the core areas of Western civilization. At first, the cattle were kept on grasslands and prairies, but later vast tracts of virgin forests in South and Central America were—and still are—cut so as to supply the meat for fast-food products like your hamburger or Big Mac.

As after centuries of exploitation, the subtropical and temperate forests were destroyed, the forests that remain are mostly in the tropical parts of the world. These are now being destroyed at high rates. But it's not only trees that are lost: it is usually only a few years before the ancient forest soil is lost too, usually well before some sustainable agriculture can be built up. Then, the fields are abandoned, and production shifts to untouched forest, where the procedure of slashing, burning, and rapid soil exhaustion is repeated. This happens not only at the small scale of supplying the needs of one family, but also on a large scale; for example, millions of hectares of primeval forest have been cut or burned so that oil palms can be grown for industrial use by international firms. Indonesia is one of the countries where this is happening. The palm oil is used in the manufacture of a range of products: margarine, cookies, Smarties candies, ice cream, petroleum jelly, lipstick, and soap, to name a few. The fast-food meat for which large numbers of cattle are used and forests destroyed, goes principally to the rich nations of the world, leaving the local people to cut forests for their own use. This also happens in Brazil to obtain ethanol from plants as a biofuel. Moreover, people in the developing world are beginning to ask for the same kind of protein-rich food, which, though justifiable, comes at the cost of extensive areas of virgin forest. If this trend continues, it is estimated that the world will have lost its virgin forests and jungles by 2040.

Meanwhile, the vast Northern Hemisphere boreal and tundra forests are being deforested for their softwood. A major use of this wood is to supply the manufacturers of milk and fruit juice cartons. Huge areas are being deforested rapidly for this, as we need millions of cartons per day. Interestingly, the printing speed of all those millions of cartons is limited by the speed of sound: at higher rates, the sound barrier holds back the throughput of carton sheets. Used cartons cannot be recycled, because they are coated with a thin layer of metal and other material. In our dumps they decompose into methane, and their coating dissipates into the rest of the waste. The deforestation of these northern forests also serves the production of your daily newspaper, your magazines, books, and the billions of paper folders, brochures, announcements, and advertisements that come through our letterboxes. The wood chips made from this softwood have many applications, from hardboard to mulch for weed control in garden borders. In Tasmania, the last primeval temperate forests are being cut down for electoral reasons: to give lumberjacks a few years more of employment before being made redundant. For whatever reason, at the end of the day, the forest is gone.

Bricks have been used for building houses in England since the Middle Ages, particularly after wooden houses and other buildings in London burned down during the Great Fire of 1666. The firing of bricks requires much wood (this function later was taken over by coal). Though bricks and roof tiles were made as early as Mesopotamian times and the Harappa civilization in Pakistan, bricks were first used on a large scale by the Romans to build five-story apartment buildings and enormously large communal buildings, such as public bathhouses. The only surviving remnants of the ancient forests of the Mediterranean and Middle East occur today around the occasional monastery. Throughout the Renaissance and thereafter, brick remained a common building material in the Mediterranean. In the little Renaissance town of Urbino, Italy, for example, the three- or four-story houses, the roads and street staircases, the huge basilica and all the other churches, a mausoleum, the castle, the Renaissance palace and a theatre, the garden walls, and the huge defense wall surrounding the town are all built of brick. Enormous numbers of bricks must have been used—and huge amounts of wood to fire them. And Urbino is only one little town out of many, one tiny speck on the map of Italy erected within one or two centuries during a couple of millennia of building.

As well as being used on a large scale for firing kilns to make bricks, pottery, and tiles, wood was also a domestic fuel. Famines do not always occur because of shortage of food as such, but also because of a shortage of wood, preventing people from cooking or baking their food. And from prehistoric times onward, roads laid out through marshes were also made of wood. In the north of Russia, I have seen the traces of a wooden road hundreds of kilometers long that once linked St. Petersburg to Archangelsk. It was constructed of long, very heavy beams, stacked on top of each other in five or six layers. As the beams underneath rotted away, new ones were laid on top.

For many centuries in antiquity, people around the Mediterranean also built their ships from wood. During the Middle Ages, too, first the Vikings, and later the Hanseatic traders of Scandinavia and North Germany, must have built fleets consisting of hundreds of wooden ships. In the Renaissance, global expansion of trade began, leading to the colonization of continents overseas by the Portuguese, the Spanish and, later, the Dutch, in their turn followed by the English. Their fleets consisted of many hundreds of ever-larger ships—merchant vessels and men-of-war. Imagine, too, how much of the Spanish forest left since Roman times was cut down to build the mighty Armada. But that fleet was scattered by storms and destroyed by the English—gone were the woods. From early on, such ships were made of hardwood, mainly oak, which soon became exhausted in the home countries, not only because of the rapid growth of the number of ships used, but also because ships soon became obsolete because of the rapid technological advances in shipbuilding. Another reason was that the wooden ships rotted away rapidly, particularly in tropical waters. Ships leaving Holland, for example, had to be dismantled after their first leg to Indonesia, after which about half the initial number were reconstructed from usable parts for their return journey. Their turnover rate was really high.

The Dutch obtained their wood from the Scandinavian and Baltic countries, parts of Europe that also remained a source of wood for the English fleet. As said, from the sixteenth century onward, the English were also already importing some of their wood from America. Yet, not that long after, the English had to turn to other regions. Cromwell is said to have colonized Ireland because of the insatiable thirst of the English for oak, and later, during the nineteenth century, their more intense search shifted to North America. By the mid-nineteenth century, the forests in America around Chicago had been completely cleared, leaving a bare landscape as far as you could see. The HMS *Beagle*, the ship that carried Darwin around the world in the first half of the nineteenth century, was built of wood plated with copper, and its onboard carpenters had to replace parts of its hull at regular intervals. Thus, shipbuilding alone was a significant factor with regard to the deforestation of first Europe and later of other continents as well, although it is only one of many aspects of a wood-based economy that is thousands of years old. The present destruction of the vast tropical and boreal forests is only the last phase of global deforestation.

Most of the wood of the forests of Earth has now turned into carbon dioxide gas. It has either been burned or it has rotted away. Once it has dissipated into the air, it is lost—unless it is retrieved at a high expense of energy. One seemingly cheap solution for retrieving some of the carbon dioxide could be to reforest part of the wooded lands lost. However, the hidden assumption here is that the newly planted forest will reconstitute a real forest, with a species composition, layering, and dynamic similar to the previous forest, guaranteeing long-term persistence. But for forest dynamics to be natural and sustainable, it is essential to reinstate the spatial and temporal composition of its plant and animal species: its trees, bushes, herbs, insects, birds, and mammals, as well as huge numbers of mosses, fungi, and bacteria. Think of the millions and millions of trees and bushes, belonging to hundreds if not thousands of species. Where would our nurseries get their seeds? Where would we situate the nurseries? They'd soon need as much area as the forests themselves. Usually, we cannot plant the seeds of forest species in bare soil, because the moisture and temperature conditions are unfavorable for germination or there is insufficient shade. And where will we get the replacement forest animals and multitude of insects from? Moreover, forest dynamics, its turnover and its shifts in species composition, is extremely slow. Forest trees, for example, often reach one, two, or a few hundred years of age before they can be replaced by a tree of the next generation. In the temperate regions of the world, a natural change in forest composition, forest succession, in which biological, hydrological, soil, and climate conditions shift gradually, takes at least six hundred years.

And usually, too, after the original forest was cleared, the humus decomposed and the mineral topsoil was desiccated or eroded by water and wind. This disrupts the water content and water flow in the soil. Thus, the hydrology of the deforested area will have been altered irreparably, making it impossible to restore the forest's composition and dynamics. Look at the TV footage of heavy machinery plowing through the wet, humus-rich forest soil, leaving behind a mess of pools, deep gullies, smoking piles of broken stems and branches. A ruined, wasted world as far as you can see. The many bacteria and fungi that not only decompose the humus but also help the trees, bushes, and herbs to germinate, live, and reproduce are gone, having dried out and burnt in the sun. The birds and mammals that used to disperse seeds have dispersed in all directions, if they haven't been killed or died out. Yet without all these, a forest cannot be sustainable in the long term. Under the altered conditions sustainable reforestation is impossible.

Many forests, moreover, are found on nutrient-poor soil, which means that virtually all nutrients available for plant life have already been used up and fixed in the forest plants. Nutrient turnover is tight and rapid: new trees contain the nutrients of their immediate predecessors. Constitutionally, they replace the old ones. In fact, already during their lifetime, their fallen fruits and shed leaves, and their broken-off branches are recycled continually by birds, mammals, insects, fungi, and bacteria in smaller and rapid nutrient cycles. Forests exist and are maintained through the continual turnover of nutrients laid down in trees, bushes, and herbs, and in a whole world of animals, fungi, and bacteria, each having its own turnover rate within that of the forest as a whole.

These are only some of the local and most obvious effects of deforestation. But cutting down large stretches of forest also affects continental patterns of cloud formation and therefore of temperature and rainfall, which are impossible to reconstitute. Also, the humus and the mineral soil of a forest work like a sponge: they take up water after heavy showers or snowmelt, and they release it later in small portions, supplying forests with water year-round, which, in return, produces more humus and protects the soil against erosion. What could have been a creek or a wild torrent tumbling down the mountainside, giving life to a multitude of insects, fish, and mammals now becomes a number of pools of standing water, some trickles and winding rivulets. Forests buffer the downward flow of water and reduce fluctuations in runoff over large regions. Deforestation, instead, usually leads to alternating flooding and desiccation. Forests, too, build up and stabilize the soil, whereas deforestation inevitably leads to erosion.

There are other consequences of deforestation that cannot be reversed either. For example, many ancient forests once occurred in mountain valleys, but valleys often have good, rich soil and contain an abundant supply of clean water from the river or from creeks and rivulets flowing off the mountains—a supply that often remains constant over the year. And the river also supplies people with water and fish and a means of easy transport. Moreover, valleys with flat bottoms are also easy to work, have more or less uniform growing conditions for crops, and can be irrigated. Therefore, fishermen and farmers have always settled them, and their settlements have usually grown into villages and towns. But those farmers and townspeople had their demands for wood, crop growth, and living space and so valleys were the first to be deforested.

As populations grew larger, too little space remained for further agricultural and urban expansion, and people turned to the mountainsides for growing their food. Thus, they began to cut the mountain forests as well, so that its trees could no longer protect the forest floor against erosion. Rainwater ran off with ever-greater force, first removing the humus and then the mineral soil and stones, washing them down into the valley. The forest floor, built up over millennia, was lost forever. When the plots on the mountainsides became exhausted, adjacent forest was cut. Erosion worsened, because now the deforested area was larger. Moreover, the humus, the mineral soil, the mud and sand, the rubble, stones, boulders of all sizes, trees, and parts of the remaining forest washed down the mountain. Debris covered the arable soil on the lower slopes and in the valley and choked the river. I once saw how overnight a fertile alpine meadow was made worthless after a flood of nutrient-poor mud and stones had come down following a cloudburst. The same story is told over and over again across the world, the ongoing devastation of the tropical mountain forests of the Philippines or New Caledonia being just two cases out of many.

So, erosion from deforestation not only removes valuable soil from mountainsides but also wastes fertile soil in the valleys in-between. And when rivers get choked with silt and detritus, they can change course, thereby destroying agricultural fields, habitations, villages, and towns. Long ago, the Yellow River in China was named after the large amounts of loess it carried, as was the Yellow Sea in which it used to end. Apart from getting its muddy color, the river itself changed considerably as well: some of the stones and silt dropped to the bottom, where they stayed, raising its bed and hence its water level. The dykes containing the river kept having to be raised, increasing the risk that they would break and that the now low-lying areas of the floodplain would be inundated and thousands of people would drown. Over long stretches, the Yellow River flows ten or more meters above the countryside. It no longer reaches the sea. Upstream is a heavily eroded landscape without vegetation, intersected by deep, steep gullies—a landscape completely and irretrievably lost for further habitation.

Deforestation has other consequences. Particularly in the tropics and subtropics, groundwater reverses its flow; instead of sinking slowly into the soil, it rises upwards, bringing dissolved salts and minerals from deeper layers to the surface. After the forest has been cut, the sun shines directly on the soil surface, heating it up and drying it out and sucking up even more water. As the groundwater evaporates, the salts and minerals from deep down are left behind and precipitate either as a hardpan just under the surface or as a salty crust. This process can be enhanced when irrigation water from rivers, lakes, or aquifers adds more salts. Most animal and plant species, including those of valuable crops, are physiologically unable to grow in salt-impregnated soil or salt crust, forcing farmers to abandon their land.

In Egypt, the salts and minerals were washed away each year by the flooding Nile. In Mesopotamia this did not happen, and therefore soil salination caused this ancient civilization to end. From those ancient times onward, salt and minerals have always come up to the surface in the previously moist, wooded valleys of the Middle East, gradually forming mineral hardpans and salty crusts, so that eventually, the soil is unsuitable even for more salt-tolerant crops. In the end, large areas of once-fertile land had to be left behind as useless desert, glistening with salt. More recently, this also seemingly occurred in Central America, contributing to the decline and demise of the Aztec civilization.

In other parts of the world, dissolved iron oxide was similarly transported to the surface layers, often precipitating at some lower level within the soil then obstructing the downflow of rainwater. In northwestern Europe, for example, it can often been found as an almost impenetrable iron pan some six to seven centimeters thick at a depth of about sixty centimeters. Then, with increasing amounts of rainwater, the land became waterlogged, forming wet heathland. Local people had to build their sod huts and farms on higher parts where they cultivated their oats and potatoes, whereas their sheep grazed the heather on the lower parts when these dried up during summer. Yet, thin layers of minerals at various levels in the sandy soil show that the area was once covered by dry forest. Recently, the resulting thin iron banks underneath heathland in northwest Europe, which cause the heath to be wet for most of the year, have been broken up by deep plowing, and can be taken into cultivation again.

Deforestation, therefore, is almost as old as human civilization; it has resulted in vast expanses of continents being denuded and irreparably degraded. Forest clearance results not only in a gradual loss of the food-growing area, but also in a loss of biodiversity. This loss will never be known; we cannot possibly reconstruct a forest on paper, or in practice by any refor-estation scheme—assuming that the newly planted trees and other species would establish under the altered conditions. They are gone, together with all other connected life-forms and with previous hydrological and climatic conditions.

The common theme underlying this brief history of deforestation and its consequences is, again, the growth of the highly demanding human popula-tion. Population growth results in vast deforestation on all continents. In Europe, first the Mediterranean forests were felled, then the temperate ones, and then the northern boreal forests and taiga. In Africa, Australia, and North America, the native people did the same. Finally, global population growth is finishing off the exceedingly rich tropical jungles. Our demands for agricultural land and wood and for timber for all aspects of urbaniza-tion are still increasing, and our tools for cutting down trees are becoming ever-more efficient and powerful. All natural forests are expected to have disappeared by around 2040, because our food supply has to double, among other reasons. We make ever-larger and heavier machinery and ever-faster means of transportation. As a result, we are rapidly approaching what has aptly been called the "timber famine": the period of permanent global short-ages of timber. Forest regrowth hardly occurs in most deforested parts of the globe anymore. Our future generations will have to make do without wood, without forests, with a bare, denuded world. They will be left with the con-sequences of our recklessness and greed.

13 THE LOSS OF BIODIVERSITY

Whenever I look at photo books or television programs on mountain life in the European Alps, the Rockies, the Andes, or the Himalayas I'm always impressed by the rich flora: meadows full of flowers forming a brightly colored blanket spread out over the mountainside. And each of those thousands of flowers seems different. Even if they are of the same red or yellow, they still have different intensities of red, different shades of yellow. Some flowers show off with their large size, whereas others stand humble and shy in-between the foliage with their tiny, greenish-yellow flowers. If you're lucky enough to get to such mountains yourself, carefully wading through this bountiful life, you'll also smell the flowers; the scent is never the same in any two flowers or any two meadows. And each flower is visited by insects flying around in huge numbers: flies, bees, wasps, and shiny beetles. They all add their bit to the low buzz surrounding you, and the crickets lower in the vegetation add their sharp chirping.

Those crickets stagger clumsily around in the turf, eating from grass stems and blades, whereas the bees sample nectar from all the flowers, flying busily between the meadow and their hive. If they are wild, their hives can be found hanging high up in a nearby tree, or, depending on the species, in-between the stones heaped up in a corner of the meadow. You might find the nests of predatory wasps dug into the ground and closed off by little pebbles and the sonorously buzzing bumblebees and beetles creeping in a tussock of grass or in their burrows in the soil. Spiders rush off hastily as you walk through the meadow; look at the butterflies fluttering silently between the flowers, sucking nectar, rolling out their tongues to reach it.

If you move slowly or sit down silently, after some time you might see some other animal life: young marmots chasing each other, mountain goats clashing heads, or deer, seemingly on tiptoe, alertly treading into the field. Little birds twitter in the air, and far away in the distance are one or two eagles approaching, circling higher and higher along the mountain slope into the sky until they disappear from sight, still looking down from their great height in search of prey.

Yet, sitting there, you know that this amazingly rich abundance of life will last only a short time; soon, it will be too cold for most of those insects to fly or for lizards or the occasional snake to rustle amid the vegetation. Plants, in their turn, will die off and rot away in the following fall rains or wither in the winds of the winter cold. Each year in spring, however, when the sun warms up the slopes, life bursts out again, and there they are back again—the flowers, insects, marmots and lizards, the clashing mountain goats, and the eagles circling in the air!

But then, after having watched life in that mountain meadow, after letting its beauty and smells sink in, go and sit down next to a little creek or ditch, or on a seaside cliff with its thousands of screeching birds coming and going, bringing food for their chicks. Or look at the mosses and trees growing in impossible places in narrow crevices in a steep rock face. In the water of some fast flowing creek, you see little, primitive insects crawling, occasionally being pecked at underwater by a bird, a dipper, feeding on them. Above the ditch dragonflies hover, hunting for their insect prey; in the water, you see small fish, sticklebacks, repairing and defending their nests in which their female had just laid a clutch of eggs. Or go snorkeling and look down into the water along a coral bank; all those fish nibbling on little algae, others waiting in their shelter for prey, or a little school of yet other, silvery fish silently trekking underneath you, unnoticed—arriving from nowhere, then sliding into a more distant nowhere again.

Wherever you look—arctic areas, savannas, or jungles—you always find an abundance of life of different forms and colors, different behaviors: animals feeding, defending, breeding, or resting for hours on end, digesting their food. All fascinating forms of life, life of an amazing variety, an immense diversity far beyond any imagination. All forms with their own history, the history of the individual organism, and each of them with the unique evolutionary history of the species to which they belong.

From whence did this exuberance come—all these colors, these sounds, this business, all those plants hugging together, yet each reaching out to show off their beauty, all trying to attract the insects, or catch the wind? Insects that day in day out sample the nectar the plants offer, eat the grass, or each other? From whence did this endless variety of life appear? This amazing world? Although scientists have ideas and theories and opinions, they really don't know exactly. What seems clear is that there seems neither rhyme nor reason in the way all these colors and sounds and movements, these differences in behavior originated; there seems no purpose behind their origin, no strict and clear line in their evolution. No cause, no purpose, no end. Evolution is a haphazard, chancy process, as is the daily life of all those individual organisms. This huge, fascinating variation, down to its weirdest outgrowths, is a product of evolution. None of all the animals knows what the next day will bring. They don't have the slightest inkling. Life is living by the day, by the hour, by the minute. Hoping for the best. Blindly changing. Evolving.

One day, the day I was writing this chapter, we saw an extremely black cloud approaching at great speed, and very soon big and heavy hailstones were battering our windows and our summer garden. Our Eden was ruined within minutes: the storm left behind broken plants, drooping flowers, and wilting vegetation. A couple of kilometers away there had been hardly any rain and only a soft breeze.

Such a fierce hailstorm, especially in summer, is a rare event where we live, and the fact that it passed over our house and garden is even rarer; it was only bad luck that our garden was hit at all, and then so hard. But such chance events are the rule in nature; "rare" events occur all the time and everywhere, although always with different intensities. One time, insects are blown off the plant on which they are feeding, another time they drown in a raindrop. A mite can wait for hours, days, weeks on end for the leg of some goat or dog to come past on which it can feed. The little, tender parachutes of dandelions, on the brink of leaving the plant to drift away on the wind to begin another generation elsewhere, stick together in the morning dew. Now, none of them will ever leave to grow out into a new dandelion; they will rot away underneath the leaves of their plant parent. The chances were against them.

What happens to those single, individual animals and plants also happens to the species to which they belong; sooner or later, they die out because the chances were against them for too long. A rare succession of dry summers or an occasional dry and severe winter might finish them off. Or perhaps they sneak through, either in the same form, or, in the longer term, as a different species. Never the same way, never in the same direction, never at the same rate. Never knowing what other species do, always following their own evolutionary path. And because of all the differences between places, differences in the number of times favorable or hazardous conditions follow each other, they are jolted about evolutionarily, eventually reaching the unsorted abundance and diversity of life-forms of a mountain meadow or sea cliff. Chance ruled their origin and existence, chance drove them together and took them apart, and chance caused the sheer endless variety of life-forms in which they evolved.

The evolution of Earth itself is a chancy process too: the billions and billions of chemical reactions since the first hesitant ones four billion years ago, the slight changes in the flow of energy, its way to degradation and decay, thereby increasing its complexity. Amazing. First, tiny, invisible organisms—bacteria—slowly became recognizable. Then, the more complex unicellular animal-like creatures and algae came along, then multicellular ones: worms, snails, crabs and lobsters, insects, fish, dinosaurs, birds, and mice. Eventually, there appeared apes, man, complex societies, architecture, boat building, trade. Then art and science, music, computers, travels into outer space, globalization. This is mind-boggling, dazzling, incomprehensible in its beauty and grandeur. And yet, it is all by chance, blind to the future, blind to the next day to come.

This is biodiversity, the diversity of life; its road to this diversity is diversification. And realize that we humans are now putting all this beauty at risk. We are making the world uniform and constant, eroding the land, and poisoning and extinguishing life. Homogenizing this staggering diversity. Almost four billion years of history. Hundreds, thousands, tens of thousands of those amazing little beings, those little bits of burning life, are vanishing each month, each year without ever having been seen. Their structure, color, and behavior vanish with them, never to return, never to be seen again. Exhausted, crushed, or poisoned. Because we don't want to see, don't want to know, don't want to wonder. We have no time to sit down in a meadow, to look around, and watch. We want to eat and earn and grow, in spite of the world. Neglecting the world.

With the increasing numbers of people, more and more species are dying out, a process conservationists are trying to stop or at least to slow down and direct. Many of them, therefore, count species over the years to see whether they are decreasing, and, if so, to see what can be done to halt this. In technical terms, these numbers represent the local species richness.

A low species richness, however, may occur equally in a field of nettles or a mud flat colonized by rarer species. It is the differences in their biology, their structure or way of living, that are important. Instead of merely arriving at a species number, we need to express the differences in biological specificity, in the diversity in identity and biological quality. This biological diversity should be expressed by some measure of diversity that represents the range of variation of life-forms present, the whole diversity of lifestyles among plants, the variety of ways of living, moving around, and behaving. All their qualities are unique; species cannot be counted together without losing what is essential for biology, essential for their own survival. When quantified, biodiversity gives biological information about an area. The diversity in an area and the nature of the lifestyles of the species found there tell us how remarkable—and, hence, valuable—that area is biologically.

What we can do is to consider a certain trait, say, the nitrogen content of the soil, and arrange various plant species along a line according to their nitrogen preference, from nitrogen-poor to nitrogen-rich. Then, we can count the number of species that prefer a certain nitrogen content, and plot these species numbers along the line, as frequencies. The range of this frequency diagram expresses the variation, the diversity of the plants with respect to their nitrogen preference or tolerance. Now we have made a measure of biological diversity among those plants, a measure of their biodiversity, which contains more information than merely the number of species on its own. We can do this for any biological, morphological, physiological, behavioral, or evolutionary trait, or for several traits combined. If the local biodiversity now decreases, this then enables us to specify which part of the diversity range is missing, which indicates the kind of biological loss. It may happen that the left-hand part of the range, say, containing the nitrogen-intolerant species, is lost; once we know this, we can take specific countermeasures to restore that part of the range.

A large proportion of our present plant species actually originated and evolved many millions of years ago under nitrogen-poor conditions, and only a relatively few specialist species are adapted to more nitrogen-rich ones. Interestingly, each time a glacial period ended during the last couple of millions of years, the first colonizers after the ice receded were nitrogen-tolerant or nitrogen-requiring plant species. But, gradually, over the thousands of years of the following glacial period (a geological instant), they have always been replaced by nitrogen-avoiding species, which continued to dominate for the duration of the rest of the interglacial period. Now, with agriculture encroaching into natural areas, nitrogen-containing fertilizers often enter into the environmental chemistry of nitrogen-deficient marshes or woodland, with the result that all the ancient life-forms of plants and animals typically living under nitrogen-poor conditions are lost. At present, northwestern Europe is the worst in this respect, having the highest nitrogen intensities and therefore the fewest wild species. Worldwide, though, the widely scattered patches of nitrogen deposits will expand; and by 2050 will fuse, leaving hardly or no place at all for the most ancient, natural nitrogen-avoiding vegetation. Thus, by extending our nitrogen-hungry agriculture, we are destroying an enormously large section of the biological diversity on Earth, the product of a long evolution from the ancient past.

If we see the biodiversity defined this way decreasing, this measure may indicate a biological loss better than the mere number of species could ever do, as it specifies what part of the biological diversity is being lost. For example, for many years, a worldwide organization has been especially concerned with wetlands as a special habitat for birds or insects or plants that are adapted to the wet conditions found there. If we lost the wetlands of a particular continent, such as Australia, or worldwide, we would lose all the species with their widely diverging properties typically adjusted to wetness and water; herons waiting for fish to feed on, the fish themselves, the plants adapted to seasonal or annual changes in humidity, and so on. Losing these life-forms means losing a whole section of our biologically evolved heritage on Earth at one stroke. Wetlands are also essential for the transequatorial long-distance migration of birds flying from the high Arctic to the southern tip of South America. These birds rely on specific marshes as refueling stations along the way. Similarly, the woods between Canada and Florida are stepping stones for the monarch butterflies that migrate south by the millions. Branches of trees have been known to break under the weight of the huge numbers of butterflies that have settled on them! If these woods are destroyed, the butterflies will no longer be able to migrate in those numbers over such amazingly long distances.

Our drainage of wetlands has already affected many species on a staggeringly large scale. During the seventeenth century alone, between ten and fifteen million beavers were trapped for their fur in North America. The wetlands created behind their dams dried up and turned into meadowland. Consequently, the aquatic insects, the fish and the frogs, the fish-eating herons, the geese and all the species of duck living in those wetlands died out over large tracts of North America. The muskrats and otters froze in their burrows; trees, once the drumming logs for the springtime mating of ruffed grouse, were gone, and so were the plants, algae, the unicellulars. All interdependent life-forms, all specifically dependent on this beaver-made environment, all gone. Meanwhile, the fur coats and hats have almost all decayed; only the occasional one survives in some county museum.

Each year we cut down thirteen million hectares of tropical forest for timber or to plant oil palms, we destroy mangrove swamps to satisfy our taste for shrimp; we have virtually eradicated a unique set of small fish in Lake Victoria because of our taste for Nile perch, and we are bringing the incredibly large schools of sardines and anchovies from the deep ocean waters of the Pacific and herring from the Atlantic to the brink of extinction, because we use this protein-rich food for nursing salmon. Apart from this, over half of the annual 180,000 tonnes of krill is fed to farmed salmon, too; this reduction in available food is one of the causes of the rapid population decline of penguins in the Antarctic. Some of this salmon we sell as canned gourmet food for cats.

First, the Sumerians; then the Assyrians, Egyptians, Greeks, and the Romans; then the Europeans eliminated lions and elephants, ostriches, hippos, wolves, bears, and martens from the Middle East, North Africa, and Europe. Then the whole of Siberia and China and North America followed suit and tigers, buffaloes, deer, beavers, raccoons, and mountain lions were wiped out, as were the hundreds of thousands of sea otters and the tens of millions of seals inhabiting the Pacific shores, the hundreds of thousands of whales commuting annually between the Arctic and Antarctic oceans, often reproducing in Australian waters. A large part of the unique and ancient Australian marsupial fauna has been wiped out, as have the gigantic moas of New Zealand. The story is the same for equatorial Africa, South America, and India. What a massacre. What remains, what we now conserve, are leftovers—crumbs of what life on Earth once was. In each case, we have lost entire sets of species, sets specific to certain living conditions, all with their unique properties uniquely evolved over millions of years. In each case, we have irretrievably lost entire chunks of the biodiversity of our world in one go. Chopping down the irretrievable tree of life.

People have always cut down forests, frequently and at regular intervals, with the result that forest turnover rate increased and older trees and ancient forests were lost. Most of the woods found today in Europe, for example, are relatively young, and you can only find the cut off, rotting, dead trunks of once impressive, massive trees as posts at the entrance of some age-old farmsteads in countries like Norway. This increase in the turnover rates implies that all the insects with long development times died out. The ones that remain are species adapted to the high turnover rates typical for young forests and arable fields; they have short generations, great dispersal capacities, and also require nitrogen-rich nutrients or food for fast reproduction. In contrast, those lacking the traits for high temporal and spatial turnover and population buildup succumb one after the other. One by one, we are replacing the flowery meadows by dull blankets of uniform green. What are left, apart from the species of our arable fields and pastures, are weeds and pests.

We did more. We homogenized the naturally diverse and dynamic conditions in our arable fields, making them uniform, constant, and stable. Thus, especially from the nineteenth century onward, we mechanized agriculture, consolidated land, and enlarged the fields. Larger, uniform fields, however, make it easier for weeds and pests to develop and survive (fields, by the way, are where they also find their preferred nutrients and living conditions). Typically, their numbers are high, helped by the fact that they often have no natural enemies, either because they had been left behind in their home country or have been killed by our broad-spectrum pesticides. Our arable fields and pasturelands are now also dominated by species from several continents, because the seed trade has achieved their long-distance dispersal. They are also transported by thousands of ships, airplanes, trucks, and cars that crisscross the world, carrying the seeds, eggs, and larvae of pest species or ornamental plants. Often, too, the wind blows their seeds, eggs, or their parasites in all directions across the uniform fields, or high up through the air to the next ones. Of course, these aliens don't always take to their new environment or don't develop into a new weed or pest straightaway, but after some years their numbers may explode. We are eradicating species by the thousands and at the same time encouraging the remaining ones to occur everywhere and in huge numbers. We are curtailing biodiversity, the biological diversity of life.

But there is more. Although the traits selected differ, the same annihilation of species is happening within and around our rapidly growing villages and towns, around harbors, industrial complexes, and roads, where large chunks of the natural biodiversity are lost, because the conditions there are different again, differently uniform and differently dynamic. This extreme environment has high concentrations of minerals and metals; consists of impenetrable brick, concrete, or asphalt; is polluted by dirt and oil leakage; has a changeable humidity; and is exposed permanently to the scorching sun. In such conditions, only a very restricted and specific set of species will survive. Obviously, insect-pollinated plants, along with their pollinators, are lacking. In some countries, foreign flowers are sown in road verges, but the native insects and small mammals are either blown away by passing cars or are killed when crossing the asphalt. Furthermore, at a regional scale, roads form barriers for the runners of plants and for all mobile animals, from deer to toads, from beetles to spiders, so that adjacent populations are fragmented and reduced in number and ultimately die out. In this way, mobility and spatial dynamics are selected against.

New broad-spectrum pesticides not only kill the predators at the top of the food chain, they also affect the sense of direction of bees. They are then unable to find either their source of nectar, or their hive. Because part of its crop depends on pollination, India may suffer a crop loss of 20 percent to 30 percent if pollinating insects are killed. In the Netherlands, this particular insecticide reaches concentrations several thousands of times higher than the legally permitted maximum. During the 1960s and 1970s, Rachel Carson had a major impact on the legislation concerning the production and application of petrochemical products in agriculture. But the effect of her revolution was only temporary; biodiversity has declined rapidly since then. Over the years, thousands of new chemicals have been developed and dispersed into the environment.

Now, after denying so many species any living space, and after altering the living conditions of others, we have also begun to reduce the genetic diversity of those that are useful to us in order to maximize crop yield and profit. This is achieved by optimizing the match between their demands and the conditions offered. Some specialized factories are producing seeds that make plants resistant to certain pests and diseases so that they don't need to be sprayed. Moreover, the yield is uniform and conforms to the market. In animal husbandry, it is cheaper to use only one bull, boar, or ram to inseminate large numbers of cows, sows, and ewes artificially, even years after the bull has died. Farmers can tick off the desired qualities of offspring on a form and order semen from a specific sire. However, the genetic uniformity and intense geographical interchange also increases the chance of pandemic diseases breaking out on farms; some of these diseases are infectious or even deadly to humans.

In this way we are losing large amounts of genetic diversity in our crops, cattle, and poultry—diversity produced in the past with great effort and which we may badly need in future. Once this genetic diversity is lost, we can never get it back again. An additional problem is that seed factories are interested in continuous income, so it is not in their interest to produce stable genetic forms of selected genes; unstable genetic combinations, hybrids, are more profitable. These fall apart in the next generation and need to be replaced by a new hybrid the following year. Of course, buying ever-new genetic types burdens the cost of agricultural production, a cost neither many farmers nor most of the consumers in poor developing countries can afford. Often, farmers aren't even entitled to use their own races when seed factories have patented them. The continual production of new genetic hybrids of crop species results in a race between diseases and seed factories, which, so far, the seed factories seem to be winning. In the longer term, however, with most of the ancient races having died out, we will run out of genes to be recombined, and then the production of our food will be at risk.

The same trend is occurring in poultry and fish, particularly now that standards of living are rising in large parts of the world. Meanwhile, hundreds of races have already died out locally or globally. Worse, now that most freshwater fish have died out in many waters of all continents, and many marine species are on the verge of extinction, diseases are also occurring in the breeding cages suspended in the open sea where fish like salmon are raised. These diseases can infect wild populations of salmon or other marine species, reducing their survival chances. Also, the genes of the farmed fish are spreading into the wild populations. Similarly, frogs kept in high densities for human pregnancy tests have contracted a deadly fungal disease which is now spreading through many frog species worldwide. As in the case of plagues of weeds and pests, endemic diseases that used to occur only rarely now easily become pandemics. They can sweep across the world from one continent or hemisphere to the other within weeks.

Thus, the amazing biodiversity of the world is rapidly being reduced, sometimes almost to zero. This is the result of all the wasting of land through the erosion, salination and pollution of the soils, the depletion of the naturally available nutrients and their substitution by a continuous supply of chemical fertilizers. It is the result, also, of deforestation and urbanization, of more efficient farming and profit maximization, and, ultimately, the result of our excessive and still rapidly growing numbers and consumption.

To feed our billions, after first having industrialized our society, we industrialized life; we industrialized species, we industrialized landscapes, we industrialized oceans. We are emptying the oceans, emptying mountains and landscapes, emptying rivers, turning them into dirty dry gutters, alien to life. And in the end, you may ask "Why?"

14 WASTED LAND

In Friesland, one of the northern provinces of the Netherlands where I spent many a summer as a young boy, the country folk have a saying that on wet days, cows eat with five mouths. They mean that when it's wet, each leg sinks deeply through the turf as the cow treads on it. This "poaching," as it's called, spoils the grass and thus wastes the cow's food. As well as trampling on their food, cows lie on it when chewing their cud or sleeping, and every now and then, they defecate on it. They therefore need to be moved to fresh pasture frequently—or be kept in barns and have harvested grass brought to them.

The cows' treatment of grass is very wasteful behavior, but don't we too waste usable, often excellent agricultural land by building our towns and industries, our harbors, distribution centers, refineries, and roads on it? In the past, small settlements of farmers or fisher folk expanded and became our large towns, occupying much more land. Look at Italy, where large and small factories are situated in fertile valleys. Or look at northwestern Europe, where large areas of fertile land left behind by the glaciations have now been built over. Look at the history of the big cities of the world, like New York, Los Angeles, or Chicago, which together are home to many millions of people. Look at the fertile coastal areas all over the world, now rapidly being urbanized, industrialized, and dissected by roads. These were the areas where we used to obtain our drinking water. The coastal fringes, riversides, and deltas of all continents were usually the best of our arable lands or lay close to the best fishing grounds. Once, the Netherlands was one large, fertile delta with rivers and lakes full of fish that was close to the breeding grounds of the fish of the North Sea; herring and cod were caught by the hundreds of thousands. Now, the country has one of the highest population densities of the world, the rivers and lakes are polluted and gray, and the sea has been overfished by beam trawling.

What is happening is not confined to the behavior of cows in their meadows or the urbanization of large areas by humans. Depleting and wasting the ground they live on is one of the reasons most species are always on the move. Of course, birds move quicker than trees, but trees also move, although their moving around can take several generations or centuries. Our hunter-gather ancestors were roaming. Present-day nomads still roam with their livestock, especially in poor and dry environments where regrowth of the scarce vegetation is slow and where their firewood is soon used up. In tropical forests, regrowth is fast, but there, too, rapid soil exhaustion forces shifting farmers to keep moving, clearing plots for their next crops by burning and cutting down trees. These slash-and-burn cultures have occurred throughout human history and are still found in many parts of the world.

At some point, people settled down permanently in villages and small towns close to their fields. There was a risk of soil exhaustion, but they discovered that adding their own excrement and that of their animals to the soil improved their crop yields grow. Occasionally, this manure contained diseases, which could threaten the survival of their permanent settlement. The early Egyptians escaped soil exhaustion—but not outbreaks of disease or pests—because each year the Nile flooded the land, leaving behind fertile silt. Thus, the farmers were able to cultivate the same land year after year in a narrow stretch of land between the river and the desert where tens of thousands of people were concentrated. But, of course, the consequence was that the upper reaches of the Nile eventually became exhausted and lost their topsoil. In contrast, in Mesopotamia the Euphrates and Tigris did not flood land and fertilize it, so instead the people built an extensive irrigation system to fertilize and water their land. Gradually, the nutrients were depleted and the region lost its fertility.

Dense vegetation reduces the evaporation of water from the soil by preventing the sun from shining directly on the surface. Moreover, without vegetation, soil dries out, and groundwater rises to the surface, bringing with it minerals from deep down. This is happening in many parts of the world, such as the southwestern United States, and large parts of Australia, which are thus wasted for crop growing and raising livestock. As mentioned in chapter 10, laterite is formed in subtropical and tropical parts of the world by the precipitation of iron oxides. Thus, soil salination and laterite formation are two kinds of human wastage of land.

Another wasting of the environment caused or aggravated by human action is wind and water erosion. In wind erosion, the top layers of the soil dry out when the wind blows away the humus and mineral soil after the vegetation has been removed. This also happened naturally in the Northern Hemisphere around the southern rim of the ice sheets during and after the ice age, which is why there are huge belts of sand and loess running from west to east across Eurasia and North America. These loess deposits are ideal for crops; they retain rainwater, allow plants to root deeply, and contain sufficient nutrients.

Wind-blown dust from the Sahara still forms large clouds that drift across Europe and the Atlantic up to Iceland. Some of this dust is also blown westward across the Atlantic to South America, or eastward up to China or beyond. What is left behind is bare rocks and fast-moving sand dunes hundreds of meters high. Interestingly, in Roman times, the Sahara was still covered by tropical and subtropical vegetation. Wild animals roamed in these areas, the game hunted by Egyptian and Mesopotamian kings and aristocracy. In Roman times they were captured and taken to Rome for gladiatorial sport in the Colosseum. Egyptian hieroglyphs, depictions of Assyrian winged lions, and the Lion Gate in Mycenae, Greece, all testify to this vanished fauna. Once, there were big lakes in the central Sahara where people fished and hippopotamuses were abundant. Beautiful rock paintings from successive periods show how the Sahara gradually dried out and the fauna changed. Eventually the region reached its present, extremely eroded state. It is said that the origin of this enormous desert is at least partly human (but this theory is controversial).

Elsewhere, wind erosion also happens as the result of an ancient way of fertilization, when after the grain harvest the farmers burn the stubble, believing that the nutrients remain in the ash and are returned to the soil during the next rain. However, just a little wind blows this ash away to forests or the sea. High winds during the mid-1930s blew away much of the topsoil in the dustbowl region in the United States. Here, farmers had plowed in the turf of the prairie vegetation, which had previously prevented the humus layer from blowing away. After a succession of dry years, dust storms removed the humus that had developed over centuries and had retained soil moisture in the soil under the vegetation. It dispersed over part of the continent and into the Atlantic, hundreds of kilometers away. What remained were often dry, barren sand dunes.

Another form of erosion, water erosion, occurs when people clear forests on mountain sides for agriculture or buildings, which usually results in a massive wash down of the valuable humus and soil. However, water erosion also happens on very gentle slopes, when runoff flows over the ground after a heavy rain shower. In the flat parts of central Australia, the river beds may remain dry for years, but after heavy rains they quickly turn into wild, muddy torrents carrying debris. When their waters subside, small trees, dead branches, and grass are left hanging at heights of three to four meters in the trees, and deep gullies have been gouged out, which soon dry up. Similar gullies occur in the eastern Mediterranean and North Africa, where they are known as *wadis*. Water erosion may be local, regional, or large scale, such as in northeastern China, where vast expanses of land have suffered water erosion that washed away their loess. Either for natural or human reasons, the big rivers of the world have always formed large deltas at their mouths from deposits of sand and mud eroded in their upper reaches. Therefore, they always contain the best soil for growing cereals, though most of the sand and silt is in the submarine part of the extending deltas. At present, however, the Nile delta is contracting: the river's flow diminishes in the water reservoir behind the hydroelectric Aswan Dam so that this reservoir is now silting up with the material once forming and maintaining the delta.

Earlier in this chapter I explained how farmers manured their fields to combat soil exhaustion, another, more gradual form of wasting of land. Fertile arable land becomes exhausted when year after year crop plants extract its nitrogen and other nutrients like phosphorus or potassium and are then harvested—which means that these nutrients are taken away from the fields. This is why we now depend for our food on fertilizer which has to be mined in only a few places on Earth, and the limits of which are in sight. However, we also depend on micronutrients, nutrients present in small amounts and which we also need in small amounts, which run out gradually as they are not present in fertilizers. Another way early farmers combated soil exhaustion was to leave the land to lie fallow for several years, during which time ruderal plants and weeds took over the land. Soil bacteria fixed atmospheric nitrogen to a form these plants could use. After the weeds had died or when they were plowed in, the nitrogen they had accumulated remained in the soil and was then available for the next generation of plants, or for the next crop.

Mainly from the eighteenth century onward, people began to understand what happened, after which controlled rotation schemes were installed. A major advance was to use different crops rather than weeds to restore the soil's nitrogen content; the advantage was that they could be harvested and used and they also helped prevent the buildup of pest populations. Nowadays, rotations are unnecessary, because farmers can use artificial fertilizers and pesticides made from natural gas and oil. However, these chemicals often kill all the soil life responsible for humus decomposition—the soil fungi and insects—as well as the nematodes and worms that are essential for the soil aeration needed for the plants' root physiology. The humus helps maintain the soil moisture content. When land is farmed without an intimate knowledge of the local soil conditions, there is a danger of large-scale soil deterioration, particularly when the "farmers" are entrepreneurs and food corporations seek to maximize profit, abandoning the land when profits drop. And if they started off by clearing or "developing" natural areas, such as forests or marshes, the result is often rapid and large-scale environmental and biodiversity loss. Fertilization and irrigation lull us into a false sense of security. Because yields are still good, we tend to forget that without these inputs, our agricultural land would be wasteland.

Around industrial-scale feedlots the nitrogen concentration from the ammonia emitted into the air from the manure and urine produced by the livestock can be very high. The ammonia is dispersed by the wind and deposited on the land and water. The nitrogen concentrations on arable land in the Netherlands can reach forty times the amount permitted, whereas 60 percent of U.S. ammonia emissions are from feedlots. Nitrogen deposited from the air is literally on top of the nitrogen added deliberately when fertilizing the fields. Yet, if we stopped fertilizing or irrigating our farmland, crop yields would fall dramatically and we would realize that it has become wasteland. Within several generations, we may be forced to stop treating the land because the fossil fuels from which chemical fertilizers and pesticides are made have run out, as have the phosphorus mines, and aquifers have run dry. Wasted farmland, and, if we can't find alternatives on time, a wasted world. Our sheer numbers and personal demands are too much of a burden on the natural world.

Soil compaction is another problem that wastes arable land. It happens when heavy farm machinery is used to plow, harvest, spread manure or fertilizer, or control pests. Compacted soil has lost the small pockets of air through which soil water could move, and so it contains too little water for the crops to grow. Also, soil life becomes impossible when nematodes and worms or insect larvae, all of which break down old plant material and humus, can no longer penetrate the dry soil, cannot get enough oxygen, or drown in stagnating rainwater. Crop growth depends on loose, crumbly, well-aerated soil. When a crop takes up nutrients, it uses energy; its roots respire and therefore have to take up oxygen.

In cold winters, soil plowed in the fall freezes and the growing frost crystals break up the clods into finer crumbs. A severe frost kills many overwintering pests, so reducing the risk to the next crop. Under a warming climate, however, soil pests have to be combated chemically, which pollutes the groundwater, and as many pesticides are often nonspecific, they harm beneficial insects and worms. Moreover, they break down in the soil only partially and very slowly and remain toxic. Herbicides too can accumulate in the soil, as can veterinary and human pharmaceuticals. Only a few plant or animal species can tolerate high concentrations of these chemicals. Such weed vegetations and animal pests also have to be treated chemically, as it is often difficult to remove them mechanically. Some weeds have even evolved seeds similar to those of the crops, so they cannot be separated before the new season and are sown along with the crop over a large region. In the United States alone, their extermination costs several hundreds of billions of dollars each year.

High concentrations of medicaments given to livestock and passed with their urine can now be found in meadows and arable land. They kill off life not only in that land and in nearby ditches, creeks, and big rivers, but also in the seas and oceans into which these water courses discharge. They can cause malformations and malfunctioning in wildlife; some marine fish are covered in pustules. Similarly, pharmaceuticals and hormones from contraceptives used by people end up in their original chemical form in sewage systems and thus pollute the water reservoirs from which we obtain our drinking water. We can thus take in a cocktail of medicaments not intended for us, and their combined load could well affect our health.

All kinds of chemical waste discharged by laboratories, factories, garden centers, and horticultural industries pollute large tracts of land and end up in the groundwater, creeks, and rivers—one large network of open sewage runoff. Dutch ditches, lakes, and rivers and even parts of the North Sea have thus become opaque and unsafe for swimming. Water containing high concentrations of detergents, bactericides, paints, and many kinds of plastics in all stages of decomposition is unsuitable for any biological life or derived human use, yet it is used for irrigating arable land or meadows.

Pollution from spilled industrial wastewater makes the groundwater unsuitable for drinking, or for irrigating the land. In some oil-producing countries, oil can leak out of pipelines, thus wasting large expanses of land or ocean for all forms of life for many years to come. The large-scale mining of tar sands in Alberta, Canada, similarly results in incredibly large areas of wasteland. Moreover, this mining costs large amounts of water, and the disruption to the regional hydrology continues long after the miners have left. The same applies for all other forms of open mining, such as stone quarries, open-cast mines for metals, such as those for gold in Nevada, or for brown coal. In Germany, for example, the groundwater of large areas around such deep, open-cast brown coal pits has fallen by many meters, well below the reach of crops. In other cases, water seeping through landfills pollutes the groundwater of the surrounding area. In China, the blood of children living near dumps containing electronic waste has been found to contain too high levels of lead, cadmium, and other metals; their health and performance at school is affected.

A related form of environmental wasting comes from spilling wastewater after metals have been chemically freed from the ore. If this water is discharged directly into rivers, it kills aquatic life. If discharged into open settling ponds these become filled with extremely poisonous cyanogens and metal cyanides. If, as once happened in Hungary, the bund around the settling pond is breached, its toxic water flows into water courses and eventually penetrates the ground, getting into the groundwater deeper in the soil, from where it can flow unseen in unexpected directions and over large distances. More recently, it happened again in the same country, when a reservoir containing waste material from aluminum extraction failed and remnants of this metal plus those of other poisonous ones diluted in sodium hydroxide with pH values up to 13 covered villages, orchards, arable fields, and the natural environment. And it still happens in Turkey where people near a silver mine die of cancer and where a large basin with twenty-five million cubic meters of potassium cyanide used for the release of silver from its rocky ore broke through its dams because of rain. Metals and other minerals and chemicals emitted from factory chimneys rain down over considerable distances around coal mines and chemical plants, endangering and sometimes killing wildlife. Many plants and animals have thus died out locally around crematoriums because of heavy metal pollution resulting from partly vaporized false teeth and metal implants. Irreversible soil and groundwater pollution also occurs after the use of very stable chemical compounds such as DDT (dichlorodiphenyltrichloroethane), PCB (polychlorinated biphenyl), dioxin (e.g., 2,3,7,8-tetrachlorodibenzodioxine), or large numbers of pharmaceuticals. Such chemicals often result in reduced fertility, high incidences of embryonic deformation, abortion, lifelong malfunctioning, or direct physical damage to the skin, eyes, intestine, kidneys, or liver.

Yet another way of wasting potentially habitable land is through warfare, particularly through large-scale destruction: spraying open water and fields with chemicals, defoliating tropical forests, deliberately spilling and burning oil on land or in the sea, firebombing villages and towns, burning land

and vegetation, and planting landmines and other explosives. In chemical and nuclear warfare, the groundwater of vast stretches of land is polluted by nondecomposable chemicals or depleted uranium.

One of the problems underlying metal pollution is that over the billions of years of life on Earth, once-abundant metals have either been used up or have been chemically bound and have then precipitated, forming sediment. Later, the sediment became rock in which they were locked up, which means that living systems had to adapt to their lowered concentrations, such as by forming pumps. Organisms still utilize metals (iron, for example) in their chemistry, so over millions of years they have developed strong pumps to extract them from their environment. When the concentrations of metals such as iron are high, the strong pumps that organisms evolved to deal with low concentrations extract too much, and too much metal accumulates in the organisms and may ultimately kill them. On the other hand, too much dissolved metal in water can hamper the working of the sideline of the more recently evolved fish with which these communicate, and which operates with low electric voltages. Not being able to communicate can affect their reproductive capacity. In contrast, in the geological past, other metals, such as calcium, became more easily available and so organisms developed mechanisms to expel them from cells in order to keep their concentrations low: calcium precipitates outside the cells, forming bone and shells. However, the bones and shells of marine animals become weak when rising carbon dioxide concentrations cause water to acidify and thus bind the calcium now necessary for mechanical reasons and for protection against predators. Acidification also releases aluminum and other metals in the soil, which are poisonous. Big changes in the availability of metals in the environment inevitably cause many organisms to die as a consequence.

An increasing amount of land is wasted because all the nondecomposable solid household refuse, industrial waste, and old cars and tires have to be put somewhere. For example, each tire in a large dump can contain rainwater in which mosquitoes breed and then swarm into the surrounding area, making it virtually uninhabitable. Mosquitoes can also breed in the water of flower vases in our ever-growing cemeteries. And every hour, the English throw away enough garbage to fill the Royal Albert Hall in London (which has a seating capacity of over 5,000). This amount is dumped in small portions all over the country, gradually filling one landfill after another or forming mountains of garbage, polluting the groundwater over a much larger area and producing an acrid stench and toxic fumes. Some incinerators for burning household refuse, however, have had to be dismantled because they released poisonous dioxins, causing malformations in animals and humans. Now, large amounts of refuse are shipped to poor countries, where people scavenge for usable bits and pieces in unhealthy conditions.

We are wasting not only the terrestrial environment but also the seemingly boundless oceans, though fortunately the discharge of chemical and nuclear waste into the oceans has been prohibited in the developed part of the world. I have already mentioned that incredible amounts of plastic, nylon rope, and other kinds of petrochemical waste are blowing or flowing into the oceans: some 80 percent of all oceanic plastic garbage is terrestrial in origin. Many albatross chicks are fed with pieces of plastic and cigarette lighters, which fill their guts; they die from malnutrition. The extent of damage to marine animals remains unknown. Parts of the once-majestic ocean gyres have turned into whirlpools of garbage. The North Pacific gyre is twice the size of the United States and is known as the Great Pacific Garbage Patch because it now contains as much plastic as the largest dump in California. In fact, it consists of two whirlpools, the Eastern and the Western Garbage Patches. Similar patches are found in the Atlantic and in the Sargasso Sea. And this is only the visible part of all the plastic—an unknown amount sinks to the ocean floor. Whether on the surface or on the bottom of the ocean, the plastic fragments into minute pieces. The result is so-called plastic soup—opaque water full of microplastic. These particles account for 95 percent of all the plastic in the ocean, and their concentrations can reach a hundred thousand particles per cubic meter. It takes about one thousand years for all this plastic to reach the ocean floor where it can be covered by other precipitates. Meanwhile, microplastic can choke the unicellular organisms on which fish and other animals feed. And it can also poison organisms, because when discharged into the oceans, poisons such as DDT, dioxins, PCB, and pesticides concentrate a hundredfold by attaching to microplastic. As the particles of microplastic are the same size as microorganisms, they are impossible to filter out without affecting those. Only the large visible pieces on the surface can be retrieved.

Petrochemical products can also poison marine life when plastic compounds dissolve in the water or directly in the tissues of an organism. This happens with Bisphenol A (BPA), which is found in plastic foil, beakers, and cutlery. A large category of plastics containing the very poisonous phthalates (plastic softeners) occur in PVC and certain insecticides. Both kinds of compounds affect aspects of our male and female reproductive systems, and both can be transferred from mother to child. Phthalates are now known to affect the behavior of boys as well, making them less masculine and reducing their sperm production. Phthalates can dissolve in seawater and will affect marine life and then land animals—including humans—who eat fish. There are several tens of thousands of other compounds that give plastic its specific properties. Alternative packaging materials such as cotton, paper and carton, or decomposable plastic based on starch or sugarcane are all made from recent rather than fossil carbon compounds and could thus diminish or solve many problems. However, to produce their raw materials, large areas of agricultural land have to be taken away from food production. Also, sugarcane is used for the production of lactic acid from which polystyrene can be made. Obviously, in addition to food production being reduced, the production of these decomposable products requires freshwater and nutrients during plant growth and processing. Ideally, biofuel would be used for the production of the energy required at the various steps, itself costing part of our agricultural land, minerals, and freshwater. We live from more than from food alone.

Water is being affected by yet another kind of pollution—heat pollution— which results from the release of cooling water and heats up canals, rivers, or even stretches of coastal waters. No life is possible in those heated waters, because warm water contains little or no oxygen. The natural, biological cleaning of water by oxygen-using bacteria therefore becomes impossible. Instead, the organic waste in these warm waters decomposes anaerobically—a process that creates bad odors. Today, much of this kind of dead surface water on and around continents is opaque, brown, and risky to use. When clean freshwater runs out, people pump groundwater from deep in the ground. (In deltaic regions in Bangladesh, West Bengal, Cambodia, and Taiwan, however, this groundwater can contain natural high concentrations of arsenic, which is poisonous because it replaces phosphorus in many molecules, thereby messing up our biochemistry.)

Much land could become wasted if climate change due to chemical pollution of the atmosphere forces cultivation to shift to areas with unfavorable soils. This may happen in the Northern Hemisphere and in most of Australia. The west-east belts of deep fertile loess soils in the United States and Eurasia currently coincide with zones in which the climate is favorable for growing crops. As the United States and Eurasia function as the bread baskets of the world, any shift in these climate belts may threaten the food of millions. A northward shift of the favorable climate into Canada, Scandinavia, or Siberia—areas once covered by glacial ice and where thin soil now covers hard rock—would be disastrous. Other parts of the world, such as the western part of South America, southwestern United States, northeastern Africa, and Australia, may dry out as the climate warms up. These areas are already prone to salination or desertification. Crop failures, along with water shortages in the cities, are expected to become more frequent in the narrow belt of habitable land along the Australian coast.

In western Eurasia, there is a large triangle of human habitation containing all the major cities, industrial areas, and farmland. It stretches from Scandinavia in the west to Portugal and Spain along the Atlantic coast and tapers in the east toward Irkutsk in Russia. Like a Russian nesting doll, this large triangle includes a smaller one of cereal crop growing, beginning from a narrower base in Western Europe along the Atlantic, roughly from the south of Scandinavia to the north of the Iberian Peninsula, extending its tip only slightly further than Moscow. Immediately north and northeast of this triangle lie the vast boreal and permafrost forests and bogs of Siberia, almost devoid of human occupation or activity. South of the triangle, it is too hot for human habitation: it is the region of savannas and large deserts. Imagine what will happen to this vast agricultural and populous triangle when climate changes.

Human habitation elsewhere in the world also exists under a relatively narrow set of climatic and soil conditions only. We, like the plants and animals we use for food, cannot live everywhere; we are restricted by ecological and economic constraints. Over 60 percent of the world population lives within a zone of sixty kilometers from the coast. Hundreds of millions of people from rural, inland populations are migrating to the coasts, attracted by the urbanization and industrialization. What makes it worse is the timing: at exactly the time when we may run out of energy, mineral sources, and cultivable coastal land, we will need more food and living space because our numbers and demands are continuing to grow.

We need to be aware of all the waste and wasteland we are making, of the fact that we are well on the way of wasting our world. We must act, and soon.

Environmental Destruction and Its Implications

During our existence on Earth, we have become increasingly dependent on a human-made environment. But increasingly, our waste is polluting that artificial, human-made environment. Our living conditions will be determined not only by our growing numbers and demands but also by the duration of their effects. The longer we pollute, overuse, waste, and destroy the environment and its biodiversity, the worse those effects will be. Moreover, as the impending exhaustion of our fossil fuels, water, rare metals, and phosphorus shows, none of our resources is renewable indefinitely; our crops, too, still coming up anew each year, ultimately depend on nonrenewable resources whose limits will soon be reached. Whatever we do, in all respects our resource use remains linear to a very large extent, with resource depletion on one end and waste accumulation on the other. What makes all this especially worrying is that many of these trends will climax all at once, partly because of the scarcity of the remaining resources, and partly because of the rapidity of the increase of their use.

It is not possible to pinpoint the date on which we might reach such a point, and we may also see some new phenomena, such as an increasing incidence of large environmental fluctuations and extreme conditions. In our attempts to make the environment more uniform and constant, we are losing natural buffers against any fluctuations, be they buffers against climate, disease incidence, or social responses to shortages and heat. Worldwide, between 1970 and 1990, 480 billion tonnes of topsoil was lost—equivalent to all the cropland in India. And during the same period, an area as big as all the cultivated area of China turned into desert—some 120 million hectares. In the face of a still-growing world population, this alarming loss of invaluable agricultural land is still going on.

TOWARD A COLLAPSE OF OUR SOCIETY?

There are two kinds of growth: the growth of our numbers and demands and the growth of the infrastructure of the society we form. These two kinds of growth are heavily intertwined, infrastructural growth being a direct result of growth in our numbers, whereas the growth in infrastructure makes the growth in our numbers possible, and accelerates it. It will be virtually impossible to stop either the growth of our numbers or that of societal structuring, and this preferably at the same time.

Many populations in the underdeveloped world of today remain relatively unorganized as they grow, which means that, in a certain way, they are overpopulated concerning their food supply, other resources, and their production and deposition of waste. The internally well-organized Western countries, in contrast, may become overorganized, which could increase the chances that they get out of balance, although still having access to resources. This difference is of increasing relevance to our future as it can push the world population as a whole out of balance.

D

Processes within the Human Population

Several processes go along with the growth of our numbers, such as urbanization or migration. Also, with increasing numbers, the number of diseases can grow, while endemic diseases turn into epidemic ones. Other processes concern a possible gradual stabilization of our numbers and demands, although it seems more likely that these kinds of gradual processes do not occur. Numerical growth is more likely to result in a sudden crash or collapse of the population when the infrastructure supporting all people destabilizes and implodes.

15   WHAT IS OVERPOPULATION?

Most people have heard about the theory of natural selection proposed by the nineteenth-century Englishman Charles Darwin to explain how biological evolution may be driven. It was already known that evolution itself—the change in properties of living organisms over the generations—occurred, but until Darwin advanced his theory, no satisfactory explanation had been put forward. Darwin suggested that in the same way that animal and plant breeders use only a select few individuals to produce pigeons, dogs, cows, sheep, and tulips with certain properties, in nature selection occurs when a few surviving individuals are left to reproduce, whereas others are not. For one reason or the other, the latter died before leaving offspring. For breeding purposes, a breeder may use rabbits of a certain fur color only—white for example—and reject any brown rabbits. Then, by consistently selecting the whitest rabbits to breed from, after several rabbit generations the breeder will have white rabbits only, which all produce white offspring. All the rabbits giving brown offspring will have gone. Individuals with favored properties therefore have a reproductive advantage over those with less favored ones.

Even in Darwin's time and slightly before, many people felt that England was heavily overpopulated, as shown by the masses of undernourished poor. Under those conditions, the weak individuals would unavoidably succumb and the stronger would survive. Curiously, this observation was turned into policy: because of increased mortality among the weak (often the physically and economically weak, that is, the poor) the quality of the English population as a whole was thought to improve to the advantage of the nation. In 1776 already, the economist Adam Smith had recommended to governments not to regulate food prices in times of famine. Later, this was also the opinion of the economist and parson Thomas Robert Malthus, who wrote about it in his widely read book of 1798, *An Essay on the Principle of Population*. He argued that without human interference (no assistance to the poor and destitute), society would itself adapt to the existing food conditions. Hands off: leave it, *laissez faire*. Darwin was among its many readers; he later applied these ideas to processes happening in natural populations of animals and plants as evolutionary driving forces. In nature, adverse living conditions would do the selection, continually changing the properties of organisms so that they better matched current living conditions. According to him, those adverse conditions would lead to competition, which then selected the weak from the strong. A contemporary of Darwin, the biologist Alfred Russel Wallace, had read Malthus's book as well, and he too came up with this idea of biological improvement, although through a slightly different process—the struggle for existence, which is broader. It acts directly on disadvantageous properties, not indirectly via competition. Actually, if the idea of competition for food as an exclusive evolutionary driving force were correct, then there couldn't be species, and maybe there would be no life because competition between individuals of the same species (those with the most similar ecological interests) is stronger than competition between individuals of different species and, hence, with different interests. Those "con-specifics" would be selected out first, as it happened in Malthus's humans, leaving the chimps and gorillas untouched.

Throughout history there have always been concerns about human overpopulation. These concerns go back to Babylonian times, and probably much earlier. The most ancient literature known mentions it. Clay tablets dating from 1800 BCE recount legends of overpopulation and of catastrophes—drastic countermeasures taken by the gods—despite the fact that there were far less humans around in those distant times. Not long after Malthus, the French mathematician Pierre François Verhulst constructed a formula, called the logistic equation, which described how a population grows from low numbers to high. If this maximum level is crossed, increased mortality might result from famines. This extra mortality connects the equation with the ideas about selection held by Smith, Malthus, Darwin, and Wallace: selection weeds out the weak. Modern ecology and evolutionary ecology at large are still based on this idea.

Competition for food operating through an increase in the rate of mortality among the weak was a typically British model for understanding a response to overpopulation. Of course, food shortage can also operate directly through an increase in general starvation, affecting all people equally, not only the poor. In contrast, in France food shortages were thought to affect the rate of fertility rather than that of mortality by delaying the age of marriage. And Russians, knowing the vastness of their deserts and tundras, objected to the model as well. Overall, a concentration on one factor only—food shortage having an impact on numbers of people—means a gross oversimplification of natural and societal processes where epidemic diseases, climate variation, reduction of soil fertility, energy shortage, industrialization, and so on, each with its geographical variation, historical changes, and migration, operate on and influence each other. In fact, up to the nineteenth century, with its increase in medical knowledge, human population had always principally been affected by epidemic diseases rather than food shortage.

A humane society, however, might decide to make the economically weak a bit stronger by enabling them to get some of the extra food produced rather than let them die. Then, not only the gods but people themselves might do something about the effects of food shortages. Slightly before 1800, for example, roughly the time that Adam Smith, Malthus and others were developing their hands-off ideas, the famous French chemist Lavoisier actually did experiments about improving food conditions in France by improving agricultural practices, with the aim of tackling the food problems underlying overpopulation. This does not mean that nothing at all happened in Britain with respect to agricultural improvement. For example, halfway through the eighteenth century already, the geologist James Hutton had introduced the iron plow from East Anglia into Scotland which meant a much higher efficiency. This approach has been followed in the West and the developing world.

It is impossible to predict developments in complex systems such as the weather or climate, and certainly in the world population. Reliable weather forecasts can be made no more than one week in advance. Similarly, estimations of climate change or our future numbers and living conditions typically end within a couple of decades from now, that is, in 2030 or at best in 2050, which means that the projections do not even cover a single—our own—generation. However, on the basis of the numbers of people in various age groups, demography studies the changes over several generations. Not that complicated, you might think. The predictions, however, refer to extrapolated trends in the numbers of people in each of those age groups, and even small mistakes in counting cause large differences in calculation results, which become larger the further into the future the prediction is. Will the number of children per family remain the same or decrease? How old will people become during a given time period: older, or will their maximum age stabilize? These estimations, moreover, are made irrespective of complicating changes in food conditions, climate, or effects of pollution. Usually, demography emphasizes the study of changes in numbers rather than their probable cause. "All other things being equal" is the general adage, but usually, those other things do not remain equal but change continually. But even under such highly simplified and constant conditions, a slight difference in estimation produces a very different answer. For example, a difference of only 0.5 in the global average of number of children per family results in a difference of over a billion people in the next generation. In fact, the difference between the low estimate of the number of people in 2100, 10.1 billion, and the high one of 15.8 billion in that year results from an average difference of half a child more per female. How precise and certain can our estimations be when over two billion people are living in virtually impenetrable slums? This uncertainty explains, for example, why the decline in global child mortality rates over 2009 is greater than calculated.

Obviously, such small differences in the estimation of the numbers in various age groups in the world population are extremely likely, especially under changing conditions. Small wonder, then, that hardly any predictions are accurate. Less than ten years ago, for example, politicians suggested that the eight hundred million hungry people in the world could soon be halved to four hundred million, but instead the number has risen to over one billion, with unknown demographic consequences.

Throughout history, the human population has been growing at varying rates, and at different rates at different places. Which of these rates can we extrapolate? Local differences occasionally led to population extinction. And after the Roman period, for example, European populations dropped substantially. Yet, the global average in numbers increased steadily, and during the nineteenth century, numbers began to rise steeply, especially in Europe and North America. These increasing growth rates were the result of better medical and food conditions and improved sanitation and hygiene that caused mortality to slow.

A further complication is that the extrapolations are not made according to a known and anticipated causal mechanism determining the numbers. Instead, existing trends from the past to the present are extrapolated into a different future. In fact, our future numbers will be determined by comprehensive mechanisms, such as the availability of food, as determined by, say, a combination of developments in economy, climate, pollution, land waste, urbanization and politics. How does such a combination determine our numbers? And what if the operating mechanism changes its behavior or breaks down? We don't know. We are driving by looking in the rearview mirror, without using the pedals and steering wheel, and without realizing how much fuel is in the tank.

Deliberate checks on further growth of our numbers are either absent or, if they occur, can hardly have any lasting effect. Reducing family size, for example, to one child per family, as in China, is not drastic enough, and would be almost impossible to carry out on a global scale and for many generations. One of its complications is a temporary imbalance in the population's age distribution, which is one reason why the Chinese are now thinking of relaxing their one-child-per-family rule. Contrast this with Russia or France, for example, which have a declining population and where reproduction is encouraged. However, such countermeasures do not always work, since, apart from financial reasons, people have become accustomed to small families and may resist incentives to increase their size.

As in the past, effects of diseases on population growth are only regional. From the fourteenth century to well into the fifteenth, plague raged across Europe, in many cases reducing populations by two-thirds, leaving behind a badly damaged society. But from a global perspective, this was a mere blip. Moreover, the effect it did have was smoothed out during the next few centuries. In Iceland, mortality from measles, another once-deadly disease, showed an irregularly undulating pattern, indicating that previous losses, however serious, were soon after compensated for. The massive 90 percent mortality of the indigenous inhabitants of North and South America as a result of smallpox introduced by the Spanish and Portuguese, and later by the English and French, was soon compensated for by new, rapidly reproducing immigrants. Even a reduction of 90 percent would prove to be too small to bring our present world population of 6.8 billion down to 500 million or so. Such a reduction is highly unlikely because of the huge absolute numbers involved. If a disaster were to reduce our numbers this drastically, deliberate measures would have to be taken to prevent compensatory growth of the surviving population.

The same applies to wars; in terms of our present numbers, they too are demographically negligible, and have no significant and lasting effect on a world scale. The devastating Thirty Years' War fought during the seventeenth century in Central Europe, for example, cost some 60 percent of the population their lives but left hardly a trace in the longer term, or at the scale of the whole continent. Similarly, according to some estimates, about eighty million people died worldwide during the Second World War, often under terrible conditions. Without wanting to belittle their suffering and that of all those they left behind, this was about 4 percent of the world population of two billion—hardly noticeable and rapidly compensated for on a world scale. Similarly, the number of deaths under the most horrifying conditions of trench warfare in France during the First World War was smaller than the number who died during the Spanish flu epidemic immediately following the war. Meanwhile, wars have become financially unfeasible and economically disadvantageous. Wars or any other deliberate measures to increase mortality rates are morally unacceptable and should never be contemplated.

Famines don't give a lasting answer, either. At the moment of this writing, there are about one billion people under the hunger line: globally, about one in seven. They live mainly in underdeveloped parts of the world: sub-Saharan Africa (except for South Africa and Botswana), India and its neighboring countries, and Indonesia and the Philippines, along with some Central and South American countries. Famines in poor countries where there is racial and religious tension complicate the problem. Mass mortality due to starvation cannot be condoned, and even if one billion were to die from hunger, again, this would be a mere blip.

In principle, we can reduce our own numbers ourselves, drastically and within a short time: for example, by an almost complete and worldwide stop on reproduction via contraception and mass sterilization. However, the measures needed are extensive and extremely deep-cutting, felt to be immoral, inhumane, and in violation of our basic personal freedoms. They are therefore impossible to carry out. The only option left is a less drastic strategy, although this will cost more time and carries the risk that our resources will run out in the meantime or that pollution, climate warming, and environmental destruction will begin to have their effect and reduce our population for us but in a much less humane way. From this perspective, our future will depend on a fragile balance. Any hopes that mortality rates will achieve the required population reduction without us doing anything are in vain. We have passed the point of no return. Given our present, still rapidly growing numbers, Malthus's mortality can no longer be of any help, while Lavoisier, supplying more food, makes the problem worse.

The combination of a reduced mortality rate and a high fertility rate has resulted in an enormous explosion in the number of people in the world. But that is not all: large-scale emigration has extended and enhanced the growth process. First, Western Europeans settled at increasing rates in North America, after which they also turned to other regions on other continents: South Africa, Australia, and New Zealand. Ethnic Russians moved into the vast emptiness of Siberia, and Han Chinese into the surrounding space south and west of China. In many cases, the colonists exterminated the local populations, whereas at present, migrants are interpenetrating those populations by the millions. This huge expansion into new territory hid the population explosion actually going on in the countries of origin, but soon the newly established populations underwent the same demographic growth.

Apart from this overflow into other parts of the world, people moved into previously empty parts of their own countries, then they deforested and drained the land. Urbanization and industrialization took off. People moved by the thousands from the countryside into the towns; they were "pushed" by the mechanization of agriculture out of the countryside and "pulled" by industrialization into the towns. This concentration of people in cities is still going on in the underdeveloped and developing countries of Asia, Africa, and South America, although the possibilities of industrialization vary greatly between countries. The decreased mortality in these countries, coupled with constant, relatively high fertility rates, explains why they have the greatest number of large cities and hypercities. When urbanization is not accompanied by large-scale industrialization, however, tens of thousands of jobless people amass in the numerous, vast shantytowns surrounding the cities; currently the number of their inhabitants worldwide approaches two billion—roughly one in three of us.

Many of us attribute the growing problems with our living conditions to the growth of the world economy or to its globalization. If only that could be stopped. However, the growth of the world economy necessarily has to follow that of our numbers. Although those billions living in slums stand apart from our economy, we all need to be fed, watered, housed, warmed or cooled, dressed, cleaned, and transported. And this personal care necessarily requires an enormously large and complex organization, itself having its own requirements and producing waste. Among the components of this system are shopkeeping, lawmaking and enforcement, insurance, office space, transport for people and freight, factories, electronic communication and banks, shareholding, mining, power plants, and waste disposal. This organization is indispensible if we are to keep going. In our present numbers we can't each be self-sufficient; this infrastructure of our society, life in the countryside, villages, cities, conurbations, and megacities without all these external inputs and facilities is unimaginable. We have reason to build those cities and this infrastructure. We have to accept that the larger our numbers, the even larger these cities and infrastructure become, and the greater the need for large-scale mechanization and automation to get all the work done. We ran out of people a long time ago; now there are simply insufficient people to maintain all people, making it compulsory to speed up and enhance agricultural production and to use all kinds of machines running on fossil energy to do all the work needed. Throughout history, technological and organizational innovations, agricultural mechanization, and societal industrialization have helped us to feed ourselves; they followed our growth. Growth in our numbers accelerates growth in economy, expenditure, resource use, and waste.

Our numbers require the organization of our life to be globalized, and therefore, national infrastructures have grown out into a superstructure of global dimensions. Globalization is like national infrastructures and the structure of our own body. And infrastructures too have their costs: all the energy each of us requires for living flows through our bodies, and it similarly flows through societal infrastructures of individual countries. A large energy input reduces the friction in the running of our global society. And we need the materials that carry this energy and shape its flow. Our numbers are too high to be checked either by ourselves or by disasters beyond our responsibility and capacity.

Does this incapability suggest that the world is overpopulated by humans? What is overpopulation, actually? The first criterion of overpopulation is that our numbers have grown beyond the amount of space, food, mineral, and energy resources available; pollution is beyond the capacity of natural decomposition and recycling processes; and environmental deterioration is beyond nature's capacity to bounce back. The shortage of resources and the amount of pollution generated are static criteria for judging whether or not our world is overpopulated. The second criterion is that our numbers are beyond our capacity of control. A third criterion is that our future is now also determined by the dynamic of the structure of the population and society itself, far out of our own reach. Together with the dynamic of the environment, that of the self-accelerating effects of the greenhouse gases in particular, such factors relate to the social, economic, or environmental stability. Any fluctuation or instability leads to social collapse.

For many people, a dynamic overpopulation model keeps alive their hope that the population problem will resolve itself. According to such a model, our numbers will stabilize when the rates of mortality and fertility become attuned to each other—not in a Malthusian way by increasing mortality, not by deliberate birth control, but by an automatic control of fertility by demographic and societal factors.

Such a process of a decline in fertility in the European populations during the nineteenth century toward a presumed balance with mortality is taken as a model, the so-called transition model, for a similar stabilization process in the rest of the world. It is questionable, however, if this is justified, that is, whether under different conditions other populations will go through the same transition process, and if they do, whether they will do so at the same rate or at a higher one (assuming that this process ever took place in Europe).

One significant difference is that the Europeans had the economic opportunity to go through such a phase: they obtained their industrial resources, income, and food from their colonies. Today's underdeveloped countries don't have such an advantage. They remain poor, because they lack a substantial industrial sector and trade with foreign resource areas and markets. Our ways of exacting low prices are still in place and therefore they can't buy food and other resources abroad as the West could and still can. And they can't build up their own industry. Without financial resources, they can't build up their own systems of health care and sanitation or their own system of social security. Instead, they have to feed their growing, jobless population in spite of industrial and agricultural competition and exhaustion from outside. Their reproductive surplus cannot emigrate like the Europeans did, leaving by the hundreds of millions from their still relatively small populations. Many developing and developed countries are closing their doors anyway. They can't afford to fight colonial wars, such as the ones fought by seventeenth-century Holland or nineteenth-century England in which tens of thousands died (another macabre demographic outlet of the time). During the European demographic transition period with falling fertility rates, fertility even had to be promoted for military reasons. Pro-natalist groups in various countries, such as the United States, still argue that keeping the population young and growing will give them an economic and military advantage relative to other countries with ageing populations like Russia, China, or many European countries.

Moreover, many developing countries, such as China, India, and Korea, until recently themselves underdeveloped, are now running short of food and energy and are leasing or buying land in Africa and South America, or are buying mines to supply their increasing demands for minerals. At the time of this writing, somewhere between fifteen and twenty million hectares of African agricultural land alone, an area equal in size to France, have been added in this way to their own crop growing areas. And this at a time that food resources are becoming stressed, if sufficient, for the Africans themselves; the resources are under stress because of population growth and desiccation due to climate warming. Moreover, roughly one-quarter of the population in these African countries consists of children younger than fifteen years who will still be in their reproductive phase at the time of an expected global stabilization of population growth, around 2050. All these factors keep countries from entering some demographic transition phase, a phase that would last at least a hundred years. Instead, they are trapped in their preindustrial agricultural phase with its heavily fluctuating mortality rates due to recurrent famine, civil war, disease, and social disruption. The poor, nondeveloping countries may thus follow a Malthusian pattern with high fertility and mortality rates. As approximately 75 percent of the global population growth comes from those countries, one may wonder whether the expected stabilization of the world population will indeed happen, and how. Moreover, for many decades and greatly due to external forces, in most underdeveloped countries, a pro-natalist policy has been in place to stimulate fertility, along with medical care to reduce mortality, and this attitude will stay. Under the present conditions, the worldwide application of the transition model seems dangerously misleading.

But does this model even apply to Europe and its colonies themselves? In the nineteenth century, the model did not apply to the countries England dominated, including Ireland, India, and Egypt. Instead, England destroyed the local industries in India and Egypt so that cotton from these countries could be processed in England, and the cotton products were sold back at a profit. Those countries were thus thrown back into the agrarian, preindustrial state from which they are still struggling to emerge. They all maintained a high net reproduction rate; they did not undergo a demographic transition phase. Because of these measures, during and after the Industrial Revolution in England, Continental Europe, and North America, industry, technology, and trade brought in the money with which capital for social security was built up, and with which conditions of feeding, health, sanitation, and education were improved. This led to a greatly increased reproductive rate during the nineteenth century there.

The transition model fares no better during the second half of the twentieth century: rather than slowing down, during the 1960s and 1970s, the annual global net reproductive rate soared to an incredible, historical all-time high of 2.02 percent. After that, it declined, though it remained at an unprecedented high value. The continuing high values challenge the transition model's contention that an increase in wealth is accompanied by a decline in fertility. Data for 2007, for example, show that the number of children per female is highest in the wealthy northwestern Europe and lowest in the relatively poor Mediterranean and Eastern European countries. Still, along with global trends, in many countries reproduction rates drop to well below the replacement rate, implying that the rates of mortality and fertility do not tune up at all as they would do according to the transition model.

In relation to the improving living conditions of the nineteenth century and thereafter, two processes are stimulating further growth. The first is the increase in longevity. Average life expectancy in many parts of Europe used to be thirty years, whereas now it is seventy-five years or more (an increase by a factor of 2.5). This means that the number of mouths to be fed has similarly increased by 2.5. In China alone, by 2050 the number of people over sixty years of age is estimated to be about half a billion, all to be fed, housed, and dressed, resulting in population growth without an increase in fertility.

Secondly, the increase in longevity was followed by a delayed decrease in the number of births, indicating that an unambiguous physical balancing mechanism was absent—as was a criterion about the location of a new optimum level. Actually, how can a birth rate respond to mortality achieved some decades later, reached under different conditions? By what demographic mechanisms could the Chinese reproduction rate tune to, say, the Indian rate? Still, this is what the transition model as applied to our global population demands. If there is any response, it is via human decisions themselves. But such decisions vary according to social and economic rules and developments. For several reasons, too, in many societies there is a fear of reducing the fertility rate—a fear counteracting the reaching of any balance. As the decrease in fertility rate was delayed, and that of mortality has been low for decades, the number of humans is increasing rapidly: worldwide, populations have been growing at staggering rates, higher than even before the Industrial Revolution. The average annual growth rate of 0.34 percent of the world population expected in 2050 would approach some presumed stabilization level, but this is still three times higher than the 0.11 percent it temporarily reached around 1600. Before then, it had always been well below that percentage, but it needs to be zero percent for there to be stabilization. Moreover, a decreasing percentage of an increasing, already high absolute number still results in a substantial increase in absolute terms. Thus, the increase of the world population would still amount to some forty million people each year by 2050, compared with the present seventy million achieved at a growth rate three times higher than forecast.

To call observations on such a complex set of processes a model overstates what we know about them: as we don't know the mechanism linking the rates of fertility and mortality, we simply have no model. Models represent process mechanisms. Also, the nondemographic processes driving changes in fertility and mortality are largely unknown, whereas suggestions made as to their joint operation necessary for a model differ for the two processes. For example, the education of women is often mentioned as a factor driving a reduction in fertility, but it does not surface as a factor determining mortality. The same applies to the significance of the wealth and degree of industrialization of a society as well as to their timing: in France the "transition" began around 1785, whereas in England, a wealthier and more industrialized country, it began only in the midst of the Industrial Revolution, that is, during the second half of the nineteenth century. Although there is an obvious worldwide decline in replacement rates, it is impossible to predict or manipulate these rates without a model unifying a heterogeneous yet largely unknown set of processes.

However, we should not aim to stabilize our numbers at the incredibly high level of nine or ten billion people anyway: we have to reduce our population to a number that is sustainable in the long term. For Earth has to support these nine or ten billion people for a long time—perhaps indefinitely. And Earth's resources are limited, and we inevitably keep making and accumulating waste. The assumption of some automatic fine tuning of the rates of mortality and fertility to achieve a numerical stabilization introduces some wishful thinking into the equation. But we clearly have to take matters in our own hand; relying on some invisible hand is of no use. The stakes are too high.

If, indeed, for whatever the reason, Earth is overpopulated, and if we are to reduce our numbers, by how much should we do that? Down to what level? What is our optimum number, given a certain persistence time of humankind and a certain minimum size of national infrastructures and a global superstructure to maintain such a number? After all, recycling, however well carried out, remains imperfect; we cannot avoid using up our resources and accumulating waste. Do we have a criterion?

We all know that we can't do without energy for a moment. As soon as the bloodstream in our body stops, we almost immediately become brain-damaged. Vital functions can be lost forever. To a lesser extent, other parts of our body are badly affected by energy shortage as well. The same holds for our national and global society. A cessation in our energy supply would mean that electronic negotiations, payments, and instructions to transporters could no longer be made; our computers would stop straightaway; telephone connections would be broken, affecting the necessary commerce and the local and global financial systems; factories would stop working and waste would heap up; heating and cooling systems would break down with consequences for food preparation and preservation; transport would come to a standstill, leaving large regions without food and other supplies; pumps for drinking water, irrigation, and sewage systems would stop working, exacerbating health problems; people would not be able to go to work; it would be dark indoors and out, which would increase criminal activity; warning systems wouldn't function. There is hardly any aspect of our personal, social, or economic life that does not require a constant supply of energy, large or small. Yet we know that our fossil fuels are running out, and there is still no reliable alternative. All the existing alternatives combined would cover 20 percent or less of our needs. A reliance on energy is the Achilles heel of individuals and society alike. In the case of many other resources we may be able to eke out what is left, find or make an alternative, or make do with less, but not so with energy. The supply of energy is pivotal to the working of any-thing, of any organization. Therefore, unless we find an alternative, artificial source of energy, our global population is doomed to sink back soon to its level from before the introduction of fossil fuels. That is the level of around 1900—or even around 1800—which is of just above one billion people at most. In that case, we would have to go back to one-seventh of our present numbers, or, if we don't take measures soon, to one-tenth. And that's without factoring in a certain persistence time for humankind on Earth. Moreover, finding one or more alternative, carbon-neutral energy sources, ones that do not affect the supply of our food, is compulsory for avoiding the worst thing of all, a runaway greenhouse process igniting on Earth.

Finding a solution for energy only, however, won't bring ultimate solace, because the same reasoning applies to almost all other resources, or to us be-ing smothered by our own waste and pollution. We would have to go back a very long time to reach a balance with the level of natural resources and the rates of nutrient cycles. If we remain sitting back, letting our numbers grow to just under ten billion within our generation, then only one in ten persons will have a chance of surviving. Who makes the choices? Are there humane ways to achieve the necessary reduction in population? Obviously, we can't leave our future to chance, to the wishful thinking of the transition model. It won't work; it has never worked. We must reduce our numbers ourselves.

Although we may not feel it personally, we need to realize that in both the short and the long term, our planet is heavily overpopulated with humans, perhaps from ten thousand years ago or more when it became necessary to hoard food, live together, and specialize. We did that for some reason. The situation is worsening year after year—growth to reduce or avoid bad effects of growth, leading to ever-faster growth. We are rushing headlong into this danger we have created. Great efforts must be made to develop new sources of energy and to find ways of rigorous and large-scale recycling of materials. Large parts of the process have grown out of hand, but there are still some buttons we can press. First and foremost is the one that drives it all—our numbers. Without tackling our numbers, other measures are partial and temporary.

Any measure to reduce population sounds immoral and inhumane, but measures we do not take ourselves will always be less—much less—humane and moral. We have gone too far, and we must turn back. There is no further way forward, no possibility of staying put. And we must do it ourselves.

16 BURSTING OUT OF EDEN

As boys, we often wondered whether we could touch the horizon. We realized that the horizon was far away and that like our shadows, it shifted with us with each step we took. Traveling by train did not help. So, touching the horizon seemed an absolute impossibility—like jumping over your own shadow. Later, it appeared that our horizon widened in all directions: the higher we climbed a tower, a hill, or a mountain, the higher and more distant the horizon became. Touching it became even more impossible. Still later, we realized that this observation could be turned around: getting closer and closer to the ground shortens the distance to the horizon. At the point your head touches the ground, your hand can indeed touch or even reach beyond the horizon! Amazing at first, and mindboggling for us boys! Clearly, the distance to the horizon depends very much on your own height relative to the surface of Earth. Everybody, short or tall, has his own personal horizon depending on their height and whether they are standing at the top of a tower or lying down.

The same applies to the view you take of the world. Some people have only a local view—a narrow horizon. One biologist referred to the narrow view of the local naturalist, by which he meant a view with a very restricted meaning of time and space: not untrue or vague, but constrained. This is viewing the world with a very clear and sharp eye but through a tiny little hole. I think that this often applies to ecologists. Their studies typically entail making observations of a very small surface area and over only a short period of time. A study has shown that most ecological investigations look at only a few square meters and have a time horizon of roughly three years (the duration of a study grant). This is a contrast to the view taken by geographers and biogeographers; by definition, they have a view of whole countries or islands, archipelagos, part of a continent, or even the whole world. And modern astronomers have a horizon of some fourteen or fifteen billion light years away. That is the horizon of the visible universe, determined by the age of the expanding universe. And their theorizing can even comprise several universes.

It is easy to see what can go wrong when we look at local processes only. Usually, the environment varies and fluctuates on a great number of scales (from local to global) and from the short term to the long term. One can focus on one scale or one time horizon, with the result that the shorter terms, for example, act as noise, while the longer terms become effectively constant. The result of this is that their effect on the processes at hand is difficult to detect. This is why scientists make such a selection: it makes it easier to see the effect of factors varying at the scale chosen. This procedure is basic to much scientific research; it allows scientists to study nuclear processes without having to study chemical, biological, and social processes at the same time, for example. In economics, one therefore distinguishes microeconomics from macroeconomics, and in historiography one can study century-long processes—*la longue durée* of the French historian Fernand Braudel—and distinguish them from short or intermediate processes. However, not being able to detect the effect of factors operating at other scales does not mean that they don't have any effect; they can be there but we don't see them. In ecology, however, the entire theoretical edifice is based on such a selective procedure from which the concept of carrying capacity—the subject of the present chapter—stems. Taking a broader range of temporal and spatial scales into account, therefore, means that this edifice crumbles, taking the concept of carrying capacity with it. Moreover, by emigrating or taking technological measures when local factors have become more restrictive, humans have always postponed surpassing limiting thresholds; they have temporarily raised these thresholds. They—we—have done so without worrying about the consequences for future generations: the price to be paid, the threat of running short. It is this that makes the concept of carrying capacity obsolete and misleading.

Many people talk about some maximum number of humans who could live on Earth without knowing exactly which factors determine or limit this number and without stating the time over which we wish to prolong our existence. They assume either that the present situation will continue indefinitely or that "the future" will solve any problem. Their home is the measure of all things, and when they feel there is more, they close their eyes.

However, as mentioned, around 1800, when historical time, the idea of our past and future, was born in our minds, some prominent thinkers in England and Continental Europe believed there was severe overpopulation. Malthus suggested that the number of people in Britain was growing in an ever-steeper, upward-bent curve—an exponential curve—whereas their food supplies were following a straight line—a linear curve. For example, assuming a constant reproductive rate of 2, then two people would together have four children, those four would have eight children, and those eight, in turn, would have sixteen. But the food supply would grow simply as 1, 2, 3, and so on. The food supplies would therefore soon fall short of the increasing demands of the faster growing numbers of British people. Inevitably, famine would follow, and the surplus number would die.

Of course, there is no obvious reason for the supplies to grow linearly, differently from the exponential growth of the number of consumers. Why can't they also grow exponentially and then even faster than the number of humans? Why can't they even decrease for some reason (perhaps because of an exponentially growing population of aphids eating the crop), following a downward bent curve? Or they could remain constant, determined by the constant size of the British Isles. Why not? In turn, population growth may also follow an entirely different process, requiring another growth equation to describe it. You can't decide which scenario is most likely without understanding the process.

Also at the beginning of the nineteenth century, Verhulst formulated an equation to do precisely this. His equation, the logistic equation, is still well known in ecology. According to this equation, populations grow until they reach a certain size but then stop because the resources are constant. The maximum—or ceiling—reached by the population (of people, animals, or plants) is called the "carrying capacity," of the area concerned because the amount of food in the area can carry only so many, no more. Actually, populations would first grow exponentially like Malthus suggested. Then, the constant supplies, specifically food, would gradually become more limiting, expressed by a decrease of the growth exponent. Eventually, the decreasing exponent equals one, implying that some maximum number of people has been reached—the country's carrying capacity. In humans too, approaching this maximum, this ceiling, would result in starvation and, hence, in extra mortality, whereas at the ceiling, the rates of birth and death would roughly balance each other.

The formulation of this equation was a major breakthrough; right or wrong, at last there was insight. But what exactly determines the height of this ceiling? Do other limiting factors apart from the available food operate as well—such as the amount of freshwater, metals and oil, degree of industrialization and trade, climate, or the length of time we want to prolong our existence? Various forms of pollution by excessive waste production or environmental destruction are other factors, but these are often neglected, although they can be decisive, particularly in our own future. How can we know the height of this ceiling? And does it always remain the same? Would we be able to push it up a little? But how?

Since the early nineteenth century, it has repeatedly been shown that neither the exponential growth equation nor the one of logistic growth fit the data. Usually, the data cover only a small part of the whole process, so it is impossible to test the applicability of any equation. In the few remaining cases, the data do not lie on the curve of either of these two equations. Unfortunately, other, more recent equations don't fit the data much better: it seems we still do not know which growth equation applies. And this means that we don't really know how population growth progresses or how it stops. Meanwhile, many remain confident that we will eventually reach some level at which our numbers will stabilize, and we can calculate when this will happen. But so far, predictions of population numbers have missed the mark; the actual numbers have always been far from the estimates. In 1997, for example, the United Nations estimated the world population would be 7.5 billion people in 2040; in 2004, they revised this number upward to 9.5 billion, some 28 percent higher. This was a difference of more than one-quarter, and only slightly less than the 2.2 billion that worried us so much during the 1970s.

Interestingly, from 1540 to 1700, English agrarian production was growing faster than the population, so food could be exported. But between 1750 and 1880, only the population expanded, whereas agricultural production lagged behind, not even growing linearly, as envisioned by Malthus and his followers. At that time, during the agricultural revolution, after fifty years of stagnation in agricultural growth, England had become an importer of food instead of an exporter. It seems that the rates of growth of the two processes—food production and population—vary, and they vary differently, although the growth of food production has to follow that of the population. But why? Moreover, the population number that was so worrying in 1800, about nine hundred million, was very different from present-day estimates of our world population, which are roughly eight times as high and seem not to worry us that much anymore. Yet, worrying or not, this difference shows that the assumed level of overpopulation or stabilization is not fixed. And as the most ancient written texts show, there has always been concern about overpopulation, even when numbers were small, compared with today.

Long ago, early humans went through a dip in population; their numbers were reduced to 20,000 or less. So, what is our global carrying capacity and how far can we manipulate it? And is there a limit set to this possibility of manipulation? Clearly, the concept of a fixed and knowable carrying capacity is obsolete. It is tied up with unknown technological, economic, and social issues and conditions. For example, to increase food availability, and stimulated by technological improvements, nineteenth-century agriculture was mechanized and industrialized. In turn, industrialization boosted trade and transportation, broadened the catchment area for food that now included large parts of the world, and allowed urbanization to take place. All these developments pushed up the supposed ceiling of the local numbers considerably. So far, Western technological and socioeconomic developments have managed to keep pace with the growth of our population.

Look again at what has happened: since the earliest times of our existence on Earth, we have extended our geographical distribution from some confined parts of East and South Africa to all of Africa, then into Eurasia and further into Australia and the Americas. We built tents and houses for shelter and storage, and we dressed to keep warm. We adapted and changed our diet, making starch and meat the core of our diet, which eventually made us obese. To that end, we modified cereals and domesticated animals, and we deforested and drained the land, plundered the rivers and oceans. Throughout human history we have organized ourselves, mechanized, industrialized, and urbanized our world. We have extended our lifetime, up to the point that either our body or our mind begins to give up. And in parts of the world, we can stop working after reaching a particular age, and the handicapped are permanently freed from looking after themselves. Finally, all sorts of machines and chemicals have released us from tedious mental and physical work; for us, unlimited amounts of cheap energy are spouting up from deep down in the earth. Everything we wanted to do, we have done; everything needed has been achieved. For us, there is no ceiling. We can reproduce and take without worry. Yes, why worry about the future, our growing numbers? We can push up our numbers even further, ever further. Why not? Because of a shortage of resources, or the overproduction of polluting waste?

Actually, recycling could prevent negative effects of growth in resource use and waste production from happening while allowing demographic growth and a prolonged stabilization. However, part of our economic growth and our growing resource use and waste production is the result of the growth of our population, part to our growing demand, and yet another part to its accompanying organizational superstructure. No type of growth can be covered by recycling, because recycling is based on a static, nongrowth model. The purpose of recycling is the avoidance of growth in both resource use and the production of waste, but in practice it only implies their reduction.

Moreover, given certain dissipation rates, the effect of recycling is limited. How much will the environment be polluted by the materials still dissipating into it despite recycling? And is there still enough left to be recycled? Throughout history, we have scattered our waste around, dissipating minerals and valuable compounds into the environment on a large scale. However, their concentrations are still very low—too low for us to be able to retrieve them in sufficient amounts for their recycling. Most of that material is lost forever. Given enough energy, for example, you can retrieve some of the carbon dioxide from the atmosphere because it still occurs there in large amounts. But it will cost much more to get it back from the oceans as well. In all those cases where the concentrations of chemicals are lower and more dispersed, or where the material has dissipated into the soil or the groundwater, their retrieval will cost even more energy so that it becomes impossible to get it back in sufficient amounts and in a reasonably pure form. So much energy that in practice you'd better forget it.

Recycling costs a lot of energy. In terms of energy, most of our waste is the end product of a running-down process; waste can only be brought into circulation again by adding a large amount of energy. Look at photosynthesis in plants. Here, two extremely stable biological waste products, carbon dioxide and water, left by animals, fungi, plant roots, and people are brought back into circulation by the plant's use of solar energy. This energy is mostly used to split up water, after which the carbon dioxide binds with the hydrogen thus released. The carbohydrates—the sugar, starch, or oils in plants—therefore don't contain energy themselves, but contain virtual energy. The real energy is obtained by the reaction of hydrogen and oxygen when they form water again. Look at the little plant material on your plate: how much you do with that amount of energy in a day. So much solar energy has therefore gone into the splitting up of water, our initial waste product.

The mere retrieval of materials to be recycled also costs much energy. As mentioned in an earlier chapter, to give iron a certain quality, we mix it with other metals to form an alloy. It will first cost energy to retrieve sufficient alloy from scattered metal-containing waste, and then it will cost more energy to separate the various metals from the alloy. The purer we want the metals to be, the more energy is required—disproportionally more. Often, we can purify only to a certain point, always leaving part of the other metals behind, because it costs too much energy to retrieve the last little bit. These very last traces left are impurities that reduce the reusability of the iron.

Think of all the minerals in our urine and excrement: each day in any given large town, hundreds of tonnes are dumped into rivers at a single outlet. There, they dissolve to lower concentrations, after which a river carries them to big lakes or the ocean. In the vastness of the ocean, they dissolve and decompose to even lower concentrations and eventually sink to the bottom. How can we ever retrieve them from that enormous volume of water, or separate them from all the sediment on the vast ocean floor? In other words, over time, dissipation sets limits to any recycling scheme. Our consumption inevitably exhausts and degrades the land. We unavoidably need a continuous input from new resources, and we will continue polluting the environment with our waste as a result. And this continuous new input and output defines the ever falling level of a postulated carrying capacity. Relative to the constant amount of resources on Earth, our use of them inevitably grows.

Recycling merely postpones the moment we must take measures to reduce our numbers; it cannot make such measures unnecessary. By recycling, we buy time; with our present and our future numbers and demands, recycling itself can't be the ultimate solution. Recycling is our present form of the old perpetuum mobile, which a long time ago has already been proven not to work.

Could the development of more efficient production methods and logistics, more efficient machines, or more economical transport be of some help? Whether through specialization or through technological improvement, all progress in history might follow a trend toward higher efficiency in individual households, factories, administrative systems, and the globalization of commerce. Everything runs more smoothly and costs less energy and material, and is cheaper as a result. Though this may be true locally, from a more global perspective it may work out differently. It is true for the replacement of an inefficient machine by an efficient one, although part of the old, inefficient machine may have to be melted down before the material can be reused. A technical problem, maybe. The real problem, however, is that we can build more efficient machines but we can't drop people. Even if we make them redundant they continue to live, have their house and family, their clothes, car, and still use resources and produce waste. If they find a new job, they often need a new machine themselves. And when we create several new, efficient jobs for one less efficient one, they all need to be equipped with machines. And they need to be organized, and their work to be coordinated, which means further growth. Therefore, globally, we use more resources and produce more waste per head, which makes the system as a whole run less efficiently. Initially, all hunter-gatherers were self-sufficient, but that changed over time. Apart from the 2–3 percent of farmers in the West needed to feed us all, at present we need a superstructure of 97–98 percent of other workers, supported by a staggering amount of machines and organizations of all sorts, plus an infrastructure of roads and buildings, all produced by and running on nonhuman energy and mostly using nonrecyclable, nonnatural materials. Overall, our efficient, global system works less efficiently than the hunter-gatherer system of the Stone Age that had no such structure.

A structure to push up the ceiling to feed our growing numbers and to keep it at an incredibly high level has its costs. Growth feeds further growth, because local efficiency feeds global inefficiency, which, in turn, has to be tackled by introducing more local efficiency measures, requiring more growth. Waste recycling, however efficient, requires even more organization and energy than we presently need, adding to the already rapidly growing demand from our growing numbers and ever-more complex society.

Throughout history, we have always worried about overpopulation, about some shortage of natural resources, but due to our organizational and technological abilities we have always been able to push up this ceiling to higher and higher levels. Through our ability to organize and develop our highly efficient, extremely diversified technology, by using animal energy and then ever-more inanimate energy, our numbers came to dictate the level of this ceiling, rather than the other way around. To this end, we use an amazing and ever-growing amount of energy, an amount essential to keep us and our local and global organizations and technology going. We can't do without all this energy, organization, and technology. And we can't do without minerals and metals, giving it its material shape. For thousands of years, since the human population began to grow, we have always overused the natural resources in our environment like any other species, refusing to share them with other species. Diseases determined our numbers. And we evaded the nutrient cycles by producing nondecomposable waste with our mud or brick houses and temples, our pottery and kilns. Basically, there was eventually no carrying capacity based on planning for the future, on efficiency or optimal use of materials and energy, neither in nature nor in human society at any one scale. Yet, in the longer term, such a system inevitably leads to destabilization and eventual collapse. What can we do if our society collapses or if our technology is no longer fed by the necessary amount of energy and materials? What will happen to our computers, telephones, sewage systems, agricultural machines, or our means of transport? To what level would our numbers fall back? Would this be possible? Our own organizations and technologies have become an integral part of our personal environment. An environment enlarged relative to our natural, biological environment. Can this part collapse, and if so, how?

Systems collapse when the course and outcome of several interdependent processes fail, even when one or two of those processes work but are not attuned. They either amplify or reduce the course and effects of each other. Even the slightest change can have disastrous effects. Think of a set of billiard balls, the first ball hitting the next in a minutely different way than intended, this next one getting a more different course or impetus than it should have had, hitting the third ball again more differently. The differences add up, or even multiply. Soon, they are hitting each other in lots of different ways, and in different numbers: balls, rolling back along entirely different courses, hitting other balls coming to lie in completely unexpected, often unwanted configurations. The initially minute difference accumulates, amplifies, with the result that the process outcome becomes unpredictable; it becomes widely different from anything intended. The effect of the combined behavior of the balls bouncing against each other becomes nonlinear, and its end result unpredictable, chaotic in the mathematical sense. Because of this amplifying effect of minute variations within the process, no two games can ever be the same, which is actually why we remain intrigued playing them. The same happens in society, where the process is even fed by increasing amounts of energy throughput rather than by diminishing amounts (as in billiards). Therefore, societal processes quickly grow completely out of hand, whereas the deteriorating system of caroming billiard balls dies. In society, small disturbances can make the whole system collapse. Anything can cause the collapse, at any time, at any place; fatal, ever-greater disturbances follow each other, faster and faster, apparently from all sides. Unpredictable and unmanageable. Interestingly, system collapse can result in processes following Verhulst's logistic equation. Therefore, even if this equation of growth stabilization does apply in natural or societal systems, system collapse can occur. The carrying capacity cannot prevent anything from happening by reducing overuse, for example, but is subject to the processes of deterioration and collapse.

When a process deteriorates because of increasing fluctuations, two responses are possible: those that dampen the fluctuations and those that increase their amplitude. In the first, some standard value should exist against which actual fluctuations can be measured and then weakened. They can be dampened by a mechanism that counteracts the deviations from this value.

The greater the deviation, the stronger the counteraction. This process mechanism is known as a control mechanism. Similar to biological control systems, two norms are therefore of relevance in industrial and societal systems: the first one determining the standard value, and the second keeping the amplitude of fluctuation close to this level. By contrast, the second response process enhances any deviations from the standard value: the greater the deviation, the even greater the amplitude becomes. The first response is called a negative feedback system, and the second a positive one. The negative feedback system, the one minimizing the deviations, stabilizes the process, keeping the fluctuation pattern tight, whereas the second, enhancing process destabilizes it, and results in often uncontrollable system behavior, up to the point that it collapses or crashes. Financial systems, for example, can crash because of the operating of some inherent self-enhancing process, and such crashes can have many effects on society, such as crashes in food supplies or working conditions. Keynesian-directed spending, in fact, concerns positive feedback loops introduced as contrary amplification process by a government in an attempt to achieve control of the positive loop of a crash, whereas avoiding governmental spending by reducing taxes exacerbates the positive feedback loop. As mentioned, these counteracting positive feedback loops leave the capitalistic, *laissez faire* economic system untouched, whereas negative feedback fluctuation control is favored by left-wing governments. Yet, in the face of collapse, we may have to take measures according to plan.

Feedback refers to some process of interaction, in fact, not between two different processes but within one and the same process. Growth in numbers or in capital can enhance itself, resulting in a positive feedback loop and therefore in increasingly greater numbers or in increasingly violently fluctuating markets (like a financial market). Or in a negative feedback loop it can reduce growth or fluctuations to a stable situation. In both cases, the greater the diversion from some initial value, the greater the enhancement effect of the loop. This is, actually, also what we call exponential growth; feedback processes are therefore exponential processes. And so are all processes of interaction within a system or organization, not only feedback processes. They enhance the effect of one process by another one exponentially. Unfortunately, with one or two positive loops within a system in which negative ones prevail, the positive, destabilizing ones tend to dominate, eventually causing the system to crash or collapse, typically a rapid and ever-accelerating, self-enhancing process.

System collapse depends on many interdependencies, often of a different nature and duration and so impossible to attune to each other. The failing of their operation can therefore be lessened by dividing the system up into a number of subsystems, each operating with a small number of constraining, negative feedback processes. A modular set-up of life processes with their check-and-balance structures within and between cells and organs, and in which duplicate mechanisms can take over functions when something does go wrong, prevents this chaotic outcome from occurring. One of the measures taken during the global financial crisis that began in 2007 was to set up barriers where the system had grown interdependent. Another measure was to take some pivotal firms and banks temporarily out of the system. However, the global superstructure gradually keeps tightening, which increases the risk of total collapse. Such a possibility of collapse, inherent to any organization, is not accounted for by the concept of carrying capacity. Still, the possibility of collapse will increasingly determine our future.

One assumption on which the concept of carrying capacity is based is that nature is harmonious and self-contained: the physical environment is constant and uniform. Temporal fluctuations, such as those in climate, changes such as those in organization or technology, and spatial variation, such as in the chemical composition and humidity of the soil, are excluded from the concept. Maximizing our yield or profit requires that we make our environment constant and uniform. And by preventing war or diseases from occurring, we can make it harmonious. We are therefore living in a homogeneous, technological, well-organized world, increasingly more detached from the natural environment. And we hope that by recycling materials and by transforming energy ourselves from an overabundance of solar energy, that our human society is becoming self-contained, even allowing further population growth. Thus, we will be free from any limitation by a naturally set carrying capacity. In this way, we have built an artificial, man-made world, but one still within the earthly, variable, and changeable world. But it is particularly those small, local and short-term instabilities that can cause our world to collapse; just as in the example of the caroming billiard balls, these relatively small disturbances can deflect societal processes from the intended course into a completely different and often undesirable direction. There is no question about the fact that we are overusing and depleting the earthly resources at a high rate, and that we pollute and deteriorate it in return. And there is no question that this results from our too-high numbers, our too-great personal demands, and our reckless behavior that lacks any planning. In a sense, you could therefore say that we are greatly overshooting the carrying capacity of the environment, using this concept in a very broad and general way. But the very generality of its use leaves enough vagueness for skeptics to avoid their responsibilities. It is the local and temporal variation and the specificity of the materials being depleted or that are polluting the environment with their waste that contain the risks to the sustainability of humankind, whereas general concepts like the one of carrying capacity hide them. Models of carrying capacity, combined with those of unlimited recycling are deceptive, keeping us from taking measures based on insight and understanding of the limitations and the capriciousness of our earthly environment and their impact.

The idea of a harmonious nature in which animals and plants keep each other's numbers below some carrying capacity dates back to Egyptian times. Biological populations and human societies were long believed to be under the effective control of the gods or their representatives, such as kings or priests. Since about 1800, it has been believed that they are under the control of a multitude of interacting factors, explained by concepts such as the French *laissez faire*, Adam Smith's "invisible hand," ecological concepts of "carrying capacity," "community," and "ecosystem" or, more recently, "the Gaia hypothesis," "the market," "the future," and "technology." These ideas concerning a supposed harmony of nature led to models in which all process elements—species, humans, or nutrient and waste levels—have their functional place and keep each other functionally within bounds that guarantee long-term persistence. Not too many lions for the number of antelopes, and not too many antelopes for the amount of grass; not too much or too little grass—or daisies—for the fixation of carbon dioxide regulating climate. Or, "eventually" the rates of birth and mortality will "balance each other out." But this is the hopeful, romantic image of the world still operating within the narrow horizon of the ancients without any inkling about what living in great numbers in a highly intricate global society and a fluctuating and changing environment entails.

There is no wise, mysterious Gaia keeping watch over us all, waving her wand, keeping us from disaster. We cannot close our eyes in a *laissez faire* attitude any longer, trusting some equally unknowable, mysterious carrying capacity for the sustenance of our life and our numbers, our love and our happiness, our money and food, trusting "the future." When our numbers grow too large, our resource use and waste production too reckless, our global organization too complex, we must take measures ourselves. With our own hands.

17 URBANIZATION

This book was conceived and partly written in a hotel in what used to be the mint of Troyes, a medieval town in northeastern France. During the Middle Ages, this town was a commercial hub in one of the more wealthy regions of France, Champagne. So, as it grew in importance, its population became sizable. As a commercial hub, it had a regionally important market and became an administrative and cultural center. For example, not far down the street from my hotel is a palace built of white stone that was owned by the French king. The many, often rich, inhabitants of Troyes, however, usually occupied huge timbered houses of four or, occasionally, five floors. Most of these magnificent medieval houses, along with smaller ones, are still standing, and are being restored.

The use of all this wood reflects the growth of markets and fairs, plus the accompanying road networks, and this growth reflected in turn the growth of the production of surpluses on the one hand, and the growth of numbers of people and demand on the other. And it also reflected the rapid growth of the hinterland surrounding the villages and cities supplying these with all that wood, as well as with the food and clothing their inhabitants needed. The numbers of people rose, the countryside filled up, and urbanization funneled off those numbers. Urbanization of a countryside like that around Troyes took place all over Europe, as well as in other parts of the world, such as the Middle East, North Africa, and China. No wonder that all of Europe—from North Africa to Scandinavia and from Ireland to far into Poland, the Baltic states, and the Russian plain—and later, large parts of North America were all deforested at a high rate. The hinterland of Western Europe expanded to other parts of the world. Now, France is a wheat-growing country with vast, open, undulating fields. Its occasional woodlots consist of shrubby young oak trees only, if the oaks haven't been replaced by fruit trees, poplars, or wheat.

Urbanization, the concentration—contraction—of a roughly uniformly distributed population into high-density towns, gave rise to a geographical network of larger and smaller towns and these differences in size to a hierarchical set-up of the local administration and economy. This hierarchical network expanded and amplified as the population grew. As a spatial saturation process—growing plus seeding out—it was the quickest for absorbing people and production surpluses and the most efficient one to maximize the use of space for living in, although this was obviously at the expense of the natural woodland area. Moreover, during the late eighteenth and the nineteenth and twentieth centuries, cities grew out from individual villages around the initially local industry, and the cities, in turn, into vast, industrialized conurbations.

All those cities and industries made specific demands on the environment, and therefore had specific impacts on their surroundings in terms of winning resources and waste disposal. Thus, as the population grew and the cities and their catchment and waste deposition areas expanded, the space of France—and indeed that of Europe and all the world—became saturated, not only with people, increasingly contracting into villages, towns, and conurbations, but also with a dense network knitting together human habitation, agricultural activities, industrialization, information and financial exchange, and the supporting mineral and energy supplies. Eventually, this network spanned the world and transcended national borders and national sovereignty. The network consisted of areas for living, agriculture, and industry, and of roads and canals connecting these areas with each other. The areas contained houses, workshops, factories, shops, storehouses, and so on. All these components are to do with the material side of the network or infrastructure of the region or country; the accompanying educational and financial systems, the social and judicial systems, and the like form the immaterial infrastructure. Over the centuries, these structures tightened and grew in space and intensity, eventually reaching their present global extent: the global, transnational socioeconomic superstructure supporting us all. It is our survival tool, facilitating billions of us living together.

Thus, as the urban centers grew, their source areas of food, water, wood, minerals, fuels, industrial products, and waste disposal areas—their catchment areas or hinterlands—grew accordingly, although not necessarily at the same rates. Water, for example, may have been available within the towns for some time, but much later it became necessary to bring it in from the surrounding areas. Nowadays it is essential to have a system for integrating specialized social functions. Think of the distances our food alone has to cover to enable us to live in megacities, or in complex federations such as the United States, or in the global community. In the—still exceptional—case of the United States, the average distance food covers from producer to consumer is about 3,200 kilometers. We saw that, increasingly, drinking water covers similar distances. And, of course, we do not live on bread and water alone: we need different forms of energy, chemicals, instruments, apparatus, machinery, and the raw materials they are made of. They all have their extensive catchment areas, and they all have their place in an urbanizing world. These areas grow with the size of cities or with the degree of urbanization of a region or country. Yet, all those spatially distributed conditions determine the living conditions of the concentrating populations, of the urbanites. They are part and parcel of the emergent system and cannot be ignored when discussing the historical background of the growing populations. The populations, their dynamics, their contractions, and their influence on their environment are one. They are part of our urbanized world.

I used the expansion of Troyes as a partly hypothetical example of early urbanization, but there are many different ways of urbanization, and these ways change over time. For example, during the English Industrial Revolution, industries concentrated in the cities, whereas at present, towns deindustrialize and the countryside industrializes, and offices follow suit. Regions of urbanization, mining, and industrial activity arose, and others declined, resulting in a perpetual spatial dynamic at continental and intercontinental scales. All the activities of these regions had to be coordinated, both within the towns and between them and over increasingly larger distances as a population, and hence, its hinterland for food production, trade, and waste disposal grew. From early on, people began to specialize as the population grew, and as specialists, they began to depend on each others' products, and therefore on each other. They had to work the same hours and had to deliver the goods at agreed qualities, prices, and times. Coordination and standardization of labor and products required a central administration and a jurisdiction following general rules and laws. Eventually, the administrations and rules began to control the behavior of people and often developed a dynamic of their own, beyond the capacity of individual people.

Work concentrated locally in areas separate from those where workers lived and slept; transportation systems were necessary to transport goods and people. Influenced by the development of megacities and globalized trade, the system controlling urban society now consists mainly of a network of interacting computers, operating at high costs of energy. The number of interactions between at first the people themselves and then between social groups and strata—craftsmen, transporters, administrators, banks, firms, industries, and corporations—to be coordinated within this network has grown too large and complex to be handled or manipulated by humans. At present, urbanization is only one, spatial, aspect of the structuring of a growing society. Others are specialization, division of labor, mechanization, and automation. All are inextricably bound to each other.

As the world urbanizes (now already 60 percent of all people live in towns and soon this will be 75 percent), their demands increase, since urbanites in particular profit from industrialization and improving conditions. Also, industries are situated within or near towns. Industries often require large amounts of water, apart from all kinds of minerals or wood. In this way, the amount of freshwater used per person is a multiple of the two to five liters a person actually needs per day (the amount varies, depending on the aridity of the climate). This is another kind of virtual water, now as seen from the perspective of humans themselves rather than from their products. Thus, in the developed, urbanized world, per capita daily water use is in the order of a hundred to three hundred liters, with exceptional cases of a thousand liters, which increases the catchment area for food and freshwater around our growing cities accordingly, or even faster. Yet, the amount of available freshwater is decreasing. For example, all those cities plus the roads that connect them occupy a certain surface area that could have been used for collecting freshwater and agricultural production.

In South America, China, and India too, there are large cities of many people—often tens of millions—living closely together in multistory flats and vast shantytowns, and these obviously require an extensive hinterland or catchment area for the products they need for living and working. All their requirements are exhausting the local agricultural fields, mines, forests, and marine fish stocks at ever larger distances. We need an enormous array of products for our existence. And, conversely, after their use, we need a large area to dispose of as waste all the used nutrients and nonrecyclable materials, and industrial products, as well as the rubble of derelict and outdated houses and laboratories, machinery, offices, factories, roads, and harbors. We produce waste, and we damage, exhaust, and pollute our environment by our existence. Unfortunately, we can't separate the area we use for waste disposal anymore from the area that supplies us with the physical space for our cities and the inputs we need for our actual urban living.

The urbanization resulting from migration from the countryside to the cities that has gone on for over a millennium is continuing and accelerating: at the moment, there are massive internal migrations in China, India, Russia, Brazil, and Africa. As most migrants are forced to leave their area of origin because of poverty and lack of work, they often end up, not as they did in nineteenth-century England in London or in the cities of the Black Country, but as squatters on the urban fringe. There, they squat on a plot of land and build a shelter of corrugated iron, planks, and mud against the heat, cold, and rain. The result is huge shantytowns, similar to—or perhaps worse than—the slums in England of one and a half centuries ago, lacking sanitation and regular supplies of water and electricity, and with streets that are gullies of mud, excrement, and stones. The children are left uneducated, and there is no medical care. Sometimes, tens of thousands of these mostly jobless squatters are evicted and their shelters cleared, although occasionally, after much misery and hardship, improved areas can be built, forming new suburban extensions of the city. One billion people, not actively taking part in our global economy, but generated by it and depending on it for their existence.

An important aspect of urbanization is rapid industrialization, including the growth of transport and extensive administrative systems. Another large sector of an industrialized society is that of the service industry, ranging from street cleaners to shopkeepers and taxi drivers to health workers. The labor gap between farmers as primary producers and urban people has broadened to such an extent that the urban part of society now lives a life of its own. In the West, a staggering 98 percent of all people form the urban superstructure of our society. In the United States, some 11 percent of the population work in the manufacturing sector, leaving a total of 2 percent + 11 percent = 13 percent of the total population earning from direct production in agriculture and industry. The remainder of all people, 87 percent, forms the organizational and service superstructure, supported, moreover, by a vast number of trucks, machines, and computers, which are, again, concentrated in the urbanized part of society.

The urbanization of society goes deeper than the growth of cities or the urban saturation of our living space alone. The initial connections between producers and the markets in the early-medieval towns has grown out into global networks of investment and financing spun between a handful of megacities across the world. Here, billions of dollars change hands on a daily basis. And there is more. Since the 1980s and 1990s, large industrial corporations dissolved into organizations buying the various components they need from other firms and assembling them into the final product bearing their name. Those components can be made in specialized firms at any place but they guarantee the highest quality to the lowest price. It has become impossible to invent, develop, and produce them all under the same roof, as this would require too much specialized knowledge, which makes the final product too expensive. Moreover, the individual shareholders have been replaced by investment managers shifting funds between corporations depending on the rise or fall of the price of the shares. Finally, the corporations themselves consist of branches in several countries, dispersed across the continents and exchanging workers, or they are owned by foreign investment firms. Independent of their location, workers are contracted on a temporary basis according to their specialization. Concrete, national corporations and banks have become abstract, transnational networks.

At present, human society is organized as a highly dynamic global, urban superstructure. This structure is indispensible and therefore helps to maintain our population. Our global superstructure has grown out to become a self-sustaining system to sustain us all, yet is still entirely dependent on a permanent and uninterrupted flow of fossil fuels. This superstructure can only exist as long as the energy and mineral supply it requires is sufficient, or, in the case of energy, if it is replaced by nonfossil energy. However, it will collapse as soon as these supplies run out. Similarly, agriculture that feeds the people and thus in the end forms, sustains and is sustained by this structure can be maintained at a sufficient level only with inputs of water and agrochemicals from fossil fuels. Our numbers, society, urban centers, and global organization ultimately express the amount of fossil energy put into our global population. Without it, there is nothing left.

18 MIGRATION

In the earliest human agricultural civilizations, people lived in more or less permanent settlements. Though these were largely farming communities, some social separation may already have taken place between farmers and the ruling classes. The early towns were congregations of farmsteads, from which every morning farmers dispersed into nearby fields and pastures, returning with their produce and livestock in the evening. Much of the non-degradable waste produced by the townsfolk remained lying around within the settlements, where it heaped up. New houses were simply built on top of the waste of the previous years and generations. In this way, the small, early towns gradually rose higher and higher above the surrounding countryside, occasionally ending up meters above the original settlements on tells.

Surprisingly, it seems that some 8,200 years ago, many of the inhabitants of at least one such tell, Sabi Abyad in northern Syria, left the town for part of the year, leaving a small population behind, presumably to look after the place. This happened during a period when average annual temperatures had dropped by about 2°C. Moreover, the region dried out, meaning that the land, which was already only marginal for agriculture, could no longer be farmed sustainably. From more recent events we know how susceptible agriculture in this part of the world is to adverse weather: during the 1920s and 1950s—periods of below-average rainfall—cattle mortality in this area was about 80 percent.

The exodus from Sabi Abyad in response to worsening environmental conditions had been preceded five thousand years earlier by a similar event at another tell in Syria, Abu Hureyra. There, a small settlement of food gatherers had lived on fruits, acorns, and pistachios, as well as on gazelle. The trees grew at walking distance from their houses, and the people developed techniques of preparing and preserving the fruits and nuts. And there were plenty of trees, so the number of people increased and some settled on the hillsides nearby. But because they had become dependent on these nuts and fruits and had built their settlement, they were no longer able to move around as freely as their hunter-gatherer forebears had done. When, therefore, about thirteen thousand years ago, the climate became drier, they got into difficulty. The trees on which they had depended could not survive the drier conditions and no longer provided the people with an assured source of food. At first, the people turned to wild cereals and asphodel, but even these ran short. After subsisting on other food plants, about ten thousand years ago the people began to grow lentils, as well as rye and einkorn. At that time, the number of humans had grown so large that migration to new, empty spaces wasn't feasible anymore, and settlement and intensive food production became necessary. Thus, as people could not or would not migrate, agriculture as a novel technological adaptation arose in the Fertile Crescent of the Near East out of necessity, a development called the Neolithic Agricultural Revolution. This new technology of food production allowed settlements to expand once more, although the poorer nutritional quality of this new diet—being mostly grain—had negative effects on human health. Body size, height, and bone thickness appear to have declined, and iron-deficiency anemia occurred. It was in this time too that people in the Near East had to adapt to gluten, and that a consequent intense selection began against human gluten sensitivity as, over the millennia, the early farmers spread into Europe. With the expansion of the settlements, the amount of work to be done increased considerably. So it seems that what triggered this development, and which had such far-reaching impacts for the progress of human civilization, was the capriciousness of climate along with the growing number of humans rather than their inventiveness.

In other times and places, however, changes in living conditions led to mass migrations: entire communities or tribes of people moved away from their birth place, their farms or their towns, occasionally covering thousands of kilometers. Sometimes too, people may have been attracted by the wealth or the abundance of food in other communities; this would explain why mountain people often descended from the surrounding regions into the Mesopotamian river valleys. Or people were chased away, as may have happened to the Jews in biblical times, migrating in and out of Egypt. Climate deterioration drove the Hittites of present-day Turkey into Syria. Successive waves of foreign tribes, such as the Dorians and the Ionians, immigrated into Greece, superimposing their own mythology and building style on existing ones. Later, Greece itself expanded along the northern fringes of the Mediterranean, thereby forming Magna Graecia, Greater Greece. The southern fringe was occupied by immigrants from Phoenicia. Large numbers of Greeks, Phoenicians, and later, Romans left their hometowns and settled abroad. Similarly, Asiatic people expanded eastward across the Pacific, settling on its hundreds of islands. More recently, although still many centuries ago, after its spectacular expansion across a large part of Europe, the Middle East, and North Africa, Rome was itself attacked by Gothic tribes roaming across Europe. Later still, hordes of nomads from the steppe regions of Eurasia and China, like the Alans and the Vandals, penetrated deeply into Europe, mixing as usual with the local people and their cultures on the way. After that, from about AD 700 onward and up to ca. 1100, the Muslims expanded in various directions, among them along the southern fringe of the Mediterranean Sea in North Africa. Then, around the turn of the first millennium, the Vikings expanded into northwestern Europe, eastern Europe, and southern Italy, and other Europeans by way of the Crusades into the Levant—usually violently—whereas during the Middle Ages, central European populations expanded eastward, first into the Baltic countries and then into Poland and beyond. In medieval Europe, foreigners also settled in other countries as agents for the firm they were working for, thereby founding German, Dutch, Italian, or Spanish trading quarters in parts of the towns allotted to them.

Then, from the sixteenth century onward, various western European countries set up trading posts in other parts of the world. Initially, the main reason they did so was not to increase their food-growing area or living space, but for commercial gain. They mainly bought luxury products, such as spices, sugar, or silk, and sold some of their own products. Often, the trading posts were no more than small forts, rather than colonies of settlers wishing to begin a new life in a new country. Thus, first the Spaniards and the Portuguese, and then the Dutch settled in both Americas, in several parts of Africa, along the South Asian coasts and India, and finally in present-day Indonesia. Later, the English and the French followed, founding their dominions even farther away, annexing parts of whole continents and thereby extending their economic hinterland on which to live. Only during the nineteenth and twentieth centuries, did large numbers of people from Europe begin to emigrate to these areas, not to trade, but to find a new life, better than in the overcrowded country they had left. Europe's growing populations were now overflowing into the surrounding world. At first they left by the thousands, but soon there were one million or more leaving each year. Most went to North America, after which the continents of the Southern Hemisphere followed: South America, South Africa, Australia, and New Zealand. Roughly speaking, people from northern and northwestern Europe went to the northern half of North America, Africa, the Far East, and Australia, whereas people from the Mediterranean countries went to the southern half of North America, Central America, and South America. Russia and China expanded similarly, colonizing the vast emptiness of Siberia and Asia: Russians went east and Chinese went west. More recently, the United States and China have penetrated former European colonies economically (and in the case of the United States, militarily) to obtain the resources and markets they need. At present, there is hardly any land left to which people can immigrate and settle, land in which overflow populations can still find a place to live.

The colonists usually invaded, colonized, and subjugated the area and its inhabitants they claimed to be theirs, raping, enslaving, torturing, murdering and exterminating as they went. The inhabitants of Tierra del Fuego were exterminated without any possibility of resistance. In New Zealand and Australia, after initial resistance, 90 percent of the aborigines were wiped out and the children of the ones remaining were taken away to be brought up by the colonists, and in North America the few surviving American Indians were confined to reservations, and so on. This has been common practice throughout the ages, mentioned in the oldest existing stories of Sumer in Mesopotamia, and in the old Roman story of the rape of the Sabine women. But the inhabitants of the subjugated, presently underdeveloped parts of the world never had any chance to colonize another country themselves.

Since time immemorial, people were also moved forcibly from one part of the world to another. The relatively recent term "slave" comes from "Slav," because so many Slavonic peoples were sold into slavery (especially during the eighth and ninth centuries, by Arabs). During the years of the Atlantic slave trade, European entrepreneurs transported thousands of African people to the Americas. Convicts were transported from Britain—at first to North America, but from 1850 onward they were taken to penal colonies in Australia. From the late nineteenth century, France transported convicts to French Guiana. During the nineteenth and the first half of the twentieth centuries, millions of people were exported as prisoners across Russian Siberia. In the 1940s, Nazi plans to relocate people from the occupied countries in Europe to elsewhere (sending the Dutch to the Ukraine, for example) did not come about because the war ended. Since 1949, however, under the government's transmigration schemes, many hundreds of thousands of farmers in Indonesia have been forced to move from densely populated areas like Java to other islands, such as Sumatra and New Guinea, but in these areas their farming has failed because of unfavorable local conditions. People have also migrated voluntarily or emigrated under duress for religious reasons (first the Muslims and Jews in Spain and then the Jews across Europe, for example) or have fled during wars or for political or economic reasons.

However, during the last decade or so, individual countries and federations have tightened their borders. The United States actively tries to prevent economic refugees from Mexico crossing the border illegally. South Africa, which has an estimated five million illegal immigrants, is tightening its borders as well. Similarly, the European Union has sharpened its rules and laws, preventing refugees from the Middle East and Africa from entering its territory; Australia had earlier done so to discourage boat people from Southeast Asia.

People feel increasingly that the world is getting overcrowded and resources are running out. In famine-stricken Africa alone, millions are on the move. When low-lying coastal areas and islands are inundated by rising sea levels, or when inland areas dry out because of climate change, the numbers of people on the move increase. And meanwhile, in large countries, such as Russia, China, India, and in Africa, there is considerable internal migration from the countryside into the slums of their exploding towns.

All these mass movements are one-way, but there have always been movements to and fro as well. From ancient times onward, for example, traders have covered long distances. One of the oldest towns known in present-day Turkey, Çatalhüyük, was at the hub of an ancient commercial network for flint. Traders in boats took copper from Cyprus, which was sold across the ancient Middle East and Mediterranean. For centuries, caravans transported salt along long trade routes through deserts or into the interior of the continents. For centuries, too, traders like Marco Polo moved over land and over sea along the silk routes linking the Mediterranean, Near East, Asia, and the Far East. Here at home I have a wooden chest hundreds of years old, one of the many left by Jewish textile traders at posts located at regular distances en route from the Netherlands into Germany and beyond.

Traders were not the only people on the move: there were also migrant workers. In Europe, many of these were gypsies. Migrant workers did seasonal work, such as haymaking, mowing grain, or digging up potatoes. Before machines took over the work, large numbers of people were also needed to drain land, canalize rivers, dig canals, and construct railroads and roads. Or there were the mercenaries: soldiers who hired themselves out. These people had no work in their villages and towns and were permanently on the move, but sometimes they stayed on after a job and mixed with the locals. On the whole, their movements were nomadic, albeit they were following work opportunities, not favorable climatic conditions. In Europe, gypsies are the only remnants of such people with a nomadic way of life.

At present, we see large movements of people within Australia following coal mining in the east and the rapidly upcoming and booming metal mining and offshore oil drilling in the west. New houses and roads are being built, and local communities are being disrupted. Throughout the world, millions of people move not as economic refugees, but for their work. They stay for months or years, or settle permanently. Their children go to local schools and may stay when their parents move on. Thus, for this reason alone, societies are becoming multilingual and multicultural. Furthermore, tourism, which often accounts for up to 10 percent of the annual revenue of a country, also brings in a large temporary influx of people: worldwide, millions of people are involved. In this way, people of different backgrounds penetrate other populations, often destabilizing them. Much of the underdeveloped and developing world lies in parts of Earth where food production and water supply are most vulnerable, and it is here that effects of climate warming will be particularly severe. Social unrest, violence, and disease, therefore, are likely. Many of the desperate will also overflow into the richer parts of the world and form streams of hungry, diseased, and infectious people. Here, resistance and social intolerance can easily lead to problems between the poor of the underdeveloped and developing world and the rich of the developed world. Whether this will result in major outbreaks of violence will depend on the tolerance of the richer areas. Being tolerant of and helpful to desperate refugees, no matter their origin, is the least we can do. Moreover, in future, the demographic surplus of underdeveloped and developing countries will be needed to supplement the shrinking populations of other countries when their reproductive rates decrease to below the replacement rate. In Germany, for example, this is already a topic of political debate.

Often, people have had to move on because their land became exhausted or their resources depleted. During the 1930s, for example, there was an exodus of farmers—Okies—from the so-called Dust Bowl in the American Midwest after the topsoil of their land blew away. In the gold rushes, huge numbers of people went to California and to Victoria, Australia, but then when the gold ran out, many left as quickly as they had come. And since time immemorial, the same regional depopulation has happened when soil nutrients ran out or when salt encrusted the topsoil, as it did many millennia ago in ancient Mesopotamia at the very beginning of human civilization. Very recently, only a few decades ago, the fishermen and cotton growers around the Aral Sea in southern Russia had to give up and leave when water as a basic agricultural resource ran short after first the two big rivers ending in this inland lake and then the lake itself dried up. The same has happened in western China with the same crop, the ever-thirsty cotton. In the central United States the fossil water of the Ogallala Aquifer will run out within decades after which a similar exodus is likely.

At present, large regions are becoming depopulated because of severe chemical pollution due to the dumping of industrial, agricultural, or military waste. Vast areas around polluting industries or nuclear plants have been abandoned. In the West, we don't feel the impact of our massive amounts of household and industrial waste that much, because we often dump it in other, poor countries, such as the Philippines, which become wasted that way. In the Philippines, as in many other parts of the world, people also are abandoning large deforested areas suffering from mud streams.

Migration and population collapse due to climatic change are well known from historical climatology. A change in the course of global air currents can affect regional weather patterns, and thus local yields. This happened, for example, around 1200 BC, when the Peloponnesian Peninsula dried out. This drought may have been so severe that it abruptly ended the flourishing Mycenaean culture in Greece, effectively depopulating Mycenae. In contrast, Athens, only a short distance away, flourished because the mountains in Greece diverted the local winds such that its climate became favorable. At the same time that Mycenae depopulated, the Hittites living to the east on the Anatolian Plateau asked in vain for help from Egypt, "lest we should starve." Pollen records suggest that drought led to severe famine, thus forcing the population to emigrate. Moreover, as said, the desperate Hittites moved to northern Syria, a region that would at that time have received 40 percent more rain than normal, and would have seen a moderate rise in temperature of 0.4–1.2°C.

Similarly, the people once living in the center of the present Sahara had to leave that part of Africa when it desiccated over the centuries after a period of wet conditions during which there were large, tropical lakes in the region, which harbored hippopotamuses and fish. The lakes dried up and steppe species moved in but in turn were replaced by desert species. Finally the area became what we know today: a huge, barren desert of migrating sand dunes of occasionally over a hundred meters high. The former human inhabitants scattered in all directions, taking their ancient cultures with them when they mixed with the local surrounding populations. We can still find heaps of fish bones left by the last people along the shores of the former lakes. The rock paintings they left show hippos and, later, the ostriches and then bovines that succeeded them.

Much longer ago, climate had deteriorated for the earliest human species living in what is now the scorching Rift Valley in East Africa, and for the Australian aborigines then living in what is now the hot and dry center of the continent, where remains can be found of fast-flowing rivers hundreds of meters wide. In Brazil, however, it was a change to wetter conditions that forced the early inhabitants who built dense networks of villages in the jungle to abandon the area. And think of the dozens of alternating glacial and interglacial phases, jolting the northern floras and faunas, along with the local human tribes, backward and forward. All the plant and animal species, the megafauna of the former grass steppe with its wooly rhinoceroses, mammoths, bison, and saber-toothed tigers, affected by these climatic changes adapted spatially through migration if they didn't die out, rather than staying put and changing genetically. Unless we recognize this huge and perpetual spatial dynamic, we will never understand biology or human history.

Urbanization is the form of recent, within-region migration producing increasingly larger conglomerations of permanent settlements. The agricultural and industrial revolutions in Britain and continental Europe resulted from the depopulation of the countryside when—if they were lucky—former farmers found work in factories in the new towns. There, they had to mix with each other whatever their social or cultural background or region of origin. Often, however, they ended in terrible slums as in many English towns, or in those of Dublin or Paris. The same happened in North America, although there the various nationalities tended to keep together: the Italians and Irish, for example. As China's east coast (home to hypercities like Beijing and Shanghai) turns into one gigantic conurbation hundreds of kilometers long, the government is developing the central and more western parts of the country. The old town of Chongqing is set to become a megacity of thirty-two million inhabitants, and ninety cities of similar size will follow. Vast numbers of Chinese are therefore flocking from the countryside and Mongolian steppes to start a new life again in the towns. Thirty years ago, the proportion of the population of China living in towns was only 18 percent, in 2010, 50 percent, and by 2020 this will be 70 percent. The same is happening in India, where, due to the Green Revolution of a few decades ago, many farmers left the countryside to seek work in the towns. All too often, however, they fail to get work or adapt and end up as squatters in shantytowns. Similar shantytowns are growing throughout the tropical and subtropical world, attracting immigrants. Within other towns, migration happens continually from the central parts outward, speeding up their expansion rate.

People lived a nomadic life for centuries, if not millennia. Only with the rise of nationalism at the end of the nineteenth century were sharp boundaries drawn, restricting people who moved with the weather or the seasons. Yet, as all the examples mentioned show, across the world and throughout history migration has been the norm rather than the exception. Accepting this is essential for preparing for our future when climate will continue to warm. As areas become drier, aquifers run out, and crops fail, famines will spread over extensive areas, forcing people to move away. These refugees have to be accepted by the native populations of the areas to which they immigrate, whatever their cultural differences and backgrounds. It will be inhumane to refuse them entry, leaving them starve to death on the other side of a border or fence. On the other hand, the immigrants should not overrun the receiving populations, as has so often occurred in the past. Their integration is a two-sided process.

Cultural adaptation and integration is always difficult—but is possible. Ancient Greek mythology shows a hierarchical set-up of their divine family which reflects the religions of the various tribes that successively immigrated into the region. Recent genetic analysis, such as that of Britain or of Europe as a whole, has revealed an enormous heterogeneity in populations where people don't have the slightest inkling about huge differences in cultural background of their past, resulting from the continuous inflow, mixing, and mutual integration of natives and new arrivals. In future we will have to rapidly accept and tolerate refugees who are unable to survive in their homeland. They have no choice, and neither do we.

Migration is more than a topic for academic dispute; it has always been essential for the perpetuation of local and regional human existence. It is the expansive counterpart of urbanization which represents population contraction, both as spatial expressions of our growing numbers. Once again, the core factor is ultimately the continued growth of the already too large numbers of people. This growth leads to secondary factors, like climate change, salination, resource depletion, waste production, environmental pollution, and eventually, migration. Narrow-mindedness is no basis for a prolonged future of humankind on Earth under the presently changing conditions. Migration complicates and intensifies all those other problems that we will unavoidably face. We must accept it as an inevitable part of human existence, both in the past and in our future.

19   THE SPREAD OF DISEASES

Until recently, thanks to its isolation, Iceland had remained free from many of the infectious diseases that were common in continental Europe, such as measles. Occasionally, a sailor or fisherman would contract measles abroad and bring the disease home to the island, where it spread from one community to the next. Then, many Icelanders would become ill and often, a large proportion died—young and old. After this epidemic had passed, measles would die out on the island because there were too few susceptible people remaining to contract the disease, and part of those who had caught it had become immune. Many years and several generations of people could go by without measles taking its toll, and therefore, the Icelanders would once again be susceptible. So, when another infected sailor disembarked, the disease was able to spread through the population once more.

Therefore, the more people in a population who are susceptible, the greater the chance that an infectious disease will wreak havoc in that population. Many Icelanders lived almost completely isolated from each other and therefore were unlikely to come into contact with an infected person. The chance of an infected sailor arriving from the continent was also small. Moreover, to be able to pass on the disease, the sick sailor would have to be in the incubation phase of the disease. In Iceland, therefore, the number of deaths from measles fluctuated hugely, irregularly and unpredictably. All this changed when the Icelandic population grew past some threshold level typical for measles. These days, because of their higher numbers plus their greater mobility thanks to public transport and private cars, few Icelanders live isolated from others. The effective pool of susceptible persons, therefore, has become larger, increasing the chance of the disease being passed from one person to another. During the twentieth century, this increased chance actually overruled the growing percentage of immune people in the ageing population. Therefore, as soon as the disease now arrives, it does not die out as before, but survives through generations, although it operates at a low key. As it always finds a few new victims, however small their number, its fluctuations are less pronounced and more regular. So the incidence of measles remains low in this larger population of Icelanders until there are again enough who are susceptible for an outbreak to flare up.

Similar to Iceland, diseases also flared up irregularly in preindustrial Europe and elsewhere where villages and towns lay isolated like islands in an ocean of forest and marsh. Such flare-ups are called epidemics. An epidemic breaks out after a period in which the disease has been absent; it rages rapidly through a population and then disappears as quickly as it came. When, in contrast, a disease remains operating at a low key, with at any given moment a small number of people infected, it is called an endemic disease. Endemic diseases are a permanent feature in large populations, like those in nineteenth-century towns and cities in England and Europe and in present-day megacities. They are particularly common in deprived urban areas. Here, the high density and mobility of the urban population favor the spread of the disease between communities. There are no major breaks or gaps within the city population that stop the disease dead in its tracks. The disease becomes more benign because, in order to stay around in the population permanently, being deadly is disadvantageous to it and is selected against.

So, in megacities such as Sao Paulo, New York, or Los Angeles, measles does not die out after some time; the disease moves around within the city, eventually returning where it started. In this way, it can close an infection cycle within a single population. It has turned from an epidemic into an endemic disease. The urbanization that accompanied trade and industrialization thus resulted in the less fluctuating pattern of mortality typical of our present society. As a result, disease mortality as a whole declines, and the population can grow faster, even independent of improvements in health care. However, this tendency is enhanced when immunity is increased artificially by vaccination—and by children going to school and becoming infected there.

Obviously, the greater mobility of the members of a population has the same effect of increasing the number of contacts on which disease transmission depends. Mobility increases with an increase in transport and trade, and with the number and quality of the canals, roads, and flight connections. Moreover, a greater number of contacts turns a general epidemic disease into an endemic childhood illness, most older, immune people still being alive. Therefore, even within one and the same disease, it can take different forms. For example, AIDS, a disease resulting from the HIV virus, occurs in East Africa along roads with much truck traffic, whereas in Thailand it is mainly found in rural communities to which young, diseased prostitutes return after having worked in Bangkok. In Western countries, it was initially concentrated in groups of homosexuals with many sexual partners, and it was dispersed along flight connections and truck routes.

In large, more or less uniform populations, endemic diseases are the inevitable result of the many contacts susceptible people have with each other. In a way, endemic diseases are unnatural and man-made. They are the direct result of our high numbers and our mobile way of living today. The same holds for childhood illnesses. When people lived in small, isolated communities, both adults and children were equally susceptible to diseases that struck occasionally. Nobody was old enough to be still immune from a previous epidemic because the disease had been absent for too long. However, this changes as the number of people grew and number of contacts within and outside the community increased. When a disease would reappear more frequently, the older, immune people were still alive. From that point on, the population consisted of immune adults and susceptible children. Now, when an infectious disease arrives, only the susceptible children run the chance of becoming ill, whereas the immune adults remain free of it. So, in growing populations, infectious diseases like measles or smallpox can become childhood illnesses. Not surprisingly, therefore, in Europe, similar to measles, smallpox first evolved into a childhood illness in cities, whereas it was still a general disease in the countryside, affecting people irrespective of their age. Similar to endemic diseases, the spread and gravity of childhood illnesses also reflect how big the population is and how fast it is growing, as this determines the chance of infection through contact.

This infection rate can also be accelerated by infecting people artificially. From the beginning of the eighteenth century onward, people were inoculated for smallpox by deliberately being infected with the disease. This worked, although it took a toll: not all people can develop immunity to the disease and so died from the preventive treatment. At this time, inoculation was done with the virulent virus. Two centuries later, a more benign kind of inoculation was done with cowpox, which is less virulent to humans and confers immunity to smallpox. This procedure was called vaccination, after the Italian word for cow, *vacca*. Smallpox then became a childhood illness (because all the vaccinated adults were immune) and finally was eradicated from the population of children.

Interestingly, since the mid-eighteenth century, the rate at which populations were growing in various parts of Europe has accelerated. Just before the
Agro-Industrial Revolution of the mid-1800s, the rate increased suddenly
and sharply. What caused this jump in the rate of growth and the steady
high rate of increase that followed? There are several explanations which are
all debated by the specialists, but I give only my own based on the epidemic
dynamics of diseases.

At the beginning of the Neolithic Agricultural Revolution some 10,000
years ago, population growth had also accelerated. Rough estimates suggest
it increased from around 0.0003 percent per year to 0.021–0.051 percent. An
enormous jump. After that time, from AD 1 to 1650, the annual growth rate
remained roughly 0.042 percent, so the world population did not grow more
than it had in the thousands of years previously. Then, according to rough
estimates (very rough for the earliest dates), the rate began to rise again:
from AD 1 to 1750, for example, a period just a century longer, it rose from
0.039 percent to 0.092 percent. However, from 1750 to 1850, it jumped to 0.52
percent and 0.70 percent, and from 1850 to 1900, it was 0.6 percent. Yet, from
1900 to 1950, it became 0.8 percent; from 1950 to 1975, 1.9 percent; and from
1975 to 1990, 1.8 percent. For a time during the 1970s it was 2.2 percent.

So what caused this jump and further increase in the rate of growth after
the seventeenth century? Did living conditions, medical care or sanitation
improve? Or could some demographic threshold have been crossed? Could
this possible threshold have something to do with disease thresholds when an
epidemic, deadly disease turned into an endemic childhood illness, thereby
taking a smaller toll? In the case of measles in Iceland, for example, calculations show that once the population exceeded 250,000, measles changed
from being an epidemic disease to an endemic one. Roughly at that time,
it could also have developed into a childhood illness. A similar threshold
may have been passed in Europe in the eighteenth century. I have explained
that roughly at that time, smallpox developed into a childhood illness in the
growing urban environments. And that due to the crossing of some threshold, not only does a disease change its character, but it also results in a sharp
decrease in mortality rate, and thus in a sharp increase in the growth rate of
the population. In fact, improvements in medical care, nutrition, and sanitation came later.

Not only had fewer people died because of the buffer formed by the large
proportion of immune people in the population, but a smaller proportion
died before or during their reproductive years. In their epidemic form, infectious diseases like measles or smallpox could affect a horrifying 90 percent of
the people in local communities, whereas after becoming endemic, measles

needs only four or five thousand victims in a local community per year to be sustained as an endemic disease. And these victims do not always die of it; they sustain it. After a reduction in adult mortality, infant mortality from infectious diseases dropped only when vaccination programs made infants immune to an increasing number of those diseases, as well. This greatly reduced child and adult mortality, therefore, had major demographic consequences, which might account for the population explosion during and after the eighteenth century.

The knock-on effect of the population explosion was that agricultural production had to be improved and mechanized, after which the growing society urbanized and industrialized. These socioeconomic changes would have absorbed much of the reproductive surplus. After that, the first waves of emigration began, absorbing another part of the surplus. Also, the local and national markets for raw materials and finished products had to be extended nationally by improving the network of roads, canals, and railways. As extra money came in from abroad, social conditions in Europe and North America improved in terms of nutrition, housing, health care, and sanitation, especially during the nineteenth and early-twentieth centuries, and particularly after the Second World War. In these ways, population growth accelerated and improved the standard of living well beyond the subsistence level, and this, in turn, stimulated further population growth. Nowadays, we take our opulence and reproductive rates to be normal—to such an extent that many of us think that technology is driving population growth, rather than the other way around. However, supply always follows demand.

In short, health conditions improved during the nineteenth and twentieth centuries through better nutrition, sanitation, sewage systems, medical care and pharmaceuticals, education, social and working conditions, and mechanization, as well as new sources of energy and materials. These improvements were made thanks to higher national income and industrialization, and therefore, ultimately, because there were more people. Meanwhile, European and global infrastructures improved, which increased mobility further, and thus the infection rates of new diseases followed suit, although most of those diseases were (and are) benign rather than deadly. Together, all this allowed for a further decrease in mortality and increased longevity, both of which stimulated population growth even more. Later, birth rates also began to drop, slowing down the high rates of population growth, although these have remained historically extremely high. Growth stimulates growth and is impossible to turn back. Infectious diseases, once of major demographic importance, are now local and negligible in demographic terms, horrible as they can be at a personal level.

As our numbers, mobility, and catchment areas have increased, some new infectious diseases have emerged. Some of the new diseases have resulted from the increasing contacts farmers have with greater numbers of animals. In the past there have been instances of "foreign" diseases being introduced via the same contact-driven process by which infectious diseases spread. Many of these new diseases started off as benign animal diseases but then evolved into epidemic diseases that were deadly to humans. Actually, we share most of our infectious diseases with animals like dogs, pigs, cows, and horses; the closer the contact we humans have had with a particular animal over the centuries, the more of their diseases we share. Conversely, we share the fewest diseases with animals remaining at a larger emotional or working distance from most of us, like those of rats, mice, poultry, and aquatic animals.

Take, however, the case of cholera, a disease of the last category, the aquatic animals. Initially, the cholera bacterium occurred in a small marine shrimp or crayfish found near the Indian coast. Every so often, a fisherman would contract the disease and possibly die from it. Before he died, he may have infected members of his family, as well, or other people in his village. However, the disease would die out with its victims because of a lack of new susceptible people (at that time, contacts between the coastal villages were too few and far between for the disease to spread from village to village, particularly as cholera kills its victim swiftly—within 24 hours—so there is little chance the victim will spread the disease over long distances). But as the communities grew bigger and the distance between them decreased, people had more contact with each other, and the likelihood of the disease jumping from one community to another increased. Gradually, cholera changed from an epidemic disease into an endemic one. Finally, when it found its favored living conditions in the sewage systems of big cities such as London and New York, it became endemic there, especially when it leaked into the groundwater that served as drinking water for many. And when infected water from upstream in the Thames was pumped into pipes for drinking water downstream, the result wreaked havoc in large parts of the city. For the transmission of an infectious disease like cholera, contact between people can therefore be both direct and indirect.

Cowpox was initially confined to the cattle kept in large herds by local communities in Asia. There, it spread from one animal to another and occasionally infected the cowherd. This disease, too, must initially have remained epidemic. Only when the contact between people in a community and then between communities became frequent enough could such a disease eventually evolve into an endemic disease, and finally, in Europe, into a childhood illness.

In the twentieth century, the same process may have happened with HIV, responsible for a new deadly disease: AIDS. The virus may have jumped over to humans from chimps kept in captivity in western Africa. The contact between the caged animals and people was frequent enough to allow the virus to infect a few humans. The infected persons infected other people, who in turn infected other victims, until the disease was common enough to be recognized as a new deadly disease. However, the disease has a long incubation time (ten to fifteen years before the virus develops into a disease detectable from outward symptoms), so by the time doctors realized its gravity, enough people had been infected to make it endemic, and it was already difficult to eradicate. In terms of the disease itself, its initial establishment remains the most hazardous time—when there are not enough infected patients to jump over successfully to new susceptible people, or if there aren't the right and frequent enough contacts, the disease does not take off. The "right" mix of conditions remains a matter of chance. For example, it seems likely that AIDS occurred in a Dutch hospital during the 1930s, but it remained confined and thus died out.

During the last few decades, we have seen similar global processes developing in diseases of pigs, cows, sheep, goats, and poultry. Here too, the frequency of contact between these animals kept at high densities, and between them and their keepers became so high that diseases have become new epidemic diseases. In order to halt the spread of these diseases among animals, only the infectious and potentially infectious animals need to be isolated from the susceptible ones, although in all the outbreaks that have occurred so far, large numbers of animals have been slaughtered as a preventive measure, regardless of their potential risk. Unfortunately, many traditional livestock breeds have been irretrievably lost in this way, significantly reducing the biodiversity of animal husbandry.

There are more complex variations on the theme of an animal disease developing from a succession of incidental epidemics into a single outbreak of permanent endemic proportions. Certain diseases that normally tick over between several species of wild animals instead of between domesticated ones are now able to jump over to the human population as well. An example is influenza. The influenza virus normally occurs in high densities in a number of migratory bird species, such as gulls, terns, or swans, and has spread from them via infected droppings to domesticated animals (pigs, cows, and horses), and finally to humans.

Formerly, a farm family owned only one or two cows, which made both the chance of the animals getting infected and that of the disease jumping over to the farmer's community negligible. But in today's intensive livestock farms, animals live in high densities of hundreds or even thousands. After many animals have infected each other, the farmer is also likely to contract the disease easily, after which he will infect others. Moreover, our increased mobility means that influenza now spreads rapidly between people: from a farm to a village to a town. An epidemic can soon break out as the disease is spread by sneezing or by direct physical contact (especially between schoolchildren). Large numbers of people packed together in the confined space of a bus, a train, an aircraft, or in trenches during a war increase the chance of transmission, so that the infection can be spread over large distances within hours or days. Influenza epidemics can thus rapidly develop into global pandemics, as happened with the Spanish flu at the end of the First World War.

As with measles and smallpox, many people can develop a lifelong immunity to influenza, but they become immune to only a certain genetic strain, whereas various strains succeed each other over time as epidemic waves. After several decades, when many people immune to a particular strain have died and enough susceptible newborns have entered the population, this strain can develop and reach epidemic proportions again. Thus, individual strains recur at regular intervals. Knowing the intervals of their recurrence, we can predict when an influenza epidemic may break out again, and then prepare influenza vaccines against the strain in question. However, the individual strains are genetically variable, so that new, slightly different strains can occasionally pop up and enter animal and human populations.

Taking various strains together, therefore, influenza as a whole remains a general, endemic disease for the population of adults and children, whereas as people get older, they are susceptible to a decreasing number of strains. The individual strains, however, recur in epidemic waves and cannot develop into childhood illnesses—even though the disease usually begins in nurseries and schools, where children in particular have a lot of physical contact with each other. The children unwittingly bring the disease home and infect their parents, who in turn transfer the disease to other people at work, on public transport, and in supermarkets.

Each species is biologically unique. This applies not only to plant and animal species but also to the microbes, bacteria, and viruses that cause our diseases. There are as many transmission mechanisms as there are diseases, and the transmitting species may be wild. Take rabies, for example, a deadly animal disease that also occasionally infects people. This virus develops in foxes and dogs, and because it affects their nervous system, the animals begin to behave unpredictably before they die. If they bite other animals—a grazing cow, for example, or another dog, a fox, or a human—the victims soon develop the same symptoms. Being mainly confined to the pool of foxes, this disease is restricted to the open countryside rather than villages and towns. However, when the density of both foxes and people is high, as it is in Central Europe or Britain (where urban foxes are a nuisance), villagers, urbanites, and pets are at risk. Fortunately, so far, rabies has been prevented from entering the British Isles.

Plague is another such deadly animal disease, which was common in Europe in antiquity, the late Middle Ages, and the Renaissance, up to 1720. More recently, during the late twentieth century, it still occurred in small pockets in the Far East. As a soil bacterium, it often remains endemic in rodent populations, which die in great numbers. Their fleas, which are also infected, can transmit the disease to humans by biting. It is estimated that in the Middle Ages plague killed one-third of the total European population, although locally mortality could amount to 60–70 percent, and was occasionally as much as 90 percent. Its sudden decline mainly during the mid-seventeenth century has been attributed to a change in the way houses were built: from then on, people began to build houses in stone instead of from wood, wattle-and-daub for the walls, and thatch for the roofs. This took away the biotope for black rats as disease carriers. Roughly around that time, too, black rats were partly replaced by brown rats, which prefer to live outside houses in burrows in the ground.

When a person becomes infected with bubonic plague, the result is terrible: once in the bloodstream, the bacteria enter the lymphatic system where they kill the defensive cells, turning the lymph nodes into painful, fast-growing boils full of pus, which eventually break open. Then, the bacteria reach the liver, the spleen, and the brain, causing hemorrhages that destroy these organs and lead to uncontrolled behavior. Finally, once in the lungs, they make the victim cough up blood. The close proximity between humans and rats and other rodents explains how the disease gets into the human population. But in the more densely populated cities, transmission by coughing is more likely and so there is more chance of catching the deadly pneumonic form of the disease.

The diseases I've discussed so far are infectious or contagious. Diseases that, for example, are spread when coughing or sneezing spray a great number of viruses or bacteria into the air. People nearby can breathe in this air, thereby becoming infected. But life isn't always that simple. Take malaria, for example. In the past, people mistakenly reasoned it was caused by the poor quality of the air near marshes and therefore called the disease malaria, "bad air." However, it is caused by a minute single-cell parasite and is transmitted not directly through coughing or sneezing, but indirectly through a mosquito as an intermediate, or vector. This insect injects the parasite into the bloodstream of its victim during its blood meal. In this case, the disease can be controlled by draining or spraying the marshes, the mosquitoes' biotope (marshes were drained anyway to reclaim land needed for crop production and urbanization). There is still a relationship between disease incidence and the number of people living close together, albeit a weaker one; when there are many people crowded together, mosquitoes need fly only a short distance to get their next meal and infect a new victim. Thus, the more people living together, the greater the chance of a mosquito encountering an uninfected victim, and the greater the chance of transmitting the disease and the disease becoming endemic. However, the insect has its own biotope and area of occurrence, so people living far from these are not at risk. On the other hand, it has sense organs and so can home in on its victims from some distance. Yet, those risks can change when, due to climate warming, the responsible mosquito species expands into densely populated parts of the world, like Europe.

Malaria is just one of a great variety of parasitic diseases, and the parasites range from single-cell to multi-cell, each with its own vector or transfer mechanism and its own preferred living conditions. It is their idiosyncrasy and that of any viral and bacterial disease that makes their medical treatment or eradication so difficult. Our medical arsenal is therefore elaborate and complex, and, hence, expensive, which makes it prone to collapse when funding runs out, or when medical staff cannot do their work properly during a war. Curiously, the better they do their work, the lower the risk of people contracting a disease, and the lower the funding.

This curious development occurs in Western, developed parts of the world that have gone through the epidemiological transition of the 1960s and 1970s, during which infectious diseases declined, because effective, petrochemical-based pharmaceuticals were applied. Instead, chronic diseases, such as cancer or heart diseases, took their place. However, the largest part of the world with most people still has not gone through this phase. As infectious diseases are subject to environmental variation and change, like climate change, serious problems may arise, particularly under socially and economically stressed conditions as they occur in the slums of the world.

To protect us against diseases now that natural control from low contact rates between potential victims no longer works, we have developed a medical control system. But it is not always tight enough. In practice, control often depends on the knowledge held by a single person or by only a handful of specialists in the whole world. When one or more of these experts retires or dies, there are no replacements who know from personal experience what to do. A few years ago this actually happened in India, where a fungal disease thought to have been eradicated reappeared in wheat, and only a retiree over eighty years old still knew what countermeasures had to be taken. In other cases, the medical system is still proving effective at keeping certain diseases at bay, such as the deadly Ebola virus in Africa. However, the existing system may break down if social unrest in the area allows this virus or another to get a foothold strong enough to develop into an epidemic that kills thousands of people within a short time.

At present, antibiotics and other pharmaceuticals protect us against many diseases. However, the disease vectors—viruses, bacteria, and single-cell organisms—can themselves become immune to these chemicals, so that our protection falls away. Of course, we can then develop other chemicals instead, but this strategy has its limits and can also be too costly. Recently, certain antibiotics that had been administered often and indiscriminately have lost their effectiveness and so the search is on for new ones. At some time in future, however, we may run out of possible substitutes to combat ever-more resilient superbugs and so will be left unprotected.

In other cases, such as when HIV is still restricted to a particular homosexual community, to prevent the disease from spreading we need to know the sexual partners the members of that community have had in the past, because the disease remains hidden for ten to fifteen years before its symptoms become apparent. Unless these individuals are monitored, HIV can spread extensively, with nobody noticing until it is too late. In the Netherlands, for example, there is a comprehensive database of the sexual partners of HIV-positive persons, which is updated continually. Conditional to the set-up of such a preventive medical control system is, of course, that the disease itself, the number of contacts of the people involved, and the contacts' identities are known. Obviously, in many other parts of the world, this condition is not met. In those cases, the actual—often not official—degree of infection can be as high as 30 to 40 percent of the total population, increasing the death rate so much that the age structure, and thereby the structure of society, is disrupted completely. In some developing countries, particularly the older, richer people of the ruling classes get infected more frequently, which can destabilize the country's organization. Then, foreign firms and factories, as well as tourists, stay away, leaving the country economically and socially underdeveloped. Yet, setting up and maintaining such a preventive system requires constant effort and adequate funding, well-trained medical staff, and sophisticated technology. With each relaxation or disturbance of the system, the disease concerned will flare up again.

As the world population keeps growing, and the number of contacts increases—not only between humans, but also between domesticated and wild animals, and between them and humans—the number of new and unknown diseases will increase. Often, these new diseases are as lethal as those that killed large numbers of people in the past (such as measles, dysentery, tuberculosis, smallpox, the plague, malaria, and cholera). Among the native populations of various parts of the American continent, for example, the death rates from newly introduced diseases sometimes exceeded 90 percent. Waiting in the wings are new deadly diseases that are currently still confined to specific areas, such as the one caused by the Ebola virus, or the one caused by the Lassa virus, both of which occur in parts of Africa.

Our rapidly growing world population, our increased mobility, and the growing number of people living under stressed and unhealthy conditions in shantytowns around the major cities make us more vulnerable to epidemic, pandemic, and endemic diseases. It will need increasing funding and human and material resources to be able to continue to protect us. Often, however, the increasing medical costs are already so high that they are either cut back or are privatized, increasing our risk.

20   THE DYNAMIC STRUCTURE
OF SOCIETY

One sad day in 1794—May 8 to be exact—the French chemist Antoine Lavoisier was beheaded. He is known as the father of modern chemistry, the first person to quantify his measurements very precisely, which resulted in his interpretation of chemical elements as basic, elementary units. He also refuted the phlogiston theory, which postulated that something immeasurable, nonmaterial—phlogiston—was involved in the burning of materials like wood. For these reasons, his execution is remembered by scientists, biographers, and historians in general. But all these authors have viewed him from their own perspective, sometimes from a close, personal distance, sometimes putting him into a broad, social or historical context. Yet, he has usually been described in isolation, perched beyond reach, like a stylite high up on his pillar, as if he hadn't been influenced by anybody, or by the turbulent society around him.

I have always wondered, however, what role Madame Lavoisier played in all his efforts; she was always at his side during his experiments—she made all the notes. And the mathematician Laplace, known as the French Newton, must also have influenced him. It is known that they were good friends and also that Laplace frequented their home at the time Lavoisier did his famous and critical experiments. Laplace was not interested only in pure mathematics, he was also an applied mathematician who emphasized the importance of the quantification of all kinds of measurements. Surely, at the Lavoisier's, Laplace must have spoken about his mathematical and quantification interests, already applied to astronomical, physical, and social phenomena, and now possibly to chemical ones as well. To some extent, the history of individual people—and, indeed, the history of all society—is always one-sided and incomplete, always biased, warped, and of the moment.

Similarly, we can look at the historical development of society from a broad array of angles, from various distances, emphasizing completely different aspects. And we always get a different picture. That is why some say that history is the never-finished interpretation of the past. We can look at history from the viewpoint of individual people—the life they led in their time—or look at the impact they had on society. Those are approaches biographies follow. Or we can look at the development of society as a whole, and do this from a cultural, military, diplomatic, economic or sociological viewpoint. Or we can focus on the history of architecture, art, or music in different countries or parts of the world; we can emphasize the role played by ideas, mindsets, or science. Similarly, we can look at the growth of the population of a country, its economic, demographic or military balancing with neighboring countries, or at the growth of agriculture and industry, at the growth of the judicial, monetary, or financial systems, the use of certain raw materials, and the production of waste. The possibilities are endless.

All such approaches single out as their subject a particular person or a certain aspect of a society, usually without an attempt to integrate the findings with those of other people or social aspects. By emphasizing individual, economic, or social processes, or those of resource depletion, it has always been tempting to consider only those as the principal driving forces, and the growth of our global population as driven, steered, or limited by them. Food or mineral production shortfalls would limit further population growth, for example. Sometimes, such shortfalls are thought to be cushioned by competition, but in the end it is still the food or mineral shortage that will cause problems, and, ultimately, lead to the demographic "solution." This is the Malthusian approach. And population growth itself would result from improved living conditions, from the successes of a new society, developing itself from human ingenuity. Yet, integrating various processes can put individual people or aspects of society into a more informative perspective that leads to better, deeper understanding, and therefore to better steering of society. To make this integrative reasoning concrete, I give just two—partly hypothetical—examples: one on the introduction of the plow, and the other on the development of the present-day communication system.

Plows are tools for turning the soil over rather than scraping it to get rid of grass and weeds by uprooting them. Why had plows become necessary, and in what social or economic context? Did plowing improve soil aeration, or did it give access to more nutrients? Had those nutrients been running out? For how long had they been withdrawn from the soil? How many people had those soil nutrients provided food for, and was that number of people increasing? Why such an increase? Also, how much energy does it cost to pull a plow through the soil and turn over the soil? Draught animals were needed to pull the plow: oxen or, later, horses. Eventually the animals were replaced by heavy tractors. Initially, plows were made of wood and required six oxen or horses and three men to maneuver them. Not until the early eighteenth century were plows with iron shares introduced, requiring a team of three horses and one man. Though plowing is done only during a short period of the year, the animals have to be fed, housed, and looked after year-round, which costs fodder, space, building material, and manpower—expenses that ultimately have to be paid by cultivating a larger area or possibly by practicing crop rotation. But what to do with the additional amount of produce; were there more buyers to justify those changes and investments? Were there large settlements, towns with markets, and roads connecting them, and weren't they there before?

The plow was invented in Mesopotamia and China almost at the same time around 2000 BC and was introduced into Europe during Roman times, but its use did not become widespread until the early Middle Ages. Why had it remained unknown for so long? Did it stimulate population growth, or, conversely, was it perhaps the growth of the population in the early Middle Ages that stimulated its use, triggering a change to a new structure of society in which the plow found its functional place? Were the towns growing at that time? How many new roads were there, how far and how large were the markets in these growing towns to sell the products? Were they built to connect the markets with each other and with the farmers' fields? And what did the farmers do with the money they earned? Could the products be preserved, for how long, how, and at what cost? In short, although known for centuries, plows only became economically feasible within a certain, new social context, within communities of a certain minimum size, and with a certain minimum infrastructure of roads, means of transport, traction, markets, and a monetary system.

More recently, a coherent system of a thousand discoveries, inventions and organizations—large and small—was needed to make long-distance electronic communication possible, beginning in the mid-nineteenth century in the form of wired telegraph and telephone connections. A century later, new developments led to concepts like information, which required, for example, the development of the entirely new, mathematical science of informatics. And these developments during and after the Second World War paved the way for the construction of computers. The principle underlying the operation of computers had been formulated back in the seventeenth century by the German Gottfried Leibniz and during the 1930s by the English mathematician Alan Turing. Leibniz's computer was mechanical, but new developments required electricity. Computers needed screens, and these were based on the ion tubes invented during the 1920s and 1930s. Later, during the 1980s, computers began to be linked to networks, first using existing telephone connections, then satellites, and at present, the transcontinental and transoceanic fiberglass connections— a development culminating in the invention of the Internet and creation of an elaborate, global financial and outsourcing network. The Internet has replaced a large part of the postal system, and has allowed electronic shopping on a global scale both for private people as well as for industries.

These developments would not have been possible without widely available and easily transportable energy: electricity, itself produced from transportable, cheaply obtained oil. Poor countries, such as many in Africa, can't afford a stable supply of electricity or the cost of rapidly obsolete computers and software and so the information gap between them and richer, developed countries is widening quickly. The life expectancy of both computers and software is only four or five years at most; many laptops crash after this time. A result is that poor countries cannot take part independently in the global financial network spun during recent decades. Therefore, they are falling further behind socially and economically.

Key to the emergence of today's global electronic communication network is the rapid increase in the need for short- and long-distance communication in the last 150 years. Once again, that need requires a critical mass of people to drive and to serve it. Our numbers drive the inventions needed, our energy sources, our organizations, and our technology, and they also make them obsolete. And it was the inventions and discoveries in the past that created the differences that have been intensified and have widened the economic gap between nations. It is the need to sustain all those billions of people that determines the highly intricate network of our present society.

It is essential to understand this hierarchy of processes: the more humans there are, the more energy we collectively need to keep going; the larger, more elaborate, and complex a society grows to sustain those numbers, the more energy it uses. The moment the energy flow stops, our societal organization is bound to collapse and we will suffer. Both we as individuals and the society we form are organized channels through which energy must flow uninterruptedly.

Our present numbers and society rely on coal, oil, and electricity; we cannot do without them or without the energy-hungry agrochemicals, plows, computers, and Internet connections. Or without trucks, ships, aircraft, or pipelines, high-tension lines, and fiberglass cables. They are as essential as houses, clothing, and food. They are the material tools to sustain our present billions, and we will need more or other tools as our billions keep growing. Those material tools are indispensable, but so are the immaterial societal structures of education, medical care, financial systems and networks of banking and money exchange at local and global scales, as well as our scientific, technological, and medical knowledge, and our national and international administrations and jurisdiction. For the longest part of our history, they have been unnecessary, but in this respect the past is irrelevant; now we have to think only of the present and about our future. Now, and even more so in future, these material and immaterial tools are crucial for our persistence as individuals, as a global society, and indeed in the longer term as humanity. We need them to survive and therefore we make, shape, and reshape them according to their function. Thus, the growth of our numbers and demands allows and facilitates further growth.

This incredibly elaborate and complex societal superstructure as a survival tool has allowed us to escape Malthus's scenario. But we run the risk of this multitude of interconnected escape routes becoming too elaborate and complex to sustain us in the long term—the risk of no longer being able to manipulate and maintain the entire edifice so that it collapses under its own weight. As a system, it requires too much effort from us to keep it running, it takes too much of our energy and mineral resources, and it produces too much pollution. In this regard, climate change may turn out to be only one of the first miner's canaries. The coming shortage of rare metals required for computer screens or for magnets to produce carbon-free energy will soon be another one. And so on. Moreover, when one or two processes fail to operate, the operation of the entire structure can seize up and break down. This risk can even be independent from inevitable shortages of energy, food, water, or minerals, independent of the pollution by all our waste, independent of the shrinking area of land that produces our food. Just because the system, our global organization, itself becomes too large and too complex. Then, the only humane strategy remaining to allow us to escape from the doom scenario of Malthus should be not an impersonally imposed increase in mortality through famine, disease, or war, but a deliberate decrease in our birth rates.

Long ago, in the first human settlements, increasingly more hands were needed to do all the work necessary to sustain the growing numbers of people. But soon, these human hands were not enough, and they had to be supplemented by other energy sources—first by tangible, material tools and then by intangible, organizational ones as well. And now, many centuries later, hardly anything is still done by hand in present-day Western societies. Our hands and many of the early tools we used were replaced long ago by all sorts of instruments, apparatuses and machines, often doing the work automatically, and all backed by a diversity of local and global organizations. Automatic pumps connected to automatic sprayers, connected in turn to a computer-steered dispenser supplying plants with exactly the amount of water and nutrients needed. Orders are picked from warehouses by computer-steered robots. An abstract, computerized world is supplementing or replacing increasingly larger parts of our familiar, concrete world; we can no longer do everything all by ourselves. One development followed another, first locally, then nationally, and finally continentally and globally. Agreed amounts of liquid gas of a certain quality flow silently and unseen through pipelines thousands of kilometers long, from the well across continents and seas to factories, offices, and homes. Computers pay for this automatically through abstract transactions in the form of electric pulses. All those developments together form this superstructure of technology and organization: a worldwide network of food production, mining, industry, financing, trade, and waste disposal spanned over all individual humans. Nobody can see what is happening, nobody knows all the flows, all the interdependencies. Nobody can manipulate this structure and its dynamic. Yet, it's all happening around, over, and for us, determining our life. Only when a war or a financial crash or a blizzard or flood cause it to collapse do we appreciate how much we have come to depend on it. In no time, such interruptions result in hunger, cold, darkness, and financial losses; in overflowing, malfunctioning hospitals and criminality; in chaos on the roads and in subways, shortages and poisoning of food, and looting; in closures of schools, offices, and factories. Imagine one of our megacities housing a couple of tens of millions of people, losing its connections with these various forms of local and global infrastructure. How would its inhabitants survive, how could such a megacity keep running when isolated from its surroundings, the outer world? We go to the store or the gas station, turn on the faucet at home, and feel we are in complete control of our life. This concrete world, however, is totally dependent on unknown people behind a machine they have been asked to operate, on energy and information flows, on abstractions. An abstract world is ruling the concrete life we live. We are components of an abstract global superstructure living its own life.

We depend on a great number of factors and processes in our natural and cultural environment, and their weight and operation depend on us and on each other, either directly or indirectly. For example, as our numbers rose in antiquity, people spread out in various directions from early centers of occupation, so that agricultural practices intensified locally and extended further geographically. From early on, the burning of forests and bushes polluted the air directly with carbon dioxide; soon the first rice paddies were emitting methane waste. Initially, climate changed only a little, but the emissions of greenhouse gases have intensified during our history and have been reaching dangerous levels since the Second World War. These levels have been particularly significant since the year 2000, as mechanical and chemical industries, traffic and other forms of infrastructure, deforestation, and agriculture—all fed by a rapidly increasing flow of energy—have supported the highest number of humans that has ever lived on Earth. Since around 2000, too, China, India, and Brazil have enhanced their living standards, which in turn has further enhanced the atmospheric concentration of greenhouse gases. Also, our increasing numbers rely on ever-increasing amounts of water used in agriculture, households, and industries, so shortages are beginning to affect the production of food, meat, and the water for drinking, cleaning, and industrial purposes. Such shortages have already resulted in international conflicts, though not yet openly in war. In turn, food shortages have always resulted in mass migrations, local revolts, or civil wars, which in turn have led to outbreaks of diseases. Wars have also been waged for mineral resources, and, during the last half century, for oil, gas, and coal. All these processes are interconnected and all ultimately relate to our numbers. The fact that they are processes means that this structure is a dynamic one. The flows, the processes, continually change in intensity, and their weights relative to each other are never the same; their interactions are always different.

Usually, we don't know how the flows and processes interact exactly: directly or indirectly, strongly or weakly, linearly or nonlinearly. They will often be nonlinear, which not only makes analysis difficult, but means that changes in any one of them can affect the stability of the structure as a whole. Worse, this often makes it impossible to predict what's going to happen, where and when. And if something does happen, we won't be able to steer or to stop the process. Out of necessity we are building a structure whose dynamics are unmanageable in principle. National and global financial systems are similar structures, and we now know that we actually go from one crash to the next, and that it is very difficult to manage such systems, to regain control. And the financial system is only one subsystem of many forming the dynamic global superstructure determining our future.

Structures, systems, or organizations are defined by interactive processes between a set of units. Structures are static when their elemental units do not change, like the cogwheels in an old watch, and dynamic when they are to do with processes. Our human society is a mix of them; soil condition, for example, is static, although, due to certain processes, such as salination, its properties can change. Diseases, however, are processes; they develop, die out, or tick over. The network of interconnected processes forms a dynamic structure; it changes continually, as history shows. It represents the natural and socioeconomic environment of humanity, although some connections vary or change in significance.

Recently, connections within the social structure have tightened and gained relevance, and they have grown out from national or continental networks to global ones. For example, the subsistence economies of the Dark Ages always had roads and regional markets, as well as long-distance traffic between regions along rivers or sea coasts. But recently, the social significance of traffic has become qualitatively different. Roads connecting production areas with the markets improved, particularly after the 1750s, and the road network was made denser, rivers were canalized, and canals were dug, sometimes to connect rivers and lakes, and enable continents to be crossed by shipping. Railroad networks were constructed. Then, long-distance, transoceanic traffic by ship was added, and recently, in the last sixty or so years, air traffic.

There are more and quite different connections. As we discussed in chapter 11, the rising atmospheric carbon dioxide concentrations led to higher concentrations in the ocean water, which increased its acidity. This higher water acidity, in turn, directly affects the survival chances of fish through its effect on their bones, and indirectly through its effect on animal plankton as their food. Also, higher carbon dioxide concentrations lead to higher air temperatures, and these to higher ocean water temperatures and this to the death of algae eaten by fish. Higher water temperatures also reduce the solubility of oxygen in water, and thus reduce the survival chances of fish in yet another way. For millions of people, however, fish is their main source for nitrogen, which helps the buildup and functioning of their tissues, particularly while they are growing up.

Similarly, for millennia, human society ran on some external, nonhuman source of energy: oxen, donkeys, and horses, or water, wind, wood, and coal.

And when one or two of them ran short, people could still switch to another one and continue working. In a way, humans did not depend on the availability of energy as such; by exploiting alternative, ever richer sources of energy, human society did not miss a beat. There was always energy in another form. This changed as humans increased in number to the present level; all alternative sources, even combined, are running dismally short. As a society, we are now almost completely dependent on fossil fuels for our food, for running our industries, for communication, for transportation, for supplying each of our villages and cities. When the world runs out of fossil energy, even if this does not happen for, say, five hundred years, our survival will be seriously at risk. Energy as a factor for human survival has become critical, because of the great number of mutually dependent processes running on it. The flexibility of society becomes constrained. Each of us as a human organism represents a certain amount of energy, a continuous flux of energy in fact. And so does our global society as a dynamic structure. Without energy, even with less of it, or when its flow is interrupted for only a short time, our existence will be critically affected.

The fate of the socioeconomic structure we have built depends on how its elements develop and how stable each of them is. Agricultural production in western and central parts of the United States, for example, depends on a large energy supply: for plowing, sowing, harvesting, and transporting, and also for pumping up fossil water from aquifers and manufacturing and using fertilizers and herbicides. Other parts of the U.S. economy, such as the car industry or the chemical industry, depend on the same fossil fuels and water. And the United States pays other countries for these fuels with money largely received from selling those agricultural and industrial products abroad. And it has to supply its own population of millions with the same products. If something goes wrong with its foreign energy supply, either because the energy becomes too expensive because of competition with other countries, or because the wells or gas fields run out, the entire dynamic structure of the United States will be affected. As said earlier, a rise in fuel costs has already led to hundreds of U.S. airports being closed either temporarily or permanently. This has had knock-on effects on industry and services near the airports affected. The United States is the largest energy user in the world and will therefore be hit the hardest. Similarly, when its water supply runs out, or when salination or soil erosion spreads, its agricultural production will be damaged severely, and the country will become dependent on other countries for its food supply. And the ripples of such developments will inevitably reach other parts of the world.

Energy supply and prices are also key in other countries and regions like

Europe, the Far East and South America, for example, although always in their own, specific ways. Australia, the dry continent, is particularly dependent on its own, scarce amount of water for its agricultural production and for supplying the large cities around its rim. In future, the money earned from its production of minerals, coal, and oil will not be sufficient to pay for this supply, particularly during El Niño years. Australia cannot possibly import additional water in sufficient quantities from abroad as has occasionally been done for some small Pacific islands. But Australia is far too large and too populous for that. And ocean water can be desalinized only at the cost of much energy. Water supply is already a matter of growing concern.

A pivotal problem is that everything depends on population growth. This growth implies a similar or an even faster growth in industrial and agricultural output, and therefore of that of resource use. This growing resource use determines the growth in the production of polluting waste, which then feeds back again on population growth. Sooner or later, all aspects of growth will meet their limits, which stops the internal working of the structure. These limits are not exclusively Malthusian, increasing mortality directly; they can also concern the rates of diminishing returns. Thus, the demands of cities and their surrounding shantytowns grow faster than the growth and the growing costs of the expanding hinterland can supply. For example, the transport system for people, resources and waste cannot keep up with the growth of the number of people in such cities; aboveground, the roads and railroads become choked; underground, the tunnels, pipelines, and cables become too complex. In some Indian cities, the construction of underground transport systems cannot possibly keep up with their growing numbers and needs. Pollution, salination, desertification, deforestation, and disease incidence, for example, hamper the operating of the system and use up all resources, space, and all organizational efforts made. These are dynamic limitations of organization, different from the static limitations of finite resources or technological constraints, or the limitations imposed by pollution or climate change. Together, all these diminish the returns from any effort to compensate for our growing demands. Locally, we gain in efficiency, but globally we lose it.

The organization of the operation of our world population is dynamic, intensifying and changing its interconnections with the inevitable increase in population size and demand. Our population is growing and becoming more complex and elaborate, requiring ever-higher levels of technology to maintain its organization on which our survival as individuals depends. Due to its complexity and elaborateness, this organizational structure is becoming inefficient. The multitude of efficiency measures taken gain a life of their own in their own growth, and grow unstoppably. All three aspects—the number of humans itself, the growing complexity of their organization, and the various forms of technology needed—ask for increasingly more energy and minerals, which are unavoidably all wasted. As the various interconnections tighten ever-more intensely, together they will make the structure more vulnerable to any weakness or instability in each of its component factors and processes.

We no longer live in a world of isolated factors and processes that can be treated individually. We cannot treat climate change independently of our growing numbers, of the urbanization of our societies, of the necessary technology or transport systems; independently of a certain supply of unpolluted water during the growing season; independently of the acidification of the oceans and the decline of their fish stocks; independently, too, of the chance of revolts or wars breaking out because of shortages of food or minerals, or of pandemic diseases developing. All these are interdependent, and ultimately driven by our numbers.

Contrary to a couple of centuries ago, all those interconnections have become relevant, and this has happened within a very short time; the network they form has to be dealt with as such. That is new. Together with its environment, our world population forms one operational unit: a structure, a dynamic system that is internally dynamic in its working, and externally dynamic in its growth and expansion. We cannot separate our environment from our population, cannot separate economy or sociology from ecology or from population growth. And, conversely, we cannot separate population growth from our natural and cultural environment. Population and environment have grown together; they are one.

Growing Past the Point of No Return

The human population ignites and feeds its own growth. Obviously, people have always realized that the faster the population grows, the sooner it will meet its limits. The acceleration of population growth in eighteenth-century Europe led to the second agricultural revolution, which in turn triggered the Mechanical-Industrial Revolution and boosted the second wave of urbanization and expansive mass migration. These two revolutions were interspaced by minor ones, by revolts, if you like, but which roughly contained the same ingredients: population growth, agricultural intensification and mechanization, urbanization and industrialization, and infrastructural intensification and geographical expansion. Recently, however, something entirely new has happened. After the Second World War, the societal chemical revolution occurred, further increasing population growth rates, inevitably leading to the communication revolution. However, mass migration and spatial expansion by colonization slowed down, and urbanization intensified accordingly. All revolutions, despite being unconscious, impersonal social responses to numerical growth have stimulated rather than slowed down the rate at which the populations were growing.

Malthusians suggest that growth may reduce or stop altogether because of the high mortality during wars, famines, or epidemics. However horrible and depressing it is to think this way, given our present numbers even this "solution" isn't going to work anymore.

We passed the point of no return a long time ago: imagine a two-thirds reduction of our present numbers, as once happened in Europe because of the plague. Given the world population in about 2000, this would mean an incredible mortality of four billion people during a war or as a result of a devastating famine or pandemic disease. Nobody would allow such a massacre to happen. And, in fact, it would only bring us back to the level of the 1970s, a mere thirty to forty years ago when we recognized the limits of the world for the first time. And even if the world population did go back to that level, at the current growth rate we would be back at the present level in another forty years. We therefore may decide on a radically different—and in the long term, more humane—demographic policy to prevent such an apocalypse ever happening. We'll then have to take measures that produce a humane and lasting solution rather than a temporary, inhumanely high and dehumanizing mortality rate. We could do so by reducing our birth rate radically and on time and by restricting our resource use and consumption, supplemented by recycling. What's stopping us from doing so?

E

Processes within the Global Society

The Future of Our Global Society

In times of social stress resulting from overpopulation, shortages of resources, and production of waste, one has to be careful not to disturb the system, as it can easily destabilize and collapse; the larger and more complex the system, the more likely this is. This is exactly the situation we are inexorably reaching. Not only are our populations increasing, our resources declining, piles of waste and streams of pollution growing, and the environment and biodiversity grossly deteriorating, but our economies will continue to grow to keep pace with the increasing demands of a growing population and its organizational and technological superstructure. Our world will get smaller and become more abstract, more unified, but it will also keep growing, which can lead to increasingly fierce competition for the last remaining resources. This inevitably increases the risk of social unrest or civil war, and, possibly, that of armed conflicts between countries. And this can lead to the collapse of the entire system.

21 FROM A CONCRETE TO AN ABSTRACT WORLD

Our eldest son is dyslexic. This means, simply, that he can't read without effort, and that he has difficulty doing basic sums; he has to construct each word and each number separately from individual letters and ciphers. When he "reads," he has to remember the string of previously constructed words to understand a sentence, after which he has to make something sensible of a section, combining the information from various sentences. After finally having read a page, everything gets blurred and gray. This costs much time and energy—too much with too little result for current-day practice.

Not being able to read and calculate easily means more than the straightforward technical problem of having to construct each word or number from its basic elements, the letters and ciphers. The problem lies deeper. Words and sentences, paragraphs, sections, chapters, and whole books represent increasingly abstract images of the world: when you are reading, you are not standing or walking in the middle of a concrete world, a world you can touch or smell. You can't touch or smell anything in the written world, you have to imagine it. And to make emotional sense of what you have read, you have to rely on previous experiences you need to recall. If you have no such experiences, as in computer games, a whole new world can develop within your head, often without any obvious connection with the world as you know it. Weird, fantastic worlds, worlds of long ago or in the distant future, faraway worlds, life on other, imaginary planets, worlds with different beings, life-forms with other ways of living. A whole galaxy within your head, within which you can travel with speeds greater than light, or far back into the past, or ahead into the future.

Numbers are no different from words. The number five in, say, five horses is an abstract concept for a group of horses you can see, and you can add this number to that of another group of eight horses. But five horses and three cows makes eight animals; the fact that five plus three equals eight is all abstract, as is the general concept of animal. So far, still all simple abstractions, but soon, these abstractions become very abstract when you leave the field of arithmetic for that of algebra and the numbers are replaced by letters. And from here you go to other types of mathematics describing relationships between unknown things that may never exist. It is then the relationships that you are interested in, not in the things they relate; those things can stand for anything. Those relationships are represented by symbols of often weird shapes. And from there you go to the operation of computers, or of networks of them. Networks spanning the world. Financial networks transferring money, shares, investments, or information. A completely abstract world. A mental map of the world, not just an abstract topographical map in our mind of concrete things—buildings, configurations of towns, roads, rivers, continents, and oceans—but a map of abstract things—regional and global flows of virtual water, money flows, flows of information. All in your head, or in its present-day extension, your computer's memory. And when your computer crashes, much or all of the information it contained is gone, just as your memories and knowledge vanish at the moment you die. Then, all those abstract worlds are forgotten—gone as if they had never existed, as if they had never ruled our concrete world of rice, chairs, or airplanes. Vanished without a trace.

Our son cannot easily form this kind of abstract world for himself by reading and doing sums. Luckily, he is highly skilled in forming spatial images; his world is much more spatial than anyone else's. It's another abstract world—another personal, mental world, one not easily penetrated by those of us who are readers and calculating beings.

Not being able to read or calculate even the simplest sums therefore means that you are not training yourself in the abstract way of thinking most of us share, a way of thinking basic to our way of organizing our human world, and basic to our technology. At first sight, this doesn't seem too bad; for most of human history, most people weren't able to do much abstract thinking anyway. They lived in their local subsistence communities, ruled by the local lord or the king. Illiteracy was widespread, but few people needed to read, or to add and subtract. Actual maps didn't exist. That wasn't their world. And, at present, aren't there still many jobs in which you really don't need those skills?

The last couple of centuries have seen an enormous change in this respect: human society has changed from one formed by a concrete, local world of real horses and cows into a highly abstract society of words and numbers, of machines and regulating instruments, of information and extremely complex, hierarchically arranged, global systems and organizations. These systems allow you to earn in future markets for unknown products of five or ten years hence or markets of global flows of virtual water. You can trade in computer software handling information more effectively, or in precisely circumscribed, yet unknown, amounts of specific, future information about something you'll never see. Or you can invest in a more efficient organization and show its greater efficiency by a number, the value of some variable; its value has now become, say, 3.6 instead of the previous 3.23, leaving people standing around you perplexed, in awe. 3.6! Unbelievable, 3.6!

Buying and selling shares, information, or packages of future travel arrangements or telephone calls are reorganization schemes that may change the concrete life of millions. There is money in your bank account you have never seen or touched; you know it's there only from a figure lighting up on your computer screen. Only one century ago, around 1900, one of the giants in physics, Ludwig Boltzmann, committed suicide because his calculations assumed the existence of atoms and molecules but many of his colleagues would not accept his findings. They urged him to keep to the concrete, the visible and tangible world. Our time is incomparably different from any previous time in human history. Enormous changes have been made over the last two or three centuries at amazing—absolutely staggering—rates. Changes making our once-concrete world of spades and spoons abstract. Books printed on paper are replaced by e-books to be read on your laptop, tablet, or e-reader. It is a world only existing in our heads—often not even on paper, but as such dismantled, coded, and scattered in our memories, or in those of our computers—one that can vanish or change owner in a split second. Like the genetic code in our DNA, our computers, calculators and brains do not even store the information of past events in a physical way as maps or images, but as rules and instructions for reconstructing them. A world of instructions determining the life of billions, yet one that can collapse, that can evaporate into nothing. Gone. No society, no trade, no nothing. It is a world that will leave behind real people who are now suddenly short of food and water because computers and fridges have stopped working during a regional blackout because of a mistake, perhaps, or a strike. People at present represented by a mere number will even lose that. Vanished from a computer memory means vanished from the world. Yet, a world that can also be built up, shaped, or changed at will, if we want to—when we realize it has got to change. It exists only in our head, anyway, so why not add to it, change it? That is the positive side of it.

What would happen to those hundreds of millions of people in the developed world of Europe or the United States if their electricity supply fell out for a month, or two weeks? It wouldn't just mean not being able to check e-mails, watch television, or go to the movies; more importantly, factories and whole industries would come to a standstill, and complex administrations, banks, hospitals, crucial laboratories would collapse. It would also mean that food would lie rotting in trucks along the highway and in warehouses, that diseases would spread unnoticed and unchecked. What happened in Beijing when traffic stopped for weeks in a traffic jam during the summer of 2010? What happened to the freight the trucks were transporting? Infectious diseases like cholera can be unleashed by large-scale inundations partly due to extreme weather events, such as happened in Pakistan at the same time.

Radical changes can also happen by deliberately changing the organization of societies. A couple of decades ago, for example, the Russian Communist system was replaced by a more capitalistic one; later, in China, a similar change occurred, although in a different way. How can such changes happen, how can the way of life of millions change at the whim of a government or even one person? What happens to those two hundred million Chinese farmers migrating from the country into a megacity of tens of millions of people because of the decision of one man? What happens to all the Chinese who used to live in the quarters of the old towns and now live in isolation on the nth floor of a tall appartment block on the fringe of a megacity? What happens to the Tuareg family, leaving the black, silent dome of their desert night sky for the jostle and stench of an overcrowded city or slum? What are they going to do, what thoughts go through their minds after they have been uprooted? What are they going to eat, how will they earn their living, what new relationships can they build up, what new resources will they use, what are they going to do with the waste they now make? From a concrete and personal subsistence existence they have landed in an urban, industrial, abstract, and impersonal society. Their way of life was based on eking out and recycling, but now they find themselves in a wasteful society churning out nondecomposable, nondegradable, polluting waste. We are living in impersonal, replaceable worlds—abstract worlds wasting the concrete world.

We are living in a tightening network of relationships, interactions between people, shopkeepers, and craftsmen, between organizations, firms, and corporations, between nations and confederations, between hierarchically arranged, highly abstract organizational or financial networks. These interactions are increasingly made through computers and transport systems. These networks are abstract, defined by agreements between large numbers of people, living and dead, supported by fluctuating currencies and laid down in customs, rules, regulations, and laws. Such agreements or laws, although on paper, are not concrete and stable; they are abstract, highly vulnerable to minor disturbances, unstable—they can change by the day. Possessions built up over years of suffering and hardship evaporate overnight into nothing. In their most concrete form, they exist on paper or as coin, but at present they mostly exist in our minds, or in the memories of our computers. In the past, when kings and popes died, the concrete life (feeding, moving around, talking, making love, having children) of ordinary folk remained largely unchanged. Now, a mere blackout can ruin our lives.

During a revolution or war, sufficient availability of food can turn into severe food shortages and famines in no time. Property rights, the value of money, shares, houses, and land are also thrown into disorder. Through the ages, it has been common practice for resistance organizations, armies or insurgents to destroy infrastructures (like roads, harbors, and railroads), and government or military headquarters. Immediately after a war, the damaged systems have to be rebuilt. During both world wars, a neutral country like Switzerland was a paradise with respect to food, medical, and financial and social conditions, whereas only a few minutes' walk away, across its borders in Germany or Austria, impoverished people fought, tortured, and killed each other under the most dire, lawless conditions, or they starved to death. Just across the border, a line drawn on a map.

Societies are at the mercy of an organized supply of resources, which can break down easily and abruptly, although the concrete needs of each of us remain. Societies collapse when their organization breaks down partly or wholly. Organizations of all kinds are not concrete but exist in our brains ("between our ears" as the saying goes), in books, or computers. They are abstract. Try to imagine all the complexity of whole worlds, the universe, existing between our ears. Yet, this is real.

But how did our present, weird world of abstractions begin? How did it develop during the last couple of centuries in particular? What aspects of our life and thinking were affected and dramatically changed? How did we get from a concrete way of living and thinking to an abstract one within such a short time? From a rural, personal life to a life in enormously big, noisy cities of reinforced concrete, housing millions of people?

Ever since the origin of civilization, economic growth has always been brought in line with demographic growth; in fact, civilization grew and served population growth and, hence, served economic growth. Larger, healthier, more active, and more mobile populations generate greater needs and form larger markets, which leads to more rapid and efficient production and trade. This, in turn, necessitates a further increase in scale of production and marketing. Moreover, economic growth facilitates the growth of the number of consumers and their individual demands. Rather than improve quality and personal care, growth in quantity for maximum profit became—in a way, had to become—the standard of success and the force driving all aspects of society. Production for personal needs or comfort or for the local exchange market was out; increase in production was in. Individual people lost control, and abstract societal factors took over. People are now reduced to abstract producers and consumers, their numbers serving an abstract world market on both sides.

To increase production output, highly sophisticated, automated factories replaced home workers. International trade took off, and national and international infrastructures—roads, canals, distribution networks, storage space, transportation firms, ships, cars, trucks, airplanes, and electronic communication and banking systems—were set up and continually improved, extended, and multiplied. Interest in money replaced interest in people; abstract quantity replaced concrete quality. Competition between people, firms, and banks set in, and factories, production, marketing, and the making of money had to grow and to keep growing in order to survive: the static, personal exchange model of the past was replaced by unlimited expansion and impersonal growth. Trade in shares and investments relates to profit and dominates the daily exchange of trillions of dollars across the world. An abstract way of relating to each other is the way we view each other and society.

New rules were set up to organize society; often opposing interest groups arose, expressed as political groups, resulting in democracy and a political structuring of society. All this followed a century-long trend toward individualization and professionalization, as well as toward specialization of industries, of banking, and of administrations. Democracy is increasingly to do with the counting of votes rather than the weighing of arguments that all too often are distant from the electorate, thus abstracting their identity and choices. Moreover, as our numbers grew, a hierarchy of democratic decision-making developed—from local to international. People and their society are expressed in numbers, in the stratification of organizations, and in the unseen and uncontrollable growth of both.

In order to maintain and control a particular number of individuals, society had to have a certain complexity, which, up to some point, could still be handled on paper. Hundreds of people in big halls transcribing payments from one account to the next, day in day out. But in a growing number of nations, this initially paper-based control system grew out of hand, and has therefore largely been replaced by a fast electronic system without human interference. Nowadays, almost everything—the making of appointments and contracts, the formulation of regulations and logistics, the arrangement of time schedules, travel, train and air tickets, hotel bookings, automatic payments of all sorts, job and house hunting—happens electronically within computers, and computer networks connect people, shops, firms, and banks instantaneously around the world. The man or woman at the ticket booth or counter is becoming a thing of the past, they are replaced by computer voices and mailboxes. Often, people themselves are deliberately excluded from taking part in a process: production, quality control, storage, automatic orders, deliveries, transportation, payments, checks, and reminders—everything is done by machines. Keeping your private life private is done electronically, but only if there is an option given to do this. Obviously, any stagnation or break in this flow of information and money, however short, costs millions of dollars and could be disastrous. Even the smallest mistake has to be avoided or rectified instantaneously. We live at the mercy of machines churning out letters and orders without considering human individuality, identity, or condition; organizations and machines increasingly control the concrete life of humans.

We have quantified and mechanized our world scientifically, streamlining any infrastructure or process, reducing all human labor and any interference to the minimum. By simplifying the complex concrete natural world, we have made our world abstract. Socially controlled village life has been replaced with living anonymously in towns, with identifying with a nation, and the nation has been replaced by the transnational "global village." Both the natural and the human worlds now operate at a distance from us, often in a way incomprehensible to all except the few specialists. Abstract comprehensibility—or incomprehensibility—became our new reality. We accept, apply, and extend that reality because it happens to work. But there is more—something deeper and with greater effects on our life and future. Quantification, mechanization, automation, and abstraction have alienated us from the concrete world and from each other.

With the growth of our numbers, human relationships became impersonal. Kings, popes, and vicars lost their personal grip on individual people, and the people themselves lost their grip on each other. Social control diminished. Impersonal property rights had to be defined, so legislation was passed, not only with respect to goods but also to people themselves: to the poor, to dissenters, and to the genders. And this led to the equal rights movements. At first, democracy pertained to men only, denying women their basic rights. Only after the First World War did mass movements organized by and for women develop nationally and internationally. A second wave followed in the late 1960s, during which all the established order was again questioned, across an even broader front than before. Part of this movement was the sexual revolution, throwing off much of the old patriarchal relationships within families and society in general. This resulted in a redefinition of the roles of men and women within families, and also gave women the power to determine the number of children they wanted to have. No husband, no state, and no church could command them. From now on, Western women could themselves decide whether or not to marry and to whom. They had equal rights to education and a career. As a result, first men and later women were no longer considered human property subjected to family or social rule. As such, the emancipation of women was not so much a result of better education but rather a change in social relationships within society in which men were involved as well, and often in the first place. Both developed different roles relative to each other within the family, and together, relative to society outside it. After a century-long struggle, both were free by law to choose for themselves, and both had to be educated to play their new roles successfully. An interesting and important aspect of this development was that it was international, although it did not always happen in all continents at the same time and at the same pace. But thanks to improved and global communication, particularly during the socially revolutionary time of the 1960s and 1970s, the movement occurred in most continents more or less concurrently. The first steps taken in those more revolutionary times were followed later and more gradually by many smaller ones.

I have concentrated here on the increasing freedom of women because this has had the greatest impact on further—and more far-reaching—social developments directly pertaining to the problems discussed in this book. And although there is still a long way to go, the steps taken so far have already had enormous consequences for the future of humankind, even in the relatively short term. As I have emphasized many times, the ultimate force driving the growth of resource use, waste production, environmental alienation and destruction, and urbanization and an abstracting of our world has been the growing number of humans. And, in turn, our numbers depend on the average replacement rate of either each woman or of that of both women and men together. In the first case, the rate should be 1 for stabilization, more than 1 for population growth, and less than 1 for a reduction of our numbers. For women and men combined, and given the present global conditions, this figure should be 2.3 (slightly higher than 2 because of dif-

ferent mortality risks, particularly of men). During the last fifty years, the
global figure has decreased rapidly from between 5 and 6 to 2.6. For most
individual countries it is even less, and it seems likely that after 2020 all
countries might have replacement rates below the stabilization level. Some
countries like Russia, with an overall replacement rate of 1.3 (it is only 1.1 in
the big cities), have a rapidly shrinking population, which is already causing
concern to their government. Other countries, like France, do not have such
a shrinking population, but for a long time their replacement rate has been
so low that they are trying to take measures to enhance fertility in order to
reach a stable population replacement rate again. Yet other countries, such as
Indonesia, are pursuing an active policy to reduce the replacement level.

A decline in replacement rate to a level below the one required for population stability seems to be in line with the predictions of the transition model.
However, the process mechanism is quite different; instead of an automatic
demographic response of birth rates relative to rates of mortality (an internal
demographic process without any obvious mechanism), the actual process
seems to be external to purely demographic factors. Much depends on contraception or sterilization and, depending on the culture in which they live,
these are to a greater or lesser degree the choice of women. This is important,
because, being largely independent of differences in the richness, population
size, demographic characteristics, or governmental or religious regime of a
country, the replacement rate can dip under the level of population stability,
following the global trend. The Philippines and Poland, both Catholic countries, have replacement rates of 3.3 and 1.2, respectively, whereas in Polish
cities the rate is as low as 1.1. Finally, large differences can still occur between
religious and nonreligious groups within countries, resulting in high reproductive rates in one group and low rates in others, but both are lowering all
the same. There can therefore be political reasons for maintaining or stimulating high fertility rates among certain groups in a population. However,
because of the unstoppable communication between hundreds of millions
of people, it seems likely that the trend toward lower reproductive rates will
continue across most of the world.

If my reasoning is correct, we are currently witnessing a centuries-long
trend that started in Europe and has spread relatively recently to other parts
of the world. It is a trend away from societies with much personal dominance
and social control toward societies with an impersonal, nondiscriminatory
and legally based sense of freedom. It is a trend toward the acceptance of the
individuality of people of different genders, natures, or backgrounds. That is
the positive side of being a number. In previous times, we knew each other

too well and were not very mobile, and so successive generations and families lived very closely—both literally and figuratively. Typically, the trend first started in towns and cities, where social control is the least, and only later and more gradually penetrated small communities in the surrounding countryside. Cities, therefore, have long been considered to be demographic black holes, not only because they have the highest mortality rates but also because they have the lowest reproductive ones, which remained so even when mortality decreased. Yet, because of urbanization and the countryside following the trend set by the towns, the difference between these two regions is diminishing. This means that because of the large-scale urbanization of the global population, a return to higher replacement rates is unlikely.

Given this trend, our population may be on target to reach a sufficiently low level relative to our rates of resource use, pollution, and wasting of the environment. However, some extra measures must be taken to further reduce our numbers, our demands, and the size and complexity of our society. We certainly need to change our wasteful system of resource use based on maximizing profits. And we need to develop new, carbon-neutral energy sources so as to be able to pay for the energy-hungry recycling of material, and we need to minimize global transport.

The revolution toward an abstract society means that our society is becoming prone to change: change coming from outside and change arising from internal tensions, disturbances and gradual change. Hardly anything needs to happen to make the system break down and collapse, and if one part of society goes down, it will pull down the other parts too. Our impressive global society is prone to collapse precisely because it is abstract in all respects.

Yet, not all the effects of the abstract world of great numbers of people living together are negative; as said in relation to women's rights, there are positive consequences, particularly that of increased and legislatively underpinned freedom. This has resulted in a substantial and demographically significant reduction in the global population replacement rate. Our abstractions turn real.

22 THE ENERGY AND INFORMATION CONTENT OF SOCIETY

We've all seen teenage boys, walking like olden-day sailors and fishermen: slowly and wide-legged, shouting, and talking loudly to each other. Just like chimps, although those swing through the jungle, roaring, shaking the trees, breaking their branches, or tearing them off as they go. Or like the rooster next door, chasing a hen, rushing at her head down and feathers ruffled, then stopping all of a sudden, looking in amazement at all the flutter around, wondering what had come over him. Then peace returns—as it does after the display by the boys and the chimps.

Whence all this showing off, all this energy being expended? Yes, whence all this energy, and where does it all go? Where does it lead to? It doesn't build up anything. Actually, what is this "energy"? You can't see or touch it, but it's clearly there in this behavior of showing off and also in all your daily occupations: sleeping, making the bed, going shopping, taking a walk, talking, going in to work. The buzzing machines too: all use up energy.

But what about setting up an organization, then operating and maintaining it? Does that cost energy? If so, how much? Can we measure that energy? Well, it takes many hours of hard work to restore organization destroyed during a thunderstorm or an earthquake or a war. It costs physical energy to rebuild the infrastructures of houses, power stations, bridges and cranes, roads, and telephone lines. Once the organization has been restored, does it then represent energy? Is that the energy spent on building? Does the act of organizing also cost energy—mental energy, perhaps? And what is that exactly? Is that mental energy different from the physical energy in machines or in the electricity network? All human society—the township, our global society, its social infrastructure—do they represent that kind of energy—

energy perhaps expended even a very long time ago—by the Babylonians or Romans formulating their legal and administrative systems? And how much energy is involved? How much energy does it cost to maintain organization? Does creating complex organization cost more energy than creating simple organization and, if so, is the difference proportionate to the size of the organization? Destroying a building or an organization does not release any energy; is that because they represent already spent energy only?

In this book, so far we have concentrated on the material side of life: our resources and waste. Possible energy shortages were referred to; we talked about carbon dioxide, oil, and gas, and about nutrients and our food. But we neglected problems of the nature of energy, its possible key role in running and understanding our society. And what of that other nonmaterial concept we use all the time: information? Where does that concept come from? Are these two concepts connected? What is the relationship of either of them to that of organization? What is organization? Should the complexity of organization be measured by the amount of information associated with it? Or by the energy content, or the information content of a certain complexity of organization? Would this concept help, not only to measure organization but also to understand how its complexity has built up and how vulnerable it is to collapse? Organizations—societies—can be built up, grow, or break down; systems can collapse. Does this possible system collapse perhaps connect these abstract concepts with the concrete and urgent problems we are talking about? A thousand questions. But let's be realistic: do we really have to go through all these problems? Isn't that a bit far-fetched for the purpose of this book? Is it practical to think about this? I think it is.

Because it is abstract, the concept of energy was the last of the fundamental phenomena that physics came to grips with, after which physicists turned their attention to atomic theory; the origin, extent, and nature of the universe; the burning of the sun; piezoelectric lighters; electronics; television; communication; and your Blackberry. Energy proved a really hard nut to crack—not in terms of talking about it or defining it, but in terms of making it measurable and manageable.

The study of energy and the mutual transformation of its various forms constitute a distinct field in physics. Energy is subject to the general laws of physics—laws that apply to everything we have been thinking about so far. That is because energy is basic to anything we do, to anything that is happening. And biologically, it's on energy we feed, nutrients only carry energy. Basically, each of us is an organized flow of energy toward its own degradation. The greater our numbers grow, the more energy is needed to sustain us; the greater our demands and productivity, the more energy we need. And the same holds for the infrastructures supporting our sustenance, though they grow exponentially relative to our numbers and demands. Ultimately, virtually all this energy, in whatever form we use it, comes at present from the fossil fuels we dig or pump up from the deep geological strata in the crust of Earth. Over the millennia, our numbers, infrastructures, and national and global civilizations have grown far more than the numbers of other species of animals and plants because we have found ways to generate extra energy on top of that derived from our food, and to organize this energy. We have used some of this energy to find and generate other, even more and richer sources of energy. Thus, our numbers have kept on growing, as has the complexity of our society.

Nothing in the world happens without the transfer of energy. Everything that happens—in the world, in your body (no matter how idle you are)—costs energy. Even thinking costs energy. The way we built our society and structured its dynamic, and the way we maintain it, requires much thinking, phoning, driving, and building and, therefore, much energy. Our society, as it is and works, expresses a staggering amount of energy accumulated all over the world during the long history of humanity. And its operation—its running and maintenance—also uses much energy. Everything has its basis in energy, its flow driving the world and shaping the form of our molecules, their interactions, our body, our evolution from the earliest stages onward, all being shaped by the one-way flow of degrading energy. The flow of energy shaped all life. It's mindboggling.

Two things appear to be relevant for us. First, like matter, energy cannot be destroyed. Therefore, the amount of material and energy used as resource inevitably turns into an equal amount of waste: waste material and waste energy. Second, energy is packaged as minuscule kernels of energy: particles. These particles are not equal but differ in their energy content—in their size, if you will. When we use energy, whether in the kitchen or by scraping off the tar sands in Alberta, Canada, the larger particles split up, so to speak, into smaller ones—up to as many as thirty-two—in a process called degrading of energy. Energy degradation always goes from large to small; its direction cannot be reversed. As in the biological and cultural worlds, the large, energy-rich particles are thus lost, dissipating as many small ones into the environment as heat. Then they leave Earth, dissipating into outer space. Gone forever.

Like any activity, recycling costs energy. In general, waste is the material that would cost too much energy to recycle. This energy either has to be extracted from waste material or we have to add extra energy to make the materials used stable. Some materials we make are so stable (plastics, polythene, or polyethylene, for example) that their recycling requires much energy, unless we transform them into a different kind of plastic. Usually, there aren't any biologically evolved enzymes that can degrade such materials for us in an energetically efficient way. Worse, mining our last remaining resources will also cost increasingly more energy precisely at the time that our energy sources are dwindling. Recycling waste into new material resources helps, but waste often requires more energy to recycle than it would cost to mine new raw materials. Hardly surprising, then, that when the financial crisis of 2007 struck, the recycling plants were among the first to slow down, while some even folded up. And although recycling plastic is cheap in terms of energy, it stopped as soon as money ran short.

Recycling the metals of cars, fridges, and washing machines, or the gold in computers and cell phones—and even their remanufacturing—costs more energy than their manufacture. Discarded appliances and devices have to be transported to countries like China, where labor is still cheap, to be dismantled for their metals, and then often to other parts of the world to be melted down and purified. Each step—collecting metal waste, transporting and dismantling it, melting down and separating it—costs energy. Plastic parts are still not recycled; they are burnt. Only when the resources become few will recycling be attractive.

We can recycle much of the material we use, but we cannot recycle energy, because then we would go the wrong direction—backward from smaller to larger energy particles, which is impossible. The energy we use by living, typing, or recycling always follows this one-way flow of degradation. Seen from a great distance—from, say, the moon—the degraded energy escaping into outer space gives Earth a warm glow of a special red light invisible to us: infrared light, heat radiation. As it escapes into space, this lost energy has to be replenished continually, because otherwise Earth would soon cool down to the temperature of the universe, and that's not what happens. This replenishment of our energy comes partly from Earth's core and partly from the sun. Together, these two sources keep the continents, oceans, and the atmosphere a mere 300°C above absolute zero, the great chill of outer space. Apart from this, plants catch a little of the sun's light energy and transform it into chemical energy, which they store in starch and oils and use to grow their bodies and fruits. When we eat, we use up the energy from this energized plant material by degrading it into waste energy: heat. As our numbers and demands, our technology and organizations have grown over the millennia, we came to use other energy sources as well, apart from the energy directly obtained from plants, but these new ones will run out soon. The heat thus released escapes back into space, creating the red glow. We cannot see this heat, yet it is the essence of our existence and of life, the essence of the origin and maintenance of the complex organization of our society, a complexity itself measured by that other, perhaps even more intangible thing: information.

Society is not an unstructured heap of humans, it is organized. Its origin, renewal, and extension have cost energy, and energy is needed for its dynamic, operation, and maintenance. This is why many of our routine jobs are being taken over by computers: they have to be done, but because of their complexity and magnitude we can no longer do them all by ourselves. Actually, the more efficiently society operates, the more energy it costs. Extra energy. For their starving yet growing populations, first Western Europe and then North America expanded their hinterlands to produce and buy food, minerals, and energy. But their populations kept growing and needed ever-more energy and ever-more material for their organization and maintenance. Seeking to meet these demands, Western Europe and North America developed their resource areas, whose populations—and demands—then also began growing concomitantly. Using ever-more resources of material and energy. Exhausting and depleting them. Generating ever-more waste and pollution. And thereby speeding up the degradation of energy.

The waste energy escapes as heat into space, but the material waste heaps up here on Earth. The process has no end—it is a downward spiral revolving faster and faster. We need this society to keep all our numbers going, but this society has its own, independent rules, its own dynamic, its own requirements and trash. To operate, it has to grow in magnitude and complexity; to operate, it uses up increasing amounts of material and energy itself.

If we can no longer build and run our machines producing and transporting food, materials, energy and information, and if we can no longer run our computers for our organization, we will soon run out of the basics for our existence: food, clothing, and shelter. As individuals and as a society, we need an uninterrupted yet rapidly increasing flow of resources, a continually increasing amount of energy. A continuous supply of energy is vital for any form of life, and this supply increases disproportionally as a society grows in complexity.

As individuals, we represent a flow of energy, and so does the society we form. Now, the upcoming societies, first India and China, and later South America and Africa, are increasingly demanding their justified share of energy. They need extra resources to build up their infrastructure of factories, roads, transport, and socioeconomic conditions, to fight erosion, land degradation, and pollution. That is, they need to build up their own national or continental superstructures, each representing part of the total energy flow.

Yet, the amazing result of this buildup of order in the midst of chaos is the perpetual drive toward an overall loss of order and organization, an increase in disorganization of matter and energy in the universe. All those countless tiny packages of energy are falling apart, degrading to heat. In the distant future, the entire universe will die its heat death, everything falling apart into degraded energy. Life—human life too—is but a short phase in this process, a local and temporary speeding up of this physical decay of order into disorder. But also a unique phase of growing insight into all we are doing, in everything happening around us in society: insight into the world from the smallest atom to the entire universe, an insight into life itself, its origin, organization, and buildup. We risk losing this unparalleled insight by not taking the right measures, by not channeling and directing the energy flows, we risk losing human civilization.

In all sorts of ways, therefore, we as organisms and our global society as an organization represent energy, a fast-running flow of energy. Therefore, understanding energy and its flow is pivotal to understanding the way we waste our world. It's essential to understand that we can't stop this flow, and that we can't stop the flow of material that carries this energy either. We can only reduce it by lowering our demands and our numbers.

How abstract life is. Life is but a stream, a stream of degrading energy. A stream keeping the wheel of the recycling of matter turning in the chemistry of life, and making it grow, thereby adding information. It has been flowing for almost four billion years and has resulted in the incomprehensibly high complexity of organisms, of humans, and of the global society we have built as our survival tool.

At what point does that other abstract concept, information, come in? It comes in when we say that a society as a complex organization expresses a certain amount of energy. The more complex the society, the more information it contains, and the more energy it costs to build that society up and maintain it. Energy and information are not the same, but to produce information, we need energy in one way or another. Information is a measure of complexity.

Like the concept of energy, information as a concept appears, actually, to have been a much harder nut to crack. It was defined even more recently than energy; as a concept with which to work it dates from the Second World

War. It was formulated in the 1940s by the American mathematician Norbert Wiener, who was a central figure in the development of computers and communication technology. And at that time, some physicists applied it also to describe the complexity of the genetic system, the complexity of plants and animals, and the complexity of society.

Think of a box of building blocks all jumbled together: wooden cubes, bricks, cylinders, and cones mixed up with wooden train cars and a little locomotive, all in separate pieces. A child takes out the cars and the locomotive and connects them in the right order into a train, pushing it backward and forward. The initial, scattered occurrence of the parts of the train in the box of toys represents a state of disorder—their random occurrence between the building blocks. The first step of separating them from the building blocks, creates some order: their occurrence on the floor separate from the blocks is no longer random. Putting the cars and the locomotive in the right sequence of a little train and hooking them together are the final two steps of ordering. The order is now fixed. All this ordering costs energy—mental and physical. Then, more energy is added, to move the little train along, and this in a certain direction. When the game is over, the child throws the cars back into the box, where they uncouple and become scattered again.

All these successive steps add order where it was absent before, and each step costs a certain amount of energy. Therefore, to generate order out of disorder, we need energy. And the child, therefore, added information, energy not in some way stored in the train but released in the working of the child's brain and muscles. Similarly, the train stores information—the energy has been spent—and what is left is a degree of organization measured by the amount of information. However, none of this energy is released by breaking down the order again at the end of the day; it even costs energy to unhook the cars, a process that is the same when dismantling organization in society. No energy or information is left because they weren't there. The energy has been spent in building up some complexity, and information is a measure of this complexity; it is not something physical, although we talk about it and even measure it as if it is. And it was such a hard nut to crack not because of the difficulty of defining information as a measure of organization, but because of the difficulty of defining it as a measure which people could use in calculations. And the results of these calculations were used in trading in information.

In this respect, terms like order, disorder, organization, and disorganization point to the definition of information. As a measure of complexity, it is, more generally, a measure of the level of organization, a measure of the degree of deviating from chaos, randomness, the expected: the degree of deviation from disorder. It is the opposite of entropy, a measure of disorder, disorganization, randomness. Disorganization is what you know from your television screen when something goes wrong: the snow. Snow or white noise results from uncoordinated electric pulses randomly coming down to us from interstellar space. They are completely unrelated to anything orderly here on Earth, like the ordered string of pulses you generate by talking on the phone. The more your earthly talk deviates from the heavenly white noise, from something completely disorganized, the more information it contains.

Entropy is a negative concept. It has to do with the degree of disorganization, whereas its opposite, information, is positive, dealing with the degree of organization of molecules, brains, or people. And a society builds up by making unorganized individuals depend on and adjust to each other, whereas before they were independent. Still, even though they are each other's opposites, both concepts depend on each other; information as a measure of organization takes the degree of disorganization as its bottom line, and reality is always a mixture of organization and disorganization. From this perspective, evolutionary adaptation in biology is the organization of matter away from the initial randomness of physical or chemical processes.

Creating order, therefore, costs energy, and it generates information. Building up an organism or a society is the same as generating order from something that was random before, and that costs energy. But the dissolution or collapse of that organism or society does not release any energy as a chemical bond may do. Now we know the relationship between these three abstract concepts—energy, information, and entropy—of which the latter two are the most abstract. Still, the ability to understand societies, their origin, growth, and stability critically depends on understanding their abstract nature.

As the child plays and the train is moved around, some cars occasionally uncouple and have to be put right. During any process, something always goes wrong. The maintenance of order also costs energy: for the warning of breakdown and for the actual repair. Special mechanisms exist for warning signals, as well as for repairing, and they all have to be designed and made, which also costs energy and material. Such warning mechanisms are well-known from biological organisms, from your own home, and from society. In complex machines or petrochemical plants, both steps are automated in the form of negative feedback systems that steer deviations back to the norm. The further they deviate, the stronger they are steered back. Many machines nowadays have such systems, as does your body. Society has them in the form of legislation; laws do not define norms; rather, they define limits to deviating in social or economic behavior relative to some agreed-upon norm. Without those limits, destabilizing criminal action and positive feedback processes would have free rein.

Those mechanisms themselves need to be checked and double-checked continually, and the larger and the more complex the organization, the more checking and double-checking are required. The checking entails coordinating the working of system components, and therefore the number of checks grows faster than the number of those interacting components. Mutual checking of two components requires only two checks, but mutually checking three components requires six. (This number increases practically exponentially as $n[n - 1]$ instead of as $n \times n = n^2$, with "n" representing the number of components.) Moreover, at a certain level of complexity, a new layer of checks soon has to be inserted to coordinate the coordination. The result is a hierarchical system of coordination in any organization, whether biological or societal, hence, the ruling and service components (or "sectors") separating out in the first societies from those of food production and manufacturing, and, as society grew to several billions of people, forming huge superstructures. And hence the increasingly larger amount of energy we need for building up an organization and keeping it in hand and, hence, for "adding information".

As mentioned before, as the number of people in Europe grew, it became physically impossible for individuals like kings or prelates to keep order, especially when a new societal organization had to be built up to suit the growing numbers of people. At that time, it seemed to people that these new organizations grew by themselves and were self-regulating, so, *laissez faire* became the new dictum. However, more people also meant the production of more food, and a large and more complex society needed more inputs—at first, more human labor, later coal, then increasing amounts of oil and now, more and more gas and more minerals as well. As the complexity generated by the societal superstructure is growing even faster than our numbers, our energy and material expenditure are growing accordingly.

As a measure of organization, information expresses the total amount of coordinating interrelationships between individual people. Money is one such interrelationship, expressing the amount of labor agreed to change hands between people and quantified in arbitrary yet mutually agreed upon amounts of shells, coins, paper dollars, or gold. It's a promise, a promise in gold. But during a financial depression or crash it can lose its value overnight. Social relationships are built up and maintained by energy, but as bits of information, they can vanish in a split second without leaving a trace—not releasing even a single particle of energy. We just forget our promises. Societies appear to be concrete, but they are not. They represent a certain amount of information. They are nonconcrete, abstract, and are prone to disintegrate into nothing. Like your money during a crash. Money was a form of promise, and the promise is forgotten. It had no substance. Similarly, societies can collapse for all manner of reasons—a shortage of energy, famine brought on by drought, a hurricane, war, or anything disturbing the usual pattern of relationships between people. And the greater the complexity of an organization, the more labile it is. All our relationships break down, leaving real individual people behind, isolated from each other, uncoordinated. Large, powerful organizations can thus cease to exist almost overnight, as happened to the leading American air company PanAm a long time ago, or more recently, Enron or Lehman Brothers. Governments and civilizations break down, along with their legal and health care systems. The descendents of the Incas in South America and the Maya and Aztecs in Central America live on, but their civilizations have gone. Being abstract, all those organizations, large and small, have been forgotten, have vanished from the space between our ears.

Yet, systems also live a life of their own, beyond the command of individual people. They cannot be derived directly from the behavior of individuals, nor from the relationships between individuals. And the more complex and elaborate their relationships, the more coherent and independent the system's behavior—behavior that is rooted in abstractions, relationships, and agreements between people, trust in promises made, whether they are conscious of them or not, yet nevertheless affecting the concrete life of all people.

One of the properties of our societal system is that it keeps growing, which automatically increases its complexity and, in turn, makes it more prone to collapse. The mechanisms that make its dynamics coherent are fine-tuned, which makes that if one process gets out of step, it derails all others. There are, therefore, strict conditions set to systems to make them work.

There are two basic conditions for the perpetuation of a system: first, it occurs in a stable external environment—and second, it is internally controlled by an elaborate mechanism of checks and balances—internal negative feedback loops. The first condition makes that the internal processes dominate and operate independently.

Often, the first condition to be met for a long-term persistence of an organization, the one to do with a stable external environment, cannot be met either. The exhaustion of fossil fuels, minerals, and soil nutrients are examples. A corollary of the overuse of fossil fuels is climate change, itself containing positive feedback processes, the self-accelerating processes mentioned. Climate change will cause local, regional, and global temperatures, precipitation patterns, and hydrology to change unpredictably: higher temperatures lead to more water loss from plants and animals, and thus to shortages of freshwater. Soil fertility and pollution affect us through food availability, though so far the soil nutrients can still be supplemented by minerals extracted from fossil fuels, but this can change when these run out. Exhaustion of resources will not happen all at once and to the world population as a whole, but regional differences in purchasing power will lead to local food shortages and these may grow into famine, resulting in mass migration, epidemics, and civil war. Thus, internal destabilizations of the system can spread as a result of a destabilization of external factors, leading to regional or global collapse of society.

The second condition of internal control by negative feedback loops is violated in, for example, stock markets. Those markets can be relatively quiet for some time, with share prices fluctuating only moderately, but occasionally a new firm or a technological innovation attracts attention. Some years ago this happened with the dot-com bubble, where excessive speculation in the Internet sector resulted in a rush to buy shares, which led to rocketing prices. As the higher prices attracted more buyers, prices were driven up even more. However, this boom was soon followed by a sharp drop in prices—a bust—in which traders tried to sell their shares as quickly as possible. The bubble burst and many people lost massive amounts of money.

Another dramatic—and classic—economic bubble was the tulip mania in seventeenth-century Holland. At its height, a single bulb could be worth a prestigious canal-side house in Amsterdam. Many of today's investors are speculating in certain fine French wines, which are so expensive that nobody can afford to drink them anymore! They are used as a trusted investment even though they may now be no more than dusty old bottles of pure acid—nobody knows. If they are discovered to be acid, even as soon as they are uncorked, then the value of the shares or the investment no longer covers their value, and no profits can be made from them. Investors will turn to other objects or to companies, and the overvaluation can be "corrected" at the level of society by a general economic slump, a crash, or a crisis. Crashes are inherent to positive feedback loops.

Occasionally, exceedingly severe crashes occur, such as the international economic crash of 1929 followed by the Great Depression of the 1930s, when the economies of countries all over the world were affected. Such a crash can be considered a collapse of the system (initially a collapse of the financial system). Individuals or organizations play by the rules as long as they trust that their investment in shares is covered by the value of the houses or firms of which they have shares. But as soon as the trust breaks down, the rules are broken, as well. As rules are abstract, they are not applied anymore and the system implodes—collapses.

Within or even between national societies, the same process can occur because of the differential earning of groups of people, organizations, or of nations themselves. For example, the income distribution of people within a national society can become skewed, because compared with the rich, the poor have little or nothing to save or invest. The rich earn extra income from the interest on their savings or investments, which they can again save or invest to improve their income even more. If this process continues for some time, the income distribution becomes so skewed that the rich can subject the poor or, occasionally, enslave them. The same process happens between firms or banks at the national and international levels. In the case of individual people, tax systems that subject the rich to a progressively higher rate can be used to redress large imbalances, and antitrust laws can be put in place to protect smaller firms and banks against the increasingly larger competitive power of their larger competitors. In this way, a society protects itself from the ultimate positive feedback destabilization through revolts or revolutions. What happens is that without tax legislation the flow of money follows an exponential system, whereas the legislation introduces cyclicity into this flow in order to prevent financial exhaustion of this system.

This process of unequal growth rates also happens on a global scale between countries, but as there is no international legislation on this, global destabilization may occur in future. All these destabilization processes depend on the same positive feedback system of differential earning. They are a potential threat to the stability of the system.

Positive feedback loops destabilize the system rather than controlling it as a negative feedback loop does. Negative feedback loops are essential to any orderly functioning of an organization, whereas positive ones destroy the order; they inevitably lead to disorganization, making the system crash or collapse. Unfortunately, one or more positive feedback loops can destabilize a system even though it is dominated by many negative feedback loops. The danger is that if the differences in income between countries are too big, the effect will be the same as that of positive feedback loops; they can destabilize the world order. The wave of governmental deregulation taking away negative feedback control and the spread of globalization without an accompanying legal system thus make the world system vulnerable to collapse.

The growth of firms and nations into ever-larger corporations and con-federations therefore has its risks. When one unit goes down, all go down since their interrelationships have become too tight, requiring a hierarchi-cal network of stabilizing cross-connections, each with its own stabilization mechanisms. One way of counteracting this potential effect of overcomplex dynamic systems is to split them up into several, more or less independent subsystems. Then, any unfortunate development remains restricted to a sin-gle subsystem, so that it soon peters out. Typically, national economies that had stayed relatively independent of all others remained unaffected by the global financial crash of 2007. Possibly for this reason, biological cells too are compartmentalized at various levels into subcellular units, whereas they are themselves united into organs and organ systems.

The larger and more elaborate and complex the organization becomes, the less efficiently energy is expended. We see this happening in large com-panies and governmental organizations. They smother themselves in an in-creasing number of rules necessary to keep the organization together and in tune. This is why big organizations and governments often compartmental-ize internally or split up completely, decentralizing until the growing need for overall control centralizes them again.

Another way of dampening the fluctuation pattern due to positive feed-back is to add another positive feedback process to the whole system to undo its effect, similar to contrast amplification in acoustics. For example, one of the measures against a crisis could be for the government to cut taxes to relieve employers from further costs, but this enhances the pattern, as no work is given to the unemployed. They have to be paid, but don't produce, and reduce consumption. This destabilizes the social system on top of the economic destabilization. The system can be stabilized by the government giving the unemployed alternative work temporarily—the deeper the crisis, the more work. Understandably, this measure is unpopular with employ-ers, because they are already in financial difficulty and are now having to pay extra tax as well. However, in the longer term they also benefit, as the crisis does not deepen any further and future crashes, closure of firms, and social unrest can be avoided. Their system based on economic freedom is left intact; it is merely made to work more smoothly. Such a solution can be implemented at a national scale between workers and their firms and also at international scale, between rich and poor countries, as suggested at the United Nations Monetary and Financial Conference at Bretton Woods in 1944. However, since then, deregulation measures have undone the Bretton Woods system of monetary management, freeing the way for globalization of positive financial feedback. This has led directly to poor countries paying in kind by allowing foreign firms to mine resources, use agricultural land for growing food, or deforest to obtain timber. This, in turn, often exacerbates social unrest and violence, thereby destabilizing the international order by social positive feedback, as well as resulting in resources being overused because of their reduced prices.

A way of avoiding increasing financial swings—and thereby avoiding the consequent overuse of the environment, which will be necessary under more stressful conditions in future—is to deprivatize many firms. Instead of earning from the exploitation of the earthly resources and of negating the cost of pollution, it may be necessary to invest in more expensive ways of processing, to which all people are contributing. Rather than overusing at minimal cost, we need to pay for eking out the remaining richness of Earth.

The least-stable societies will be those where several such unfavorable conditions coincide (for example, tropical countries, like those in Africa, with the highest risk of food shortage due to climate change, poor soil conditions, high rates of reproduction, and a lack of sufficient independent national income). Moreover, poor countries can ill afford to buy food, energy, fertilizers, or materials on the world market, and therefore have hardly any chance of building up a stabilizing national infrastructure. Instead, positive feedback loops will dominate; for instance, when richer countries begin to fight for their natural resources such as minerals or agricultural land. Epidemics and famines will hit first, and here too mass migrations and violent civil wars will begin and spread.

Thus, many external, concrete processes add to the risk of abstract, internal processes in societies dissolving into nothing, that is, to the risk that societies collapse. Then, without the support of an organizing superstructure, people will have to fend for themselves under conditions of severe deprivation. This means that many will succumb and die.

23   CAN OUR WORLD
POPULATION COLLAPSE?

Ring a ring o' roses,
A pocketful of posies.
A-tishoo! A-tishoo!
We all fall down.

For this happy English nursery rhyme, children hold hands to form a circle, and then dance around, singing. Nice for a birthday party. At the end, they all fall down, laughing. However, many people believe this happy, innocent little song easily remembered by young children refers to the dreaded plague that killed hundreds of thousands all over Europe; at times, two-thirds of a community would perish. The "A-tishoo! A-tishoo!" may refer to the sneezing during the pneumonic phase of the disease that can develop after the initial, bubonic phase, known for its feared red spots and boils. The first phase alone led to tens—even hundreds of thousands suffering an awful death. The frightening, painful deaths of the plague victims in the Middle Ages and in subsequent epidemics (notably the one in London in 1665) soon disappeared from the collective memory.

The worldwide wave of concern caused by the book *The Limits to Growth*, a mere forty years ago, wherein the Club of Rome warned that our earthly resources are limited, seems to have suffered the same fate. It was soon forgotten. But if that concern was justified, by pushing it out of our mind, haven't we lost much valuable time that could have been used to tackle the problem? The Club of Rome's warning in 1972 did not have the immediate consequences of the plague; its consequences are longer term and will be felt by future generations, but they will have a much larger impact: the suffering and death of hundreds of millions of people. Of billions perhaps. What have we been doing since 1972? How can we have forgotten? Instead of reducing our numbers and resource use, we have stimulated them deliberately. Since the 1970s, our reproductive rates have reached unprecedented heights, and per capita consumption has multiplied, particularly in the West. During the last ten thousand years, our numbers, demands, and reproductive rates have never been so high. Who worries?

We are now entering the second wave of concern about resource limitations, one to which concerns about energy supply and climate warming have been added. Other concerns haven't penetrated the media that deeply yet, and discussing the increase in our numbers seems to be a taboo subject that has met with resistance. Meanwhile, the tone of the debate is more positive and optimistic than it was during the 1970s. For example, worldwide, ways of reducing the amount of carbon dioxide expelled into the air to fight climate warming are widely being discussed: how this can be done by more economical energy use, by replanting forests, or by burying the carbon dioxide. And we seem satisfied by the forecasts that our numbers will stabilize around 2050, not realizing that our resource use and its consequent waste production are not connected into a perfect cycle, but are linear. Stabilizing the world's population while maintaining resource use at such an incredibly high level cannot but lead to rapid exhaustion and overpollution. We are getting closer and closer to those limits, and during the last thirty years the network of interactions has tightened into a fyke. Our higher numbers and demands give us less time to maneuver away from disaster.

In *The Limits to Growth*, Dennis Meadows and others concluded from one calculation that the number of humans could crash suddenly rather than stabilize gradually. But none of the other calculations showed this effect; their results suggested that the numbers of humans on Earth had to be reduced gradually, and with them, the overuse of natural resources. It seemed that this single result was anomalous and could be ignored, although its cause remained unclear.

Twenty years later, however, in their 1992 follow-up book *Beyond the Limits*, on the basis of calculations using data from the intermediate years, the authors reported that such crashes were no longer exceptional but had become the rule. Results without a population crash had become exceptional; crashes appeared to be normal and seemed not easily avoidable. This was a very different story. Without knowing the underlying causes, population crashes were now being attributed to delays in the fine tuning of interactions within the system and to the exceeding of limits of irreversible degradation.

But why a crash instead of a slower slide? What had changed in those twenty years? It remains difficult to understand. There are some known processes that produce such results and that can operate in concert, enhancing their effects.

Here's a riddle: a pondweed like duckweed is assumed to grow in a pond, doubling its surface cover in a single day. As the surface area of the pond is finite, the duckweed can cover it completely in thirty days. So, how much of the pond has the duckweed covered by, say, the twenty-seventh day? The answer is astonishingly low: 12.5 percent. Hardly anything. Yet it's true. Most people find it difficult to do the calculation, although the difficulty lies not in the calculation itself, but rather, in three ways of thinking that are unfamiliar. First, it is often difficult to calculate backward. It is easy to calculate the plant cover on day one, then on day two, and so on. That's how we learned to solve such problems at school. In this case, however, you start by calculating the plant cover on day twenty-nine, which is 50 percent of the total cover on day thirty, then use that figure to calculate the cover on day twenty-eight, which is half of 50 percent, or 25 percent, and then do this again for day twenty-seven, which gives the 12.5 percent. Not difficult at all—just unfamiliar.

Second, growth is usually assumed to continue indefinitely, whereas in this case there is a definite end: the 100 percent cover of the pond. The growth process, therefore, is limited, just like our resources, the amount of forest, metals, and energy are limited, or like the amount of waste we can dump into the environment. By contrast, there is no maximum temperature for Earth's atmosphere. With enough carbon dioxide in the atmosphere, here on Earth it could become as hot as on Venus, where surface temperatures are 450°C. But like us, plants can tolerate temperatures only up to some point, beyond which they wilt and die. Neither plants nor humans could live on Venus. The same holds for the amount of waste the environment can tolerate before it becomes irretrievably polluted for plants or animals; we all know that there is some such threshold, but we don't know how far pollution can proceed—we don't know its limits. In any case, the limits vary depending on the pollutant. Usually, we have no idea about the limited availability of our resources either. In principle, we can squeeze out more of a metal from its ore or from an alloy, but the amount of energy needed to do so ultimately limits us: it becomes increasingly more expensive in terms of energy and, hence, of money. Finally, our world population can in principle continue growing almost indefinitely as well—up to 10^{16}–10^{18}, according to one physicist. This criterion shows us the limits and maxima, after which we can make the backward calculations from there, like from the 100 percent in the riddle.

Finally, we are not used to thinking in terms of doubling or tripling times. That is, most of us are not used to thinking in terms of exponential growth or reduction. We can add and subtract, and we can multiply once but not several times by the same number, which is what you do when calculating exponential growth. Moreover, from personal experience, we know that a family grows by adding children: one, two, or three. But when calculated over a population, later these children have families of their own and so the process changes from an additive to a multiplicative process, resulting in exponential growth. An average of three children per family produces a total of nine children in the next generation, then twenty seven in the third. This change in thinking from a concrete additive process to the abstract one of exponential growth can be difficult. But thinking in terms of exponential growth is essential in order to be able to understand many of the processes of resource use, the growth of industries, the spread of diseases, and the complexification of society.

To get a better feel of the term "exponential," think of a bacterium dividing into two every twenty minutes. This period is called doubling time. After the first twenty minutes, the bacterium has divided into two bacteria, which become four after forty minutes, and eight bacteria after sixty minutes. With each passing twenty minutes, these numbers grow faster and faster, although the rate of growth, the doubling time, remains the same. Therefore, by adding minutes, you are multiplying the number of bacteria. And each time, you multiply the previous number by the same number, two: 1; $2 \times 1 = 2$; $2 \times 2 = 4$; $2 \times 4 = 8$, and so on, which you can also write as: 1; $2 \times 1 = 2$; $2 \times (2 \times 1) = 4$; $2 \times [2 \times (2 \times 1)] = 8$. Therefore, in the second generation, you count a single two in the equation; in the third, two twos; in the fourth, three twos. This can also be written as superscript numbers: 2^1, 2^2, 2^3, and so forth. These superscript numbers are called exponents, the number of twos in the multiplication series, the twos themselves are the doubling times. In principle, we could also use tripling or quadrupling times, using threes or fours, but the custom is to use twos. Accordingly, the growth of your population of bacteria is called exponential, as is the growth of a firm, or indeed that of the human population.

However, in those bacteria you knew the doubling time, twenty minutes, but in the human population, you don't. In fact, that's exactly what you want to know. You do know, however, the average number of offspring per time unit: the population growth rate. Now, with a trick, you can roughly estimate our doubling time by dividing the empirical number 72 by this known growth rate. For example, during the 1960s, the average annual growth rate of the world population was almost 2.2 percent, making the doubling time $72 \div 2.2 \sim 33$ years. In reality, because those rates declined somewhat over the years, this period proved to be slightly longer. In Africa and India, the growth rates are higher than average, 3.0 percent or more, giving doubling times of 24 years or less, which also holds for individual countries in Africa like Nigeria or Rwanda.

Given a population doubling time of about forty years, in each doubling period we need to expand the villages, towns, and megacities by a factor of two and add at least twice as many roads, hospitals, and schools, or appoint at least twice as many judges. In fact, we need more than twice as much, because the exponent of their growth rate relative to the growing number of humans is greater than 1. This applies to most aspects of modern society: they grow faster than the number of people does; they grow exponentially relative to our numbers. That is because we have to organize ourselves as a society, that is we have to interact. And as we have seen, those interactions grow exponentially relative to the growth of our numbers—as do the resource use and waste production of this organization on top of those of the exponential growth of our numbers. Similarly, over the last decade, the growth rate of the Chinese economy was about 11.0 percent, which means that every $72 \div 11$, or 6.5 years, its economic potential doubled. In the same way, we can calculate the growth of firms or banks in terms of doubling times from their annual growth rate as reported in their annual reports.

Because the number of workers—humans—is itself growing exponentially, the total output itself grows exponentially on top of this. We call this double exponential growth: a growth process in which the exponent itself does not remain the same (as in the case of the growing numbers of bacteria), but itself grows exponentially. Apart from this, the output per worker can also increase exponentially because a machine (a truck, computer, or robot) has taken over part of their work.

Our average personal demands in the West have increased considerably, especially since the Second World War, and our numbers have increased as well, both growth processes being exponential. Other changes in society have also been important. Some women entered the labor force, and various mechanical and administrative tasks were taken over by increasing numbers of machines and computers. In the last sixty years, the exponentially growing populations of China and India, both rapidly industrializing and urbanizing countries, have been consuming increasing amounts per head, a growth itself having large growth exponents on top of the exponentially growing populations. As mentioned, the Chinese economy has been doubling every

6.5 years. Accordingly, their resource use and waste production are increasing at a staggering rate.

These are predictable exponential processes, but are all processes predictable? The pond in the riddle has a particular size, and we also know roughly how much is still available of each of our resources on Earth. That is predictable, although sometimes new resources can be found. Recently, for example, natural gas, tar sands, shale gas and oil, and deep drilling in ocean fringes were added to coal and oil as sources of energy, giving us a little more leeway. Can we also predict what might happen to the interactive societal network—how it will grow or when various functions will drop out? And are we certain in what order they will drop out, and their various depletion rates? Or, in the case of soil salination, do we know the saturation rate? Can we predict what might happen when a pivotal resource is depleted suddenly rather than gradually? Unfortunately, often we can't. Why not?

According to Hubbert's curve (see chapter 7), there is at first a rapid, exponential increase in oil production, up to a peak, after which production drops in the same way but by another, often larger exponent. This curve applies to the production from one single well or a single country, as well as to that of the world as a whole. What happens is that first the most profitable fields are found and exploited because of their large reserves and because of the ease of oil extraction. As they begin to run out, smaller and less productive fields have to be found and taken into production, and so on until eventually only a few, barely economically viable fields are exploited. What makes this worse is that the smaller, less productive ones cost most to detect and explore and are the most numerous. In fact, the cost of mining mineral, metal, or energy sources will increase rapidly—exponentially. Similarly, farmers have always first cultivated the few richest and most extensive soils (for example, in the river valleys), and only later do they turn to the many poor ones that remain (such as the forests and mountain slopes). Fishery is a similar story. First, the species like herring and cod, with the largest schools and the largest individuals, are exploited, and only when these run out are smaller species in smaller schools fished. Thus, although the seas and oceans still appear to contain sufficient supplies of fish, the situation is serious: the stocks are rapidly approaching complete depletion, and many rare species are on the brink of extinction because they are easily overfished. And although supplies still seem abundant, the crash has already set in, a phase in the process that can prove to be unstoppable. The fish stock of the northeastern Atlantic Ocean, for example, has already been so badly affected that it is unlikely ever to bounce back to acceptable levels. There is no way that stocks of cod, plaice, and halibut in the North Sea will be able to rebound to yieldable levels before 2015, as hoped.

It is important to be aware that when a new resource begins to be exploited, an initial rapid rise in production is followed by an accelerating decline. This curve differs from that of normal growth: it is humped rather than straight or bent upward. The straight and bent curves are called linear and exponential, respectively, whereas the humped one is called nonlinear. Nonlinear curves have at least one inflection point, the point at which a rising curve becomes a falling curve, for example. Think of plants and insects, or of your computer, not working at low temperatures, better and better at higher ones until they reach a top, their optimum working value. After this, at even higher temperatures, their performance declines again, up to the point that they give up due to overheating. In the case of limited resources, like fossil fuels, we have so far usually been dealing with nonlinear curves, whereas the curves of waste disposal are exponential (rising). The problem with limited resources is that after some time, such rising curves reach an inflection point, thereby revealing their nonlinear nature. These nonlinear curves are particularly likely to cause instabilities when they occur in a network of interactive processes. This is because small changes in their location or shape have major consequences for the fine-tuning of those processes. For example, a shifting curve may rise and then decline, whereas other, stable curves keep rising; their previous tuning breaks down and destabilizes the system.

We don't know how or when such a crash develops, in what corner the problem will arise, or how fast it will go. It may be triggered by some minor problem, the effects of which amplify, pulling other sectors of society down, thereby rapidly, exponentially worsening the growing disaster. The only thing we do know is that many trends in society point in the wrong direction, making it increasingly more prone to collapse. Several of these trends have been discussed in this book, such as our growing numbers and demands, the declining biodiversity, climate change, and the rapid depletion of sources of energy, nutrients, and water. What makes this worse is that we increasingly depend on an economy based on growth, growth considered both the cause and the cure of our problems. And we know that all of these trends are interdependent. For example, sustaining our growing numbers and demands depends on a regular and unlimited supply of energy—a supply that, nevertheless, is limited. Approaching its limits means that we need to develop ever-more efficient technology to squeeze out the last remains of the fossil fuels from Earth, which, in turn, depends on ever larger investments. However, investors want to see their money back which proves increasingly more difficult the closer we get to the point of depletion. They have to invest more and more, but as less fuel can be mined, the returns become less and less. Approaching the point of depletion, investors will gradually draw their money back in order to reinvest it into something more profitable: at this point, a positive feedback loop starts up, more and more investors withdraw their money, this loop destabilizing the societal system as a whole. Energy shortage will destabilize this system because energy is the main constituent of our body, our numbers, requirements, and infrastructural organization. In fact, the decreasing trend in the energy returns on investment was already apparent in the early 1990s, a trend which continues to the present day and which may develop into the feared financial and economic positive feedback loop. Food will be more expensive to produce, leaving the poor in jeopardy. And so on. Similar trends in other basic requirements either occur already or are imminent. Before energy shortage may tighten, they can therefore also develop in the supply of nutrients or water, the extraction, recycling, or desalination of seawater, which will require increasing amounts of energy. And this, in turn, can push us faster in the direction of energy depletion, and into that of the positive feedback loop of lesser energy returns on investment. And so forth. We know the trends, we know where they will lead and how, but we don't know which of them will trigger the others to join into the one positive loop or the other, and when.

Some may reason that shifting investments away from mining energy, metals, or other minerals into the direction of, say, urbanization and industrialization, could be counteracted by later investors or by the government. But investing money is not only something personal, done by individual investors or banks trying to get most out of their money. The financial system is also part of our societal infrastructure, an abstract and dynamic part of it, and it therefore relates to other parts of that financial infrastructure as well as to the infrastructure at large. Some sectors need extra money and therefore draw investments from elsewhere, in this case from those in the mining industry. Money has to be spent in those sectors needing it most. As the money to be spent on energy, metals or minerals increases, industrialization, urbanization, and agriculture get less and come to a standstill, which causes other problems. Decreasing returns from investment in energy therefore cannot easily be stopped without affecting other parts of society in their development. Over time, their degree of overlap or tuning decreases and the processes begin to work against each other, whereas previously they enhanced each other. All this happens in many complex systems, which makes them unstable: all sorts of relationships working with or against each other in different trajectories of their fluctuation pattern, making the system break down slowly or suddenly, after which it crashes and stops working. Therefore, the more relationships there are, the more likely it is that they are diverging, that they are out of tune, either temporarily or permanently, making the system crash. And the number of relationships depends on the complexity of society and therefore ultimately on our numbers. What is typical for collapsing systems is that the interactions between their components, intended to keep the system stable and within bounds, can suddenly turn against the system causing it to destabilize and crash.

Most of us know from school how markets work; we learned the simple, two-factor systems of supply and demand. According to Adam Smith's "invisible hand" of the late eighteenth century, prices will balance each other. In such systems, supply and demand balance each other through the price asked for a product. A greater supply leads to lower prices, which increases the demand. Greater demand, in turn, increases the price, which, again, increases the supply, which brings us back to the first step, a lowering of the prices. This results in an endless, wave-like process, undulating into eternity. The same idea can be found in ecology where two competing species or a predator and its prey would similarly balance each other; or in selection theory, where two parties would compete, although here the fittest eventually wins. And in James Lovelock's Daisyworld, according to which over geological time producers and users of carbon dioxide would cause climate to fluctuate regularly around one stable level. In fact, the occurrence of a stable level is basic to the concept of carrying capacity, where abundances are also assumed to fluctuate around a certain stable level. This would also happen to the human population after 2050, when it is presumed to remain stable at the level of nine or ten billion people. All these concepts consider short-term or local dynamics underlying a long-term or global stability, never a collapse of the system. Such systems would be in equilibrium, or could even have two equilibriums at different levels from which it is difficult to escape. Fish stocks, for example, could switch to a lower equilibrium level after having been overfished for some time. It would then be almost impossible to switch back to the previous, higher level.

Although these processes do occur, reality is usually more unruly and complex. Usually, there are more than two system components interacting, and this is a point where the real problems arise. We can derive mathematical equations for the dynamics of two-component systems, but this is theoretically impossible for systems consisting of three components or more, such as in economic ones in human society, in ecologic ones in nature, or among the few planets rotating around the sun. For example, we can calculate the simplified course of Earth circling the sun, or that of the moon circling Earth by treating them as simple, inanimate globes of enormous size and weight. Yet, this calculation is not feasible with the same exactness for those three celestial bodies together: the moon circling Earth circling the sun. We can only approximate the courses they run. Indeed, when you look carefully, the planets in our solar system do not perfectly orbit the sun as explained in school books—their courses are wiggly, making the solar system a bit of a jumble. Yet, together, the planets keep each other reasonably well in orbit. Even Newton had to admit this.

Because they are much of a jumble as well, societies can crash or collapse. Such crashes not only develop rapidly, but their cause, course, and timing are unpredictable. Mathematicians call this field of study deterministic chaos: unpredictability reigns, even when nothing happens by chance; chance within the process only gives additional unpredictability. Imagine, therefore, what happens when such systems contain an element of chance as well.

"Deterministic" means that the course of the process is not subject to chance—from its beginning its outcome is completely determined and can therefore be predicted without any uncertainty—and that in deterministic chaotic processes, the unpredictability comes from a slight chance variation in the starting values of the process, just before the deterministic process begins. Therefore, repeatedly beginning the calculations with a particular single value in these models will always give the same result, but when you take a slightly different value, the result can be very different. But now think of balls rolling down a pinball machine, for example: how differently they roll, although they come from the same point of release. Some exceedingly small differences at their point of release—differences you really can't see and are due to chance—result in those huge differences in the course followed by the balls and the place they end up. Still, once the balls have been released, everything goes with predictable precision. Obviously, this is not how society works; it is dependent on huge variation in weather patterns (both in time and in space), uncertainties in economic triggering and development, nonlinear processes and on chance variation in their timing, and so on, not only at the beginning but throughout the process. All of this makes it prone to collapse or crash. It is as if the pins randomly change position all the time, and that their flexibility keeps varying from rigid to soft. Crashes can develop from anything: legislation affecting the price of houses or an explosion in a pipeline resulting in large amounts of oil leaking into the sea or on land. Recently, the first kind of crash resulted in a financial crash of global dimensions, and the second resulted in the disasters in the Gulf of Mexico and in a river flowing into Lake Michigan. Those are all chance events happening all the time but each with far-reaching effects on society. So, how does chance work, and does chance depend on the number of people making up society and its complexity? If so, does the chance of societal collapse increase over time as our numbers and their resulting societal complexity grow? Have our living conditions changed (gradual soil salination, or a sudden rise in the price of food due to drought in Australia or Russia, for example)?

Think for a moment of a die: what is the chance of throwing, say, a five? A die has six sides, each with the same chance of turning up. The chance of throwing a five is one in six, or 17 percent. Conversely, the combined chance of throwing any number other than five is five in six, or 83 percent. But how great is the chance of getting a five in two consecutive throws? That chance is obviously twice as large, or 33 percent, and the chance of getting any other number is 67 percent. Therefore, the more throws, the greater the chance of getting your preferred five at least once. And the chance of missing it reduces accordingly. The same reasoning applies to, say, the chance of some explosion happening in an oil pipe, though in this case you are interested in the chance of the event *not* happening. Now the chance that some disaster will not happen is made as small as possible, say, one in 10,000, and the chance of an explosion occurring is only one in 9,999. Obviously, these chances also depend on the length of the pipe, on the number of pipes, on the number of welds, or the number of pumping and control stations, that is on the complexity of the pipe system, and also on the length of the period the system is operating: the longer the pipes and the more there are, the greater the complexity of the system they form and the longer the period of operation, the greater the chance of something going wrong, resulting in an explosion.

Moreover, all these mistakes and disasters have different chances of happening, and all these chances are superimposed. You can try out for yourself what happens by throwing different kinds of dice, the normal one with six sides, then one with four, eight, ten, twelve, twenty, and one with thirty sides. The result is a very wiggly line when you add the outcomes of these sets of dice for a number of throws together for each point on this line. Each new point is different from any of the previous ones and therefore is impossible to predict; it was already impossible to predict the outcome of one single dice. Still, this curve resembles the real world in many respects where also many chance events occur, the one adding to another and each with a different chance of happening. In reality, moreover, the chances have different and varying weights relative to the total process as well, and they interact both linearly and nonlinearly, which we all kept constant and independent when we threw our seven sets of dice. How can we predict the future of society but in general terms of depletion and pollution rates? These are our certainties, but we really can't predict in detail what will happen and when as a social or economic result. For these societal effects we can only say that the chance of collapse increases with an increasing complexity of society, as well as with increasing stress from resource depletion, pollution, and social inequality. As all this may be rather abstract, I explain the same chance process more down to earth in terms of social processes.

By themselves the number of welds, or the number of pumping and control stations in a pipeline do not indicate certainty; maybe nothing will happen during the period concerned, or an explosion happens twice or more often. You don't know for sure, because chance means that you don't understand the process fully and therefore that you can't predict its outcome. You don't know what made you throw a five, and you therefore can't influence it happening. That's the fun of using dice in certain games, and it's also the worry that chance gives to the manipulating of processes in society. Our interest in chance is that with the growth of our numbers and, hence, the complexity of society, the chances of matters going wrong increase. As these numbers and societal complexity increase exponentially, this chance also grows exponentially. That is, it happens independently of the exact time of us running out of resources or of climate gradually getting too hot. This means that the chance that our global society will collapse at some point increases over time, and with the growth of our numbers and the complexity of society. Moreover, similar to the dice where you don't know when exactly the five will turn up, you can't know when something will happen in society that could result in system collapse; it may be some time away, but it can also happen soon. And, in fact, you don't know where it will begin, or how fast it will go, if there is a point or period after which it will accelerate, or whether the collapse will stretch out over a year, over decades or a century, beginning in some as yet unknown corner of the world and spreading out unstoppably from there. Think of the decline of ancient Rome, which took centuries; nobody knows why it declined; we have more explanations than authors. Because of the great influence of chance in all aspects of society, whose behavior is unknowable and, hence, unpredictable—manageable only up to some point, after which further developments grow out of hand.

Why the reason for a crash such as the decline of Rome is also unknowable, and why its crash unmanageable is also that people usually look at only one process in isolation, such as the invasion of the Gothic tribes or the general poisoning of people by lead in the water pipes. In many cases, however, a disaster is triggered by the coinciding of a number of different events or processes, not by a single event or process. How great is the chance of two dice simultaneously rolling a five? In that case, we can't add up the 17 percent chances of the two dice as we did before, but we have to multiply them, which gives the much smaller chance of 2.8 percent. This tells us that when several chance processes combine, as happens when the processes are complex, and together these lead to a certain result—a crash—the chance of predicting the crash is really getting small. And because we don't usually know which events will happen and which will combine, nor what their relative weights are, precisely because they are chance processes, we can't manipulate or manage them when the crash comes. Therefore, as our numbers continue to grow exponentially, the size and complexity of society increases exponentially relative to those numbers. Consequently, the predictability of a particular crash developing from the occurrence of a certain combination of chance events or processes decreases. On the other hand, the number of combinations increases exponentially as well, which makes the system increasingly less stable: anything can happen at any time. It therefore becomes more certain that anything can happen, but it becomes less certain from which direction problems arise and when—particularly if the chances of some event or process occurring are unknown, and if, on top of all this, they differ. Such combined chance processes based on different chances coinciding follow a so-called fractal process which is well known.

In such fractal processes, several superimposed chance processes occur, each operating on a different timescale, the one taking longer than the other. This is different from the market models of supply and demand, which can balance each other because they are thought to operate at the same timescale, also responding to each other without delays. But most processes in society differ greatly in response time, a change in birth rate taking effect only after about seventy years, whereas exponential growth in energy and mineral use, or in industrial production and innovation, for example, act within decades or years. The recovery of fish stocks in the ocean require yet different time periods, and a reduction of atmospheric carbon dioxide to acceptable levels will take many centuries. Moreover, resource depletion, pollution or erosion are cumulative, and their effects cannot be undone. Democratically elected governments having to take the risk of introducing drastic, unpopular measures can be replaced after four or five years, well before their measures can take effect. As they don't like to be unpopular, they don't take any measures, however good these may be. This means that the assumption of a self-adjustment of society under changing conditions through internal fine tuning is not valid. Society has never been in equilibrium because it has been able to push up any limitation set to our living conditions, and it won't reach such an equilibrium when we can't push up those limitations any further. Because of differences in response times, chaos is the first thing to develop, followed by collapse.

Moreover, because many factors can be interdependent, a crash in one sector pulls others in its wake, making it a general crash in no time and also making it more difficult to manipulate or manage. Crashes of our socioeconomic system will therefore become more frequent and less easy to control. Between 1972 and 1992, therefore, the chance of such a system collapsing may have increased so much that this is the reason behind the contradictory results of the calculations mentioned above.

One more word of caution: usually, as models are developed they tend to include increasingly more variables in order to make the expectations derived from their results more realistic and reliable. Indeed, they do become more realistic, but their results may become less reliable the more variables they contain. The reason is simple: uncertainties of estimation of the value variables take, or those due to chance, multiply and thus increase with any new variable being added. For example, when the certainty of one variable is 60 percent and that of a second in a chain, say, 40 percent, the certainty of the modeling result is 60 percent $\times$ 40 percent = 24 percent. Therefore, the calculation results of very complex—realistic—models can be disappointing, although their degree of reliability does not show up; this has to be estimated from different calculations based on different information. Therefore, instead of predicting some phenomenon, such as a demographic development, a financial crash, or societal collapse, from information on basic variables following a bottom-up approach, one should rather follow a top-down approach. In that case, you look at information on the input—resource availability and depletion—and output—waste production—of a processing unit, after which you estimate which variables dominate within this unit. Using information on only the minimum number of those variables gives the most reliable results. One cannot predict with high precision.

I think that the collapse of the present human population, its numbers and quality of life, is likely, and also that the most humane way to weather this period is to design a strategy and follow it ourselves rather than sit back and wait complacently. Unfortunately, the time for old customs and cultural traditions or of long-held beliefs and trusts is over. As the latest calculations from 1992 by Meadows and colleagues in *Beyond the Limits* showed, our world can collapse, and this can happen even before any resource has definitively been depleted; collapse may come at any time and out of nowhere. It's an inevitable, unavoidable result of the behavior of an oversized, complex, nonlinear system in which interdependent chance processes dominate.

The wave of large-scale deregulations because of the globalization of the last thirty years have only made this worse by allowing more positive feedback loops into the system.

How Likely Is It that Our World Will Collapse?

Nobody knows exactly how likely it is that our societal system will collapse or when. We know that this is theoretically inescapable, because all the local and national infrastructures and the global superstructure are based on abstractions. Moreover, system collapse follows from almost any simulation experiment based on relatively recent data—data that are now already twenty years old and are therefore too optimistic. In those twenty years, it has become even more likely that the conditions theoretically leading to system collapse will occur.

III

THE PERSISTENCE
OF MANKIND

The world is getting more organized but at the same time increasingly out of balance. In particular, that part of the world with the greatest population growth rates is the least organized and has the weakest infrastructure supporting those numbers, whereas the part of the world with reducing numbers is the best organized, its infrastructure extending across the whole world for its own good. And the poorer part, which has the greatest needs (especially with regard to basic food supply), is supplying the richer part with its own food and minerals. Also, the people there suffer the worst from pollution of their soil and freshwater and from us polluting the atmosphere with waste gases heating up their already hot and dry climate. Moreover, being poor, they usually lack sanitation and basic health care, especially in the thousands of slums surrounding their overcrowded, fast-growing megacities. The gap between those two parts of the world is widening rapidly by the year. Still, the rich part is extending and defending its resource use and denying the growth in basic rights of the poor. It may only be a matter of time before the poor are overrunning the rich in desperation.

The trigger may be resources that fall short because of their limited availability, inhibiting both the production of food and the further development and maintenance of global infrastructure and technology. And the stability of the rich part depends on the need of regular income from a still growing industry. A regular income is the price tag of richness; maintaining the huge and complex infrastructure is costly. Shortage of money and inefficiency in evenness of its spending causes instability within the rich, industrialized part of the world. And the rising prices of food, because of limiting availability of energy sources, will destabilize the countries in the poor part of the world: many of the poor are spending already most of their income on a single, meager meal a day, and they can't pay out much more.

Both between and within countries, instabilities will increase and make themselves felt. This can easily result first in shifts in and then the collapse of the whole system. And still we stick to an economic system based on growth, a system intended to cure growth with further, increased rates of growth. A system promoting uneconomic use and wastage of resources, a system of reckless, hidden pollution. And we still don't recognize the cause of it all, our growing numbers, still hoping for automatic stabilization of our numbers. Trusting that remaining stable at that high numerical level can last forever. Trusting even that our demands can keep growing while staying at that level, trusting that recycling and improving technology suffices. When do we give preference to the right to survive rather than the right to get children and grow? When do we see that we are trying to do the physically impossible? That our personal thoughtlessness means unimaginable suffering for billions of others? That it can even put the persistence of mankind, of all life on Earth severely at risk?

F

Another Future for Our Human World?

The foregoing chapters described processes potentially putting humanity into jeopardy. Chapter 24 surveys the history of humanity in exceedingly broad lines and extends these into the next decades of the present millennium. The underlying idea is that we are in a phase of development of a long process having lasted some ten to twenty thousand years or more. The problems we met during the last half-century and are going to meet in the coming years are difficult to stop or to turn back; they are rooted very deeply in our history and mentality. Therefore, compared with the problems of the past when our numbers were still relatively low, the difficulties we are going to meet seem insurmountable. On the other hand, we must solve them in order to persist as humanity and to prolong life on Earth. This is an exceedingly tall order, but we have no choice any longer, given the fact that we are reaching the total exhaustion of the resources of several essential minerals on which we ourselves and our supporting global infrastructure depend, and given the fact that the degree of pollution of our terrestrial, aquatic and atmospheric environment is beginning to destroy life both in the natural world and in our own.

24 THE ROAD WE TOOK, AND THE WAY FORWARD

Many years ago, when we still had freezing-cold winters in the Netherlands, I was invited to go on a skating tour with a friend of mine and his parents. They were Frisian, people from the province of Friesland, where most of the best medal-winning long-distance skaters come from. My friend's parents skated in tandem, one behind the other (the father behind), holding a short stick, just like on an old Dutch painting. We boys laughed at them and cheered as we overtook them, of course skating much faster. And better! But the father laughed, and called after us, "Yes, now you have tailwind, but wait until you have to skate into the headwind, then we'll see who the good skaters are!" And indeed, after a while, we had to turn from one of those magnificent, smooth, and reed-lined frozen canals onto a lake where we had to skate into the headwind and on rough ice. Which caused us the predicted—and feared—difficulties, and we were soon overtaken by my friend's parents.

I have always remembered the lesson learned: it is in difficult times—not when we're being pushed along by a tailwind—that we can show our real strength. Later in life, this proved to apply to many other events as well. And when composing this final chapter, I again thought back to that skating episode. Demographically, it has almost always been really tough in human history; but on the whole, so far, humanity has experienced tailwinds, sometimes stronger, often weaker. This was expressed as a steady growth (at first slow, then rapid) of the average number of people in the world, until the recent population explosion. And it was also expressed by a gradually more comfortable lifestyle for many; increasing demand always led to some technological solution to adjust the supply. So far, solutions have always been found for the most serious problems of food, energy, or material shortages.

Only in absolutist societies that have not been able to adjust the supply to acute shortages in demand did technological solutions fail to develop. Moreover, during the last couple of centuries we have experienced increasingly stronger tailwinds; in fact, in the last fifty years we came to believe that a strong tailwind was normal and that our technological inventiveness could last forever. Surely our still growing understanding and technological skill can never been lost, can it? It can only improve further.

Our population has grown increasingly rapidly, and so has consumption per head. We have not come to grips with the limitations set to resource use, and we have not realized that the amount of waste we are generating is soaring concomitantly. Moreover, the socioeconomic and administrative overheads have been growing exponentially relative to our exponentially growing numbers. And still we seem to be unaware of the wind behind us, and if we do feel it, it seems to be reassuringly strong, pushing us along. So we continue to throw away perfectly edible food, machines, computers, cars, even ocean steamers and airplanes. Life is good and gorgeous.

But what if there is still a sharp turn ahead—perhaps even a U-turn? Will we be able to retain our speed, are we strong and inventive enough to face the difficulties we might find ourselves in suddenly? Will we soon be so busy struggling against a headwind that we fail to notice the warning signs along the route for rough or thin ice ahead?

From its earliest stages, life has always been faced with resource shortages. At first these were simple chemicals. At that stage, the life processes must have been going on within the confined spaces of the most primitive cells, with little if any interchange with the outside world. The resulting shortages plus the accumulated waste could easily have killed off those pristine processes, and in most cases they must indeed have done so. In one or a few cases, though, the first waste products must have happened to fall apart easily, thus serving as a source for the next processes. Unintended and unnoticed, the first recycling process will have begun within the confines of this proto-cell, within the bounds set by the limitations of material interchange, bounds set to unlimited use of limited resources. Biochemical cycles were born out of necessity; recycling allowed early life to continue and to develop further, despite limiting supplies from outside. Life as a whole was shifting its limits upward as if denying them.

Some of the recycling processes that initially happened only inside the cell boundaries were also beginning to happen between the cells, involving other cell types in the recycling of matter which had now become limiting in the environment. They used up the little waste expelled as their resource, thereby starting up the large biospheric nutrient cycles. However, chemical energy did run short, as this could not be recycled. At that point, external energy obtained from the sun began to flow through this elaborate system of material recycling, keeping both internal and external cycles turning. From early on, therefore, the processes of life were based partly on the recycling of materials and partly on the linear flow of degrading energy. This linear flow kept the material cycles going, like a stream keeps waterwheels turning.

Recycling with the help of solar energy was the most fundamental invention of life, but it was followed by other amazing ones. Meanwhile, under the exhausting and polluting influence of life over billions of years, the basic chemistry of the environment was changing. Always, new solutions developed, inserted within previous ones that had to be preserved to keep the system working. Accordingly, the living systems became more complex at every step: proto-cells developed into bacteria, bacteria fused into the present-day cells, and those cells together formed multicellular organisms, first jellyfish and slugs, then fish, frogs, reptiles, birds, and mammals. And plants developed from unicellular algae into multicellular ones, these into ferns and horsetails, and these, in turn, into the present vegetation of our marshes, prairies, meadows, and dense forests. During its course, life suffered setbacks every so often when devastating mass extinctions occurred, but it arose and blossomed again, sometimes falling back to simpler forms or progressing to ever-more sophisticated complexity. Always developing new solutions. Always making do with roughly the same materials, which were constrained by the same limitations. Using increasingly more energy, which continued to channel its way through the ever-more intricate systems. And the number of ways—and therefore systems—increased as well. New species evolved, always new forms, always working differently, but always according to amazingly simple physical, mechanical, and chemical principles. Most of those species remained small, although occasionally unbelievably large individuals—sequoias, dinosaurs, and whales—evolved. Life kept pushing its natural limits, the level of its recycling of nutrients, eking out on the relatively little material of the incredibly thin top layer of Earth in which it mostly occurs.

This did not change with the ascent of humans. Humans too pushed the limits of their own material resources, using ever-more material and energy, eradicating other species to satisfy their own needs, growing in numbers, setting up organizations supporting their needs. Famines due to environmental exhaustion and instability, diseases, and wars kept checking the numbers of people until humans became able to reduce instability and combat disease, and by so doing started using increasingly more energy to collect more materials and minerals. Wars took an ever-larger toll in absolute numbers, although the percentage of fatalities relative to the total population dropped. Now, our demands continue to grow, life expectancy is increasing, and we live to an old age in which we merely consume and produce nothing. Resources are scraped together from all corners and depths of Earth; solar energy obtained directly by crops is supplemented by energy stored over geological time as fossil fuels in Earth's crust. All this is amazing, but we have never really perfected recycling ourselves—certainly not to the same level, sophistication, and efficiency as other living systems; our system of material processing mostly remains based on the linear one of slashing and burning. Overall, like any individual species does, we have been overusing the environment, exhausting it, deteriorating it, simplifying it, polluting it.

As environmental instability and the impact of physical exhaustion decreased, as famines and diseases reduced in number, intensity, and impact, and wars took ever-smaller percentages of the population, the dynamic of society humans built for themselves became based on unabated growth through competition and strife, and therefore through unabated environmental exhaustion and pollution as well. Gradually, the causes of the previous numerical setbacks were forgotten, replaced by theories of growth. Even ecology, the biological science of the living conditions of organisms, even theories of biological evolution are dominated by models of demographic growth through the exploitation and suppression of other species. Other species had to adapt to the whims of a capricious environment, but humans adapted the environment to their own needs. The problems of the concrete, physical basis of life, of a continuing process of trial and error, have receded in importance and have been replaced by those of societal growth, financial organization and gain, political stability and unification. They have been replaced by abstractions, by an abstract environment for most of us.

The abstract socioeconomic superstructure supporting us all assumes that this structure operates without internal friction and that external conditions are unlimited, constant, and stable. This is why this world of abstractions can collapse under its own weight: because of internal frictions about limitations inherent to nonrenewable resources and unstable, changeable living conditions. Our world is ultimately shaped by our concrete numbers and increasing concrete personal demands. However, as soon as our world of abstractions folds up, the concrete world of old will make itself felt again. It is still there, outside our human shell. Then, the stretched limits of resource availability and pollution will contract, but we don't know how far. In a way, Earth has been overpopulated by humans ever since they built their first settlements of tents and first domesticated their plants and animals, ten thousand years ago or longer. Perhaps it was as long ago as we learned to speak to each other, thereby making the first abstractions of the world around us and allowing social organizations to grow, arranged according to the first rules. Primitive as they were, these organized settlements of tents and farms and their first rules were our first tools of survival, our first weapons to fight problems of limitation. In some—but not all—parts of the world such a low level may be reached again. There will be large-scale migration, and wars and diseases will break out, complicating the picture.

This collapse is likely because part of our present economic growth is fed by the growth of our world population, another part by the growing standard of living, and yet another part by the growing size, complexity, and abstract nature of our structures organizing this all. Even if the standard of living remains unchanged, resource use and waste production will grow, partly because of the expected three billion people extra on Earth, and partly because of the growth of the structures needed to keep all ten billion or more alive and going. If we want to raise our overall standard of living as well, we need even larger and more complex supporting structures. We will need exponentially larger amounts of material and energy at a time that we need to recycle most of our materials, and we will need to generate our even increasing demand of energy ourselves.

New techniques have to be developed for energy production. Much depends on whether we can introduce new ways of producing energy and new ways of producing materials and of avoiding waste worldwide in time. If our timing fails, humanity will be faced with having to reduce its numbers deliberately by at least 90 percent, and this within not one single generation, but as soon as the downturn begins. If we don't do this ourselves in a way of our choosing, this incredible reduction will be forced upon us externally in the most inhumane ways. The time of living for free is over; times of tough and far-reaching decisions are lying ahead.

The obvious and unavoidable question, therefore, is what size population would ensure long-term sustainability? The pundits who suggest that our numbers are too high and need to be pushed back to some lower level rarely—if ever—give criteria for estimating a sustainable population level. Moreover, a possible stabilization level, such as the one forecast to be reached in 2050, is often equated with a level of long-term sustainability. However, these are obviously two very different things in a society based on a linear process of resource processing: resource use continues unlimited whereas our resources are limited. And pollutants keep accumulating. On the other hand, we keep using and producing, and this at still increasing rates, whereas

Earth's unpolluted, usable resources decrease. The figure below shows this process graphically. The central part of the figure is the same as the one given in figure 4, but now it is depicted within the framework of the usable part of Earth. The dashed line around the central part is expanding, whereas that of Earth is contracting. The space in between those dashed lines at the bottom indicates that the usable resources are still sustainable, their overlap at the top shows that we are overusing them.

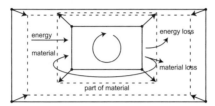

One possible criterion could be the population that could be supported when one or a few of our presently essential resources have run out. For example, in a scenario with no more oil and gas to supply us with energy, the population should be the size it was before we began to use these resources, that is, the population around 1900, which was about one billion people (roughly one-tenth that predicted for 2050). Around 1900, however, there was still plenty of wood, as well as nutrients like phosphorus and nitrogen, but these are now in increasingly short supply. Therefore, it may well be that a population of that size will still prove to be too high. In that case, we need to go back in time even further, to an even lower figure, for the population to be sustainable under conditions of shortages of energy and essential chemical elements. Another problem is estimating the degree of environmental instability permissible in order to avoid problems of temporary shortages of energy, food and materials. And with regard to sustainability, an additional criterion is the length of time *Homo sapiens* is expected to persist, given that resources are limited, and given a certain rate of recycling waste materials. If we were to halve the number of people on Earth, this time can be roughly doubled, not taking into account the unknown savings gained by not having to run such a large regulating superstructure. To be able to do this calculation we need to define the length of time we think our species should remain on the planet, the rates of consumption and pollution, and the amount of resources left.

Furthermore, it may well be that under impoverished conditions, we need to keep part of our social complexity and dynamic intact to sustain a population level under better living conditions than those present in 1900. For example, under times of severe shortages, we may need to organize and keep an extensive transport system running, along with its financial and judicial backing, to get the few remaining resources to where they are needed. And we may want to keep the same level of medical knowledge and our present technological sophistication and networks, along with the scientific infrastructures in which medicine and technology are embedded. We may also need to maintain and improve techniques of recycling and energy production continually. This organization will itself cost a certain amount of material and energy and this will have to be incorporated into the calculations and will further reduce the number of humans Earth can physically sustain.

The problem is not only when we can begin formulating and taking measures for reducing our numbers ourselves; it is also how to find the most effective and the least inhumane way of doing so. True, all human measures in this respect will be inhumane, but the issues they raise are moral and religious, whereas wars, diseases, famines, floods, and large-scale fires add unacceptable physical stress and suffering on top of them. Our physical problems can become unimaginably worse, and we can't easily exaggerate their effects and they themselves raise new problems of morality. And those problems will be far worse under those conditions. Under conditions of war, famine, thirst, or deadly pandemics, what will be left of our moral values or religious ideals? The moral and religious problems will remain, even when conditions worsen. What about our living standards and human rights? What will be left of those? What moral values are we actually defending? Which ones will we have to give up if we don't take steps to avert the worst-case scenario? And if some unimaginable catastrophe does wipe out most of the human race, bringing it back to a "safe", sustainable level, we will need to change such values anyway to prevent the same processes of growth, overuse, and overproduction of waste from recurring.

Moreover, past events suggest that unless we alter our ethos fundamentally, catastrophes will produce only a temporary set-back—a blip—in the inexorably rising line of growth. And natural disasters can't adjust the population to the most favorable level for sustainability at a chosen time period, given the availability of certain resources or a particular recycling efficiency, and so on. As natural and social disasters do not provide an alternative solution for achieving long-term sustainability, the effect of their inhumanity will be wasted. If we reduce our numbers ourselves we will at least have more control over the reduction process and can aim for the long-term level required. We are the only ones who know what we want. The measures we need to take are inhumane, but they are the least inhumane of all.

The only option open to us is a reduction in birth rate rather than an increase in the mortality rate. Deliberately reducing the birth rate is inhumane, but less so than increasing the mortality rate, certainly at the scale necessary. As a one-child-per-family policy, the Chinese example is the minimum that can be followed whilst keeping to the principle that each family should be allowed to reproduce. However, it will only roughly halve the population per generation, as two parents will be replaced by one child, and this may not prove a fast-enough reduction, particularly given the increasing life expectancy. A more rigorous measure may be necessary. The measures remaining are various methods of contraception (from the decreasing replacement rates in most countries, it is clear that contraception is already widely and successfully practiced in the world) and the voluntary sterilization of part of the world population. It is not easy for me to write this, thinking of the two children my wife and I love, and thinking of all their friendships, hopes, and plans for their futures. It seems unthinkable to deny the joy of children to others. But given the present trends, there may be no other alternative. Once more, all other measures will be far more inhumane. Alternatives accepting an increase in mass mortality cannot be morally condoned. Without alternatives, we have no choice.

Apart from the grave problems to be solved at the personal level, at the socioeconomic level it will also be difficult to overturn the now deeply engrained ideas of growth as a basic driving force in society, and replace them with those of quality standards as the main economic purpose. Although part of our present growth is to do with improvements in quality, the idea central to the Continental or Rhineland economic model, the main emphasis is on increasing quality of production, whereas the Anglo-Saxon model emphasizes increase in quantity through competition. The central concept of competition in trade and commerce, as well as between nations, though, must be supplanted by that of collaboration for improving quality and sustainability, and to eke out what is left. Under the threat of shortages, this sea change in attitude has already begun. Organizations and nations can continue to compete but not principally for money, but instead to achieve another purpose: production in the qualitatively best and materially most efficient and sustainable way. In principle, this purpose can still be achieved through private enterprise, but within a new framework in which neither energy and materials nor waste disposal are free, and in which there is optimal recycling and planned mining, use, and disposal. Large-scale recycling means in principle replacing our present economic model based on growth of profit—and thus dependent on growing resource use and generating increasing amounts of waste—by an adjusted growth model based on quality and sustainable products, and on recycling speed and efficiency. The problem remains of who is going to pay the recycling costs: the initial producer or the second user recycling the waste of the first?

All this is still insufficient if we don't succeed in replacing our socioeconomic positive feedback approach by a negative feedback one. This means that we will lose much of our freedom, though not necessarily all of it. In times of severe shortages, we have to eke out our existence, which means that we have to oversee what is left and how we can use this most economically. *Laissez faire* belongs to the past.

Apart from still basing themselves on an adjusted growth model, most developing societies in the world will replace the remains of their subsistence economy by one based on urban industry and industrial agriculture, as Western countries did from the mid-eighteenth century onward. Many subtropical and tropical countries have already entered that phase and are changing their political system accordingly into one based on constitutional democracy. However, as a global development, a new socioeconomic organization will unavoidably lose part of its chance processes basic to open markets, linear processing, and unbounded reproduction. These are deep-cutting losses, but it is the price future generations will have to pay for our thoughtless overuse and spoilage of Earth's richness.

Yet another consequence of a decline in the national populations or that of the world as a whole is that the distribution of age groups will be less favorable than that of the present growing populations. For example, in growing populations, a relatively large number of young working people can support the elderly, but this changes when the elderly begin to dominate in number. This is becoming worse now that longevity is increasing sharply at the same time. Life, therefore, will become much more expensive, especially because of an increase in medical care, so that both productivity and the age of retirement will have to increase considerably. In other words, our physical demands have to increase: by living longer, our personal demands continue for a longer time, partly canceling the effects of a lower fertility. Moreover, a continuation of our demands requires a large infrastructure to meet them, even if these demands remain the same per person. In fact, the total demand on which our resource use and waste production depend has to diminish, which means a larger reduction of our numbers and possibly the consequent breakdown of the national and global infrastructures.

All this bears the risk that care for the elderly shifts back to the personal level where it always has been. This means that many families decide or feel forced to have more children again, which counteracts present trends. If this would happen, all hopes will be gone that our numbers might stop growing at the level of ten billion.

The hardest changes of all remain at the personal level where much of our basic freedoms to choose and reproduce will inevitably be curtailed. We have to jettison many of our present, ancient, biologically determined habits, ideals, and existing moral and religious values. In principle, this is feasible if we realize that it is the least inhumane way to achieve the necessary reduction in our numbers to prolong our existence on Earth. The price is high, but it is the lowest price we can pay. Out of the sheer impossible choices available, it is the only and best choice we can still make ourselves. Objections will no doubt be raised against drastic sterilization schemes, such as about who will be exempted, whether the age distribution might be dangerously disrupted, and so on. All true, grave, and realistic considerations and arguments. But the longer we keep doubting and deliberating, the higher the reduction factor must be, the faster and the more rigorously our measures will have to work, and the less humane they will be. And the measures to be taken to curb our numbers, our resource use, and waste production will become more devastating. During the 1970s, many people felt any demographic measure was unacceptable, although taking such a measure then would have been incomparably lighter and more humane than it is now. Now we can't look the other way any longer as we did then.

We Are Wasting the World

We are wasting the world. Not only the natural world, the geological and the biological world, but our human, cultural world as well, a world that is astoundingly beautiful, a unique extension of this amazing geological and biological world. It is a world unique in the universe. It is unrealistic to cherish the hope that at some unknown place in the universe some other form of life unknown to us might still exist. And if ever we do discover other life somewhere else, this is still no argument to waste here our own. We are wasting all the amazing potential of life and humanity, all we have built up over thousands of years at the end of a long biological development, itself already almost four billion years long. We are wasting our future and, with it, the future of the natural world.

We are wasting it by asking too much, and we are wasting it by expelling too much waste in return. Waste that cannot be brought back into circulation by any present-day biological cycle, and hardly by recycling it ourselves. We scatter all this useless and polluting waste around, on land, in the groundwater and the rivers, in the oceans, and in the air. We are blinded by immediate profit, by wealth, opulence, and comfort, by improvement and quantitative growth, ignoring the consequence: an impossibly impoverished, desolate, and overheated world. We were warned some forty-odd years ago, but we paid no attention and accelerated the process instead. We know that we cannot make the same mistake twice without avoiding disaster, but we seem not to realize that we are now heading toward it at full speed. We seem not to realize that we must act.

We refuse to see the trap at the end of the fyke, we won't see that our room for maneuver is becoming ever-more restricted, won't see that we really have to stop if we are to avert unprecedented and unimaginable disaster. Deliberately, we keep our blinders on. We don't want to see.

EPILOGUE

THE EMPEROR'S NEW CLOTHES

There is a famous story by Hans Christian Andersen, "The Emperor's New Clothes," in which an emperor shows his newest clothes to the public. This had already often been done, and every time the show attracted more people. However, by now the tailor had run out of ideas and decided that the nicest dress would be the emperor's skin itself. This time, a record number of people came together and they were all amazed by how well the new dress looked. They agreed that this was certainly the nicest dress he had ever worn—until a little girl said that the emperor really had no clothes on at all, after which a wave of disappointment and hilarity went through the masses. And then, all people left with astonishing speed, much faster than they had during previous times when there had been fewer people. Just like in a gold rush or a financial boom, both the buildup as well as the decline of the numbers followed an exponential process.

A similar rise in numbers has happened to humanity; over the centuries it has uncontrollably grown out of its natural bounds, in this case, trying to escape the consequences of growth by stimulating further growth. Demographic growth stimulating agricultural and organizational growth and this, again, stimulating demographic growth, and so on in a rapidly growing spiral of growth. Similar to the growth of onlookers in Andersen's story, this is a typical self-accelerating process. To allow for this growth, adding ever-new resources to be tapped, superimposing ever-new layers of organization, all of them requiring their own resources, and each having to grow and requiring new organizations that allow them to operate. Growth on top of growth, exponential growth relative to a process growing exponentially itself. Double exponential growth, the growth exponent itself growing exponentially. Then treble exponential growth, and then multiple. Faster, ever faster.

Thus, in all civilizations, the rate of our reproductive growth led to agricultural improvements, and this to industrial growth and urbanization. This, in turn, led to an expansion of the economic catchment area of the industrializing world, and this, again, to our seeding out to other parts of the world by emigration, and this to repeated agricultural and industrial growth in those parts. In the nineteenth century, in the core area of the West, the resulting increase in income was invested in better methods of sanitation and health care, and therefore, again, to enhanced population growth. All this means that the overall rate of resource use and waste production of the processing unit got higher and higher, the one rate even enhancing the next.

To keep the wheels turning, we shifted from one source of energy to other, always easier kinds of energy, to more abundant and more energy-rich ones. From human energy to wind to wood to coal to oil to gas to nuclear power to electricity, at each step becoming less and less efficient in the actual use of the energy because of conversion losses as well as losses due to the ever-growing size, number, and layers of organization required on the way from the source of energy and its eventual use. We grew from local subsistence economies without money to global markets based on large-scale outsourcing and computerized financial exchange. We were filling one hole by digging the next. Meanwhile, we were trusting the future, trusting technology, trusting our global organization and banking. Trusting that eventually, all this growth would smoothly stop by itself, that eventually some point of stabilization, some equilibrium, would be reached and forever maintained. Not knowing how to stop it ourselves, growth grew out of hand, out of reach, beyond our capacity. But surely, this growth would outgrow itself.

We believe in the existence of some point of stabilization—in automatically reaching a level of equilibrium—whatever this might mean for the societal process concerned and, in the end, to the individual people themselves. So far, we can't see any humane alternative to solving the problems growth inevitably gives by enhancing further growth. Not growing any further means disrupting the socioeconomic system with all its dire effects: our linear system of resource use due to entropic decay requires a minimum of growth in all sectors. Not growing means losing in competition, competition between shops, between firms, corporations, nations and, finally, between continents. And the effects of this disruption are worse when we need to reduce our numbers and personal demands. But not doing anything ourselves in a humane way inevitably means further growth, and also that other, nonhuman factors take over, and that the processes of stabilization or reduction unavoidably become inhumane. Not wanting to lose out in our competitive battles, we grow, and we deplete our resources, pollute our fields, the oceans and air, and are warming the climate. We are denuding Earth from its green blanket, we are starving and killing each other. And this not only means the end of our civilized human society, it in fact can mean the very end of life here on Earth. Where is this little girl saying that the emperor is naked? When do we realize that our trust in an automatic, humane stabilization process leading to exactly the right, sustainable equilibrium level is unfounded?

Central to this all is our blind trust in equilibriums, in equilibrium population numbers, equilibrium climate conditions, or the equilibrium of supply and demand. That is, it's our trust in negative feedback processes at work while at the same time we are actively opposing their operation—voting for positive feedback instead. Negative feedback mechanisms are defined by the existence of some physical norm of reaction as well as by physical mechanisms that reduce any deviation from this norm, ones that eventually bring the process back to the norm. Think of how you keep your body temperature constant, a typical negative feedback mechanism. First, there is the temperature norm of 37°C typical for mammals, a temperature that could also have been different, such as happened in birds, where it is 40°C. When we get too hot, too close to a temperature higher than 37°C, we begin to sweat and we send blood to our hands and feet to cool off. And when we get too cold, we shiver and get goose bumps. Those are all physical mechanisms that evolved biologically to guarantee that the biochemistry of each of our cells operates optimally and constantly.

Positive feedback mechanisms, in contrast, are defined by the absence of such mechanisms. First, the growth process goes unbounded, at some point reaching an unsustainable level, which leads to an uncontrolled decline, the crash of the system. Then, after some period of recovery, growth resumes and the process reaches a similar or another level, after which once again the inevitable crash follows. Overall, we see wild, uncontrolled fluctuations happening, progressing beyond our capacity, all self-enhancing, self-accelerating exponential processes, positive ones before the crash, and negative ones after it. This is what happens in economic crashes, as well as in the ecological processes of growth that occasionally happen in populations of animals and plants. Then, in those cases of overshoot, the demand cannot be satisfied any longer by the supply at some high level of fluctuation; demand overshoots supply. And though an average level of fluctuation can be calculated, there is no physically set norm of reaction. Nor are there mechanisms for reducing the deviations to some acceptable degree. It doesn't matter whether the supply concerns the economic value of some firm and the demand the total value of shares or the ecological footprint some species has on its resources, be it an animal or plant or our own human species. In none of these instances do any of the physical mechanisms exist for making the process a negative feedback process; they are therefore all positive ones. Eventually, the positive ones all lead to a crash of some sort. In fact, the various socioeconomic growth models we have embraced in history are usually inherently opposed to some norm of stability, and mechanisms making to keep to this norm are obviously missing as a consequence. Anti-crash measures the economist John Maynard Keynes once proposed don't cure this; they only prevent the troughs in the fluctuation pattern from getting too deep while keeping the overall positive feedback loop intact. Similarly, we don't have any optimal level for a long-term persistence of human life on Earth, dependent on a certain set of minerals and energy still available. Instead, we trust that we will reach some level of stabilization soon and trust that this will be sustainable. However, we have to take adequate measures ourselves.

If we want to get our growth processes under control, it is therefore compulsory to formulate norms of total resource use in relation to resource availability, as well as to the inevitable production of all forms of waste. These norms must take the time period into account, over which we want to make use of the various resources needed. When some period of stability lasts too long, we are still depleting the resources. Given a certain time period chosen, we may need to reduce our total demand—our numbers times the demand per head—and the longer the period, the more drastic will be the reduction.

This means that we have to reduce both our personal use as well as the number of humans making use of the resources. And secondly, we need to install mechanisms that counteract any deviation from reaching and maintaining the norm. This needs to be done on a world scale, independent of history, power, or sovereignty. Any other solution will fail so that sooner or later our numbers get out of control and will crash. As soon as we have chosen either the positive feedback system or the negative one, it will run its inevitable course. This has nothing to do with economic, political, moral, or religious preference or choice; rather, it simply and unavoidably results from the process at hand—no more and no less.

By deciding what norm we might use in case we change to an eco-economic negative feedback system, we must realize that in terms of the ecological footprint we have already overshot the resources of Earth. Given our present resource use and the amount of waste we produce, we need 1.4 Earths, which may soon amount to 3 if we refuse to take measures. This means that we are already on our way to the inevitable crash; our time is running out. In those cases where there is no limit given, this will lead to a crash. This is happening with climate warming, where the increase will continue unbounded, then resulting in an overheated Earth.

We cannot excuse ourselves any longer by placing our blind trust in technology or automatic feedback mechanisms that might lead us smoothly and humanely towards some sustainable solution. We have to do it ourselves, and we have to begin soon.

ABOUT THE AUTHOR

These years, as in the early 1970s, large numbers of books have been pub-lished on problems concerning our common future. Yet, these books, how-ever valuable, always concentrate on specific problems, such as climate warming, waste disposal, or the end of oil. It is important to integrate these various problems into a single one, showing how they relate to and depend on each other. Such an integration requires a notion of the most important living conditions of mankind, a notion of human ecology.

Within biology, I have worked in ecology, biogeography, invasions re-search, and in biogenetics (the study of life's origin). I have therefore worked in all subjects pertinent to the present book. *Ecology* is concerned with the spatial and temporal variation in living conditions of organisms of all types, whether animals or plants, fungi or bacteria. *Biogeography* studies organisms of various species as they are distributed over the continents or between islands, which is determined by their living conditions, such as climate and soil conditions. *Invasions research* is concerned with the way a species spreads and influences native species when it is introduced into another con-tinent. Their spread often proves to be similar to the way diseases spread and tick over so that I came to know that part of the scientific literature as well. Finally, *biogenetics* concerns the processes that could have led to the beginning of life several billions of years ago. Understanding how life may have originated makes you also understand how it developed, how it has been maintained, and, again, what environmental factors were relevant in its development and maintenance. And it makes you understand how life transforms food into waste. Life is a mechanism that turns resources into waste, a principle central to this book. The often mathematical principles of life's development, operation, and maintenance can also be applied to the development and dynamic of socioeconomic systems. My interest in history and economics took me from biology to the ecological history of society.

Combined, these fields of study gave some understanding of the conditions with which mankind has always been confronted, as well as of changes in those conditions that will confront us in future. As a scientist, I felt the responsibility to tell what seems will threaten human society this coming century.

Also by the Author

Dynamics of Biological Invasions (Chapman and Hall, 1989)
Dynamic Biogeography (Cambridge University Press, 1990)

ACKNOWLEDGMENTS

I feel rich for having received much interest from so many people around me, and thank them all for what they have given me in that way. What they gave was substantial, because it hasn't always been easy to write this book, its subject weighing heavily on me. First of all, of course, I thank my immediate family, Claire, Eke, Sytse, and Ilona, who listened to and knew my concerns and fears for the future. David Goodall, one of the referees and an intimate friend for years, always asked about the progress I made, and making valuable suggestions. Then, there was this mysterious referee 3, whose understanding right from the beginning has been of great value throughout much of the writing. Of my friends at Leiden University I want to single out Marijke en Kees Libbenga, with whom I discussed many aspects of this book. I feel privileged to have had Christie Henry as my first contact at the Press; she led me through the whole process of designing and writing the book. She knew what I wanted and was always encouraging when things got stuck in one way or the other. Many deeply felt thanks to her. And I thank the various pleasant and able people around her at the Press who have done their bit to get the book right, in particular, Mary Gehl as a very able subeditor. Last but not least, I thank Joy Burrough, who, as corrector, native speaker, and friend, went through the text to turn my Duglish cleverly into idiomatic English, at the same time checking and improving its contents. Our son Eke prepared the text figures.

SELECTED BIBLIOGRAPHY

The literature on the subjects broached in this book is vast—often too vast to master, even for a specialist. This bibliography, therefore, contains a personal selection of titles, intended for further reading. I have tried to give both recent and older literature, as well as some broad-audience and specialized references.

Chapter 1

Ward, P. *The Medea Hypothesis: Is Life on Earth Ultimately Self-Destructive?* Princeton, NJ: Princeton University Press, 2009.
Ward, P., and D. Brownlee. *Rare Earth: Why Life Is Uncommon in the Universe.* New York: Springer, 2000.

Chapter 2

Braungart, M., and W. McDonough. *Cradle to Cradle: Remaking the Way We Make Things.* New York: North Point, 2007.
Packard, V. *The Waste Makers.* New York: David McKay, 1960.

Chapter 3

Boserup, E. *The Conditions of Agricultural Growth: The Economics of Agrarian Change under Population Pressure*. Chicago: Aldine, 1965.

———. *Population and Technological Change*. Chicago: University of Chicago Press, 1981.

Cohen, M. N. *The Food Crisis in Prehistory: Overpopulation and the Origins of Agriculture*. New Haven, CT: Yale University Press, 1977.

Commoner, B. *The Closing Circle: Nature, Man, and Technology*. New York: Random House, 1971.

Ehrlich, P. R. *The Population Bomb*. New York: Ballantine Books, 1968.

Ehrlich, P. R., and A. H. Ehrlich. *One with Nineveh: Politics, Consumption, and the Human Future*. Washington, DC: Island Press, 2004.

Gregg, P. *A Social and Economic History of Britain, 1760–1970*. 6th ed. London: Harrap, 1971.

Grigg, D. B. *Population Growth and Agrarian Change: An Historical Perspective*. Cambridge: Cambridge University Press, 1980.

Kerridge, E. *The Agricultural Revolution*. London: Allen and Unwin, 1967.

Paige, J. M. *Agrarian Revolution: Social Movements and Export Agriculture in the Underdeveloped World*. New York: Free Press, 1975.

Post, J. D. *The Last Great Subsistence Crisis in the Western World*. Baltimore, MD: Johns Hopkins University Press, 1977.

Smil, V. *Enriching the Earth: Fritz Haber, Carl Bosch, and the Transformation of World Food Production*. Cambridge, MA: MIT Press, 2001.

Chapter 4

Ashton, T. S. *The Industrial Revolution, 1760–1830*. Oxford: Oxford University Press, 1968.

Bluestone, B., and B. Harrison. *The Industrialization of America: Plant Closings, Community Abandonment, and the Dismantling of Basic Industry*. New York: Basic Books, 1982.

Foster, J. *Class Struggle and the Industrial Revolution: Early Industrial Capitalism in Three English Towns*. London: Methuen, 1974.

Gregg, P. *A Social and Economic History of Britain, 1760–1970*. 6th ed. London: Harrap, 1971.

Habakkuk, H. J. *Population Growth and Economic Development since 1750*. Leicester, UK: Leicester University Press, 1971.

Hall, P. A., and D. Soskice, eds. *Varieties of Capitalism: The Institutional Foundations of Competitive Advantage.* Oxford: Oxford University Press, 2001.

Headrick, D. R. *The Tools of Empire: Technology and European Imperialism in the Nineteenth Century.* New York: Oxford University Press, 1981.

Kemp, T. *Historical Patterns of Industrialization.* London: Longman, 1978.

Landes, D. S. *The Unbound Prometheus: Technological Change and Industrial Development in Western Europe from 1750 to the Present.* Cambridge: Cambridge University Press, 1969.

Rosenberg, N., and L. E. Birdzell. *How the West Grew Rich.* London: LB Tauris, 1986.

Rosenberg, N., R. Landau, and D. C. Mowery, eds. *Technology and the Wealth of Nations.* Palo Alto, CA: Stanford University Press, 1992.

Wrigley, E. A. *Energy and the English Industrial Revolution.* Cambridge: Cambridge University Press, 2010.

Chapter 5

Galbraith, J. H. *The New Industrial State.* Boston: Houghton Mifflin, 1967.

Hobsbawm, E. *The Age of Capital: 1848–1875.* London: Weidenfeld and Nicholson, 1962.

———. *The Age of Empire: 1875–1914.* London: Weidenfeld and Nicholson, 1987.

———. *The Age of Revolution: 1789–1848.* London: Weidenfeld and Nicholson, 1962.

———. *Nations and Nationalism since 1780: Programme, Myth, Reality.* Cambridge: Cambridge University Press, 1990.

Kindleberger, C. P. *The World Depression 1929–1939.* Berkeley: University of California Press, 1973.

Landes, D. S. *The Wealth and Poverty of Nations: Why Some Are so Rich and Some so Poor.* New York: Norton, 1998.

McNeill, W. H. *The Rise of the West: A History of the Human Community.* Chicago: University of Chicago Press, 1963.

Polanyi, K. *The Great Transformation: The Political Economic Origins of Our Time.* New York: Farrar and Rinehart, 1944.

Post, J. D. *The Last Great Subsistence Crisis in the Western World.* Baltimore, MD: Johns Hopkins University Press, 1977.

Reich, R. B. *Supercapitalism: The Transformation of Business, Democracy, and Everyday Life.* New York: Knopf, 2007.

Robin, M.-M. *Le monde selon Monsanto. De la dioxine aux OGM, une multinational qui vous veut du bien.* Paris: La Decouverte, 2008.

Shiva, V. *The Violence of the Green Revolution: Third World Agriculture, Ecology and Politics.* London: Zed Books, 1991.

Van der Ploeg, J. D. *The New Peasantries: Struggles for Autonomy and Sustainability in an Era of Empire and Globalization.* London: Earthscan, 2008.

Worster, D. *Rivers of Empire: Water, Aridity and the Growth of the American West.* New York: Pantheon Books, 1985.

Youngquist, W. L. *GeoDestinies: The Inevitable Control of Earth Resources over Nations and Individuals.* Portland, OR: National Book Co., 1997.

Chapter 6

Anonymous. *The Future for Renewable Energy.* London: James and James, 1996.

Campbell, C. J. *The Coming Oil Crisis.* Brentwood: Multi-Science Publishing, 1988.

Campbell, C. J. *The Golden Century of Oil 1950–2050: The Depletion of a Resource.* Dordrecht: Kluwer, 1991.

Commoner, B. *The Poverty of Power: Energy and the Economic Crisis.* New York: Knopf, 1976.

Deffeyes, K. S. *Beyond Oil: The View from Hubbert's Peak.* New York: Hill and Wang, 2005.

Dennis, K., and J. Ury. *After the Car.* Cambridge: Polity, 2009.

Evans, R. L. *Fuelling our Future: An Introduction to Sustainable Energy.* Cambridge: Cambridge University Press, 2007.

Gever, J., R. Kaufmann, D. Skole, and C. Vorosmarty. *Beyond Oil: The Threat to Food and Fuel in the Coming Decades.* Cambridge, MA: Ballinger, 1986.

Heinberg, R. *The Party's Over: Oil, War and the Fate of Industrial Societies.* Forest Row, UK: Clairview Books, 2003.

———. *Powerdown: Options and Actions for a Post-Carbon World.* Gabriola Island, BC: New Society Publishers, 2004.

Jevons, W. S. *The Coal Question: An Inquiry Concerning the Progress of the Nation, and the Probable Exhaustion of Our Coal-Mines.* 1906. Reprint ed. New York: Kelly, 1965.

Klare, M. T. *Blood and Oil: How America's Thirst for Petrol Is Killing Us.* New York: Henry Holt, 2004.

Leggett, J. *The Carbon War: Global Warming and the End of the Oil Era.* London: Penguin, 1999.

Lovins, A. B. *Soft Energy Paths.* Harmondsworth, UK: Penguin, 1977.

Pahl, G. *Biodiesel: Growing a New Energy Economy.* White River Junction, VT: Chelsea Green, 2005.

Rifkin, J. *The Hydrogen Economy: The Creation of the World-Wide Energy.* Harmondsworth, UK: Penguin, 2002.

Roberts, P. *The End of Oil: The Decline of the Petroleum Economy and the Rise of a New Energy Order.* London: Bloomsbury, 2004.

Simmons, M. R. *Twilight in the Desert: The Coming Saudi Oil Shock and the World Economy.* Hoboken, NJ: Wiley, 2005.

Wagner, H.-J. *Energy: The World's Race for Resources in the 21st Century.* London: Haus, 2009.

Chapter 7

Borgstrom, G. *Too Many: An Ecological Overview of Earth's Limitations.* New York: Macmillan, 1969.

Cole, H. S. D., C. Freeman, M. Jahoda, and K. L. R. Pavitt. *Models of Doom: A Critique of the Limits to Growth.* New York: Universe Books, 1973.

Commoner, B. *The Closing Circle: Nature, Man, and Technology.* New York: Random House, 1971.

———. *The Poverty of Power: Energy and the Economic Crisis.* New York: Knopf, 1976.

Forrester, J. W. *World Dynamics.* Cambridge, MA: Wright-Allan Press, 1971.

Heinberg, R. *Peak Everything: Waking Up to the Century of Decline in Earth's Resources.* Gabriola Island, BC: New Society Publishers, 2007.

Karplus, W. J. *The Heavens Are Falling: The Scientific Prediction of Catastrophes in Our Time.* New York: Plenum Press, 1992.

Meadows, D. H., D. L. Meadows, and J. Randers. *Beyond the Limits: Confronting Global Collapse, Envisioning a Sustainable Future.* White River Junction, VT: Chelsea Green., 1992.

Meadows, D. L. *The Limits to Growth.* New York: Universe Books, 1972.

Mesarović, M., and E. Pestel. *Mankind on a Turning Point.* New York: Dutton, 1974.

Romm, J. J. *The Hype about Hydrogen.* Washington, DC: Island Press, 2005.

Schmidt-Bleek, F. *The Earth: Natural Resources and Human Intervention.* London: Haus, 2009.

Simmons, I. G. *The Ecology of Natural Resources.* 2nd ed. London: Arnold, 1981.

Williams, R. J. P., and J. J. R. Fraústo da Silva. *The Chemistry of Evolution: The Development of Our Ecosystem.* Amsterdam: Elsevier, 2006.

Youngquist, W. L. *GeoDestinies: The Inevitable Control of Earth Resources over Nations and Individuals.* Portland, OR: National Book Co., 1997.

Chapter 8

Boyden, S. *Western Civilisation in Biological Perspective.* Oxford: Oxford University Press, 1987.

Byles, J. *Rubble: Unearthing the History of Demolition.* New York: Harmony Books, 2005.

Girling, R. *Rubbish! Dirt on Our Hands and Crisis Ahead.* London: Transworld, 2005.

Montgomery, D. R. *Dirt: The Erosion of Civilizations.* Berkeley: University of California Press, 2007.

Packard, V. *The Waste Makers.* New York: McKay, 1960.

Tal, A. *Pollution in a Promised Land: An Environmental History of Israel.* Berkeley: University of California Press, 2002.

Tammemagi, H. *The Waste Crisis: Landfills, Incinerations, and the Search for a Sustainable Future.* New York: Oxford University Press, 1999.

Turner, B. L. *The Earth as Transformed by Human Action: Global and Regional Changes in the Biosphere over the Past 300 Years.* Cambridge: Cambridge University Press, 1990.

Vaughn, J. *Waste Management: A Handbook.* Santa Barbara, CA: ABC-CLIO, 2009.

Williams, R. J. P., and J. J. R. Fraústo da Silva. *The Chemistry of Evolution: The Development of Our Ecosystem.* Amsterdam: Elsevier, 2006.

Chapter 9

Atkins, P. *Four Laws that Drive the Universe.* Oxford: Oxford University Press, 2007.

Braungart, M., and W. McDonough. *Cradle to Cradle: Remaking the Way We Make Things.* New York: North Point, 2007.

Harold, F. M. *The Vital Force: A Study in Bioenergetics.* New York: Freeman, 1986.

Lovins, Amory B. *Openpit Mining*. London: Earth Island, 1973.

Rifkin, J. *Entropy into the Greenhouse World*. New York: Bantam Books, 1980.

Chapter 10

Connell, D. *Water Politics in the Murray-Darling Basin*. Sydney: Federation Press, 2007.

De Villiers, M. *Water*. Toronto: Stoddart, 1999.

Glennon, R. *Water Follies: Groundwater Pumping and the Fate of America's Fresh Waters*. Washington, DC: Island Press, 2002.

Hoekstra, A. Y., and A. K. Chapagain. *Globalization of Water: Sharing the Planet's Freshwater Resources*. Oxford: Blackwell, 2008.

Ma, J., N. Y. Liu, and L. R. Sullivan. *China's Water Crisis*. Translated by N. Y. Liu and L. R. Sullivan. Norwalk, CT: EastBridge, 2003.

Pearce, F. *When the Rivers Run Dry*. London: Transworld, 2006.

Postel, S. *The Last Oasis: Facing Water Scarcity*. London: Earthscan, 1992.

Shiva, V. *Water Wars: Privatization, Pollution and Profit*. Cambridge, MA: South End Press, 2002.

Chapter 11

Archer, D. *Global Warming: Understanding the Forecast*. Oxford: Blackwell, 2007.

———. *The Long Thaw: How Humans Are Changing the Next 100,000 Years of Earth's Climate*. Princeton, NJ: Princeton University Press, 2008.

Beerling, D. J. *The Emerald Planet: How Plants Changed Earth's History*. Oxford: Oxford University Press, 2007.

Broecker, W. S. *How to Build a Habitable Planet*. Palisades, NY: Eldigio Press, 1987.

Burroughs, W. J. *Does the Weather Really Matter? The Social Implications of Climate Change*. Cambridge: Cambridge University Press, 1997.

Cox, J. D. *Climate Crash: Abrupt Climate Change and What It Means for Our Future*. Washington, DC: National Academic Press, 2005.

Davis, M. *Late Victorian Holocausts: El Niño Famines and the Making of the Third World*. London: Verso, 2002.

Fagan, B. M. *Floods, Famines, and Emperors*. New York: Basic Books, 1999.

———. *The Long Summer: How Climate Changed Civilization*. New York: Basic Books, 2004.

Gelbspan, R. *The Heat Is On: The High Stakes Battle over Earth's Threatened Climate*. Reading, MA: Addison Wesley, 1997.

Glantz, M. H. *Currents of Change*. Cambridge: Cambridge University Press, 1996.

Gore, A. *An Inconvenient Truth: The Planetary Emergency of Global Warming and What We Can Do about It*. London: Bloomsbury, 2006.

Gribbin, J. *Hothouse Earth: The Greenhouse Effect and Gaia*. New York: Grove Weidenfeld, 1990.

Hooper, M. *The Ferocious Summer: Palmer's Penguins and the Warming of Antarctica*. London: Profile Books, 2007.

Hoyt, D. V., and K. H. Schatten. *The Role of the Sun in Climate Change*. New York: Oxford University Press, 1997.

Kump, L. R., J. F. Kasting, and R. G. Crane. *The Earth System*. 2nd ed. Upper Saddle River, NJ: Prentice Hall, 2004.

Leggett, J. *The Carbon War: Global Warming and the End of the Oil Era*. London: Penguin, 1999.

Lomborg, B. *Cool It: The Skeptical Environmentalist's Guide to Global Warming*. London: Marchal Cavendish, 2007.

———, ed. *How to Spend $50 Billion to Make the World a Better Place*. Cambridge: Cambridge University Press, 2006.

———, ed. *Smart Solutions to Climate Change: Comparing Costs and Benefits*. Cambridge: Cambridge University Press, 2010.

Lynas, M. *Six Degrees: Our Future on a Hotter Planet*. London: Fourth Estate, 2007.

MacKay, D. J. C. *Sustainable Energy—Without the Hot Air*. Cambridge: UIT Cambridge, 2009.

Nordhaus, W. D. *A Question of Balance: Weighing the Options on Global Warming Policies*. New Haven, CT: Yale University Press, 2009.

Pearce, F. *The Last Generation: How Nature Will Take Her Revenge for Climate Change*. London: Transworld, 2006.

———. *Turning Up the Heat: Our Perilous Future in the Global Greenhouse*. London: Bodley Head, 1989.

Philander, S. G. *Is the Temperature Rising? The Uncertain Science of Global Warming*. Princeton, NJ: Princeton University Press, 2000.

Reay, D. *Methane and Climate Change*. London: Earthscan, 2010.

Rosenzweig, C., and D. Hillel. *Climate Change and the Global Harvest: Potential Impacts of the Greenhouse Effect on Agriculture*. New York: Oxford University Press, 1998.

Ruddiman, W. F. *Plows, Plagues and Petroleum: How Humans Took Control over Climate*. Princeton, NJ: Princeton University Press, 2005.

Schneider, S. H. *Global Warming: Are We Entering the Greenhouse Century?* San Francisco: Sierra Club Books, 1989.

Schneider, S. H. *Laboratory Earth: The Planetary Gamble We Can't Afford to Lose*. New York: Basic Books, 1997.

Schneider, S. H., and R. Londer. *The Coevolution of Climate and Life*. San Francisco: Sierra Club Books, 1984.

Singh, H. B., ed. *Composition, Chemistry, and Climate of the Atmosphere*. New York: Van Nostrand Reinhold, 1995.

Sivakumar, M. V. K., and J. Hansen, eds. *Climate Prediction and Agriculture*. Berlin: Springer, 2007.

Smith, K. *Nitrous Oxide and Climate Change*. London: Earthscan, 2010.

Solomon, S. et al., eds. *Climate Change 2007: The Physical Science Basis*. Cambridge: Cambridge University Press, 2007.

Svensmark, H., and N. Calder. *Chilling Stars: A New Theory of Climate Change*. London: Icon, 2007.

Chapter 12

Bonnicksen, T. M. *America's Ancient Forests: From the Ice Age to the Age of Discovery*. New York: Wiley, 2000.

Cronon, W. *Nature's Metropolis: Chicago and the Great West*. New York: Norton, 1991.

Harrison, R. P. *Forests: The Shadow of Civilization*. Chicago: University of Chicago Press, 1993.

Williams, M. *Americans and Their Forests: A Historical Geography*. Cambridge: Cambridge University Press, 1992.

———. *Deforesting the Earth: From Prehistory to Global Crisis, an Abridgment*. Chicago: University of Chicago Press, 2006.

Chapter 13

Beinart, W., and P. A. Coates. *Environment and History: The Taming of Nature in the USA and South Africa*. London: Routledge, 1995.

Browder, J. O., and B. J. Godfrey. *Rainforest Cities: Urbanization, Development, and Globalization of the Brazilian Amazon*. New York: Columbia University Press, 1997.

Carson, R. L. *Silent Spring*. Boston: Houghton Mifflin, 1962.

Clover, C. *The End of the Line: How Overfishing Is Changing the World and What We Eat*. London: Ebury Press, 2004.

Crosby, A. W. *The Columbian Exchange: Biological and Cultural Consequences of 1942*. Westport, CT: Greenwood Press, 1972.

———. *Ecological Imperialism: The Biological Expansion of Europe, 900–1900*. Cambridge: Cambridge University Press, 2004.

Di Castri, F., and T. Younes, eds. *Biodiversity, Science and Development: Towards a New Partnership*. Wallingford, UK: CAB International, 1996.

Dovers, S., and M. Shirlow. *Our Biosphere under Threat: Ecological Realities and Australia's Opportunities*. Melbourne: Oxford University Press, 1990.

Ehrlich, P. R., and A. H. Ehrlich. *Extinction: The Causes and Consequences of the Disappearance of Species*. London: Gollancz, 1981.

Eldredge, N. *Systematics, Ecology, and the Biodiversity Crisis*. New York: Columbia University Press, 1992.

Elvin, M. *The Retreat of the Elephants: An Environmental History of China*. New Haven, CT: Yale University Press, 2006.

Falk, D. A., M. Olwell, and C. I. Millar, eds. *Restoring Diversity: Strategies for Reintroduction of Endangered Plants*. New York: Island Press, 1996.

Flannery, T. *The Eternal Frontier: An Ecological History of North America and Its Peoples*. Melbourne: Text Publishing, 2001.

Goldoftas, B. *The Green Tiger: The Cost of Ecological Decline in the Philippines*. New York: Oxford University Press, 2006.

Graham, F. *Since Silent Spring*. Boston: Houghton Mifflin, 1970.

Grove, A. T., and O. Rackham. *The Nature of Mediterranean Europe: An Ecological History*. New Haven, CT: Yale University Press, 2001.

Huston, M. A. *Biological Diversity: The Coexistence of Species on Changing Landscapes*. Cambridge: Cambridge University Press, 1994.

Komarov, B. *The Destruction of Nature in the Soviet Union*. London: Pluto Press, 1978.

Low, T. *Feral Future: The Untold Story of Australia's Exotic Invaders*. Ringwood, Vic.: Viking, 1999.

McKibben, B. *The End of Nature*. London: Viking, 1990.

McKinney, M. L., and J. A. Drake, eds. *Biodiversity Dynamics*. New York: Columbia University Press, 1998.

Melville, E. G. K. *A Plague of Sheep: Environmental Consequences of the Conquest of Mexico*. Cambridge: Cambridge University Press, 1997.

Nelissen, N. J. M., L. Klinkers, and J. van der Straaten, eds. *Classics in Environmental Studies: An Overview of Classic Texts in Environmental Studies*. Utrecht: International Books, 1997.

Pretty, J. *Agro-Culture: Reconnecting People, Land and Nature*. London: Earthscan, 2002.

Richards, J. F. *The Unending Frontier: An Environmental History of the Early Modern World*. Berkeley: University of California Press, 2003.

Shiva, V. *Biopiracy: The Plunder of Nature and Knowledge*. Cambridge, MA: South End Press, 1997.

Solbrig, O. T., H. M. van Emden, and P. G. W. J. van Oordt, eds. *Biodiversity and Global Change*. Wallingford, UK: CAB International, 1994.

Swanson, T. M. *The Economics and Ecology of Biodiversity Decline: The Forces Driving Global Change*. Cambridge: Cambridge University Press, 1995.

Whitney, G. G. *From Coastal Wilderness to Fruited Plain: A History of Environmental Change in Temperate North America, 1500 to the Present*. Cambridge: Cambridge University Press, 1996.

Wilson, E. O. *The Diversity of Life*. London: Penguin, 1992.

Worster, D., ed. *The Ends of the Earth: Perspectives on Modern Environmental History*. Cambridge: Cambridge University Press, 1988.

Chapter 14

Blackbourn, D. *The Conquest of Nature: Water, Landscape, and the Making of Modern Germany*. New York: Norton, 2006.

Bonnifield, M. P. *The Dust Bowl: Men, Dirt, and Depression*. Albuquerque: University of New Mexico Press, 1979.

Browder, J. O., and B. J. Godfrey. *Rainforest Cities: Urbanization, Development, and Globalization of the Brazilian Amazon*. New York: Columbia University Press, 1997.

Byles, J. *Rubble: Unearthing the History of Demolition*. New York: Harmony Books, 2005.

Davis, M. *Late Victorian Holocausts: El Niño Famines and the Making of the Third World*. London: Verso, 2002.

Foster, J. B. *The Vulnerable Planet: A Short Economic History of the Environment*. New York: Monthly Review Press, 1999.

Girling, R. *Rubbish! Dirt on Our Hands and Crisis Ahead*. London: Transworld, 2005.

Grainger, A. *The Threatening Desert: Controlling Desertification*. London: Earthscan, 1990.

Grove, A. T., and O. Rackham. *The Nature of Mediterranean Europe: An Ecological History*. New Haven, CT: Yale University Press, 2001.

Hickman, L. *The Final Call: In Search of the True Cost of Our Holidays*. London: Transworld, 2007.

Hillel, D. *Out of the Earth: Civilisation and the Life of the Soil*. London: Aurum Press, 1993.

Laity, J. *Deserts and Desert Environments*. Chichester, UK: Wiley-Blackwell, 2008.

McNeill, J. *Something New under the Sun: An Environmental History of the Twentieth Century*. London: Lane, 2000.

Meyer, W. B. *Human Impact on the Earth*. Cambridge: Cambridge University Press, 1996.

Montgomery, D. R. *Dirt: The Erosion of Civilizations*. Berkeley: University of California Press, 2007.

Pretty, J. *Agro-Culture: Reconnecting People, Land and Nature*. London: Earthscan, 2002.

Smil, V. *China's Environmental Crisis: An Inquiry into the Limits of National Development*. Armonk, NY: Sharpe, 1993.

Tal, A. *Pollution in a Promised Land: An Environmental History of Israel*. Berkeley: University of California Press, 2002.

Worster, D. *Dust Bowl: The Southern Plains in the 1930s*. Oxford: Oxford University Press, 1979.

Chapter 15

Borgstrom, G. *The Hungry Planet: The Modern World at the Edge of Famine*. New York: Macmillan, 1965.

———. *Too Many: An Ecological Overview of Earth's Limitations*. New York: Macmillan, 1969.

Brown, L. R. *Who Will Feed China? Wake-Up Call for a Small Planet*. New York: Norton, 1995.

Clark, C. *History of Population growth*. London: Macmillan, 1967.

Cohen, J. E. *How Many People Can the Earth Support?* New York: Norton, 1996.

Cohen, M. N. *The Food Crisis in Prehistory: Overpopulation and the Origins of Agriculture*. New Haven, CT: Yale University Press, 1977.

Dando, W. A. *The Geography of Famine*. London: Arnold, 1980.

Ehrlich, P. R., and A. H. Ehrlich. *The Dominant Animal: Human Evolution and the Environment*. Washington, DC: Island Press, 2008.

———. *The Population Explosion*. New York: Simon and Schuster, 1990.

———. *Population, Resources, Environment*. San Francisco: Freeman, 1970.

Forrester, J. W. *World Dynamics*. Cambridge, MA: Wright-Allan Press, 1971.

Gilbert, G. *World Population: A Reference Handbook.* Santa Barbara, CA: ABC-CLIO, 2001.

Hahlbrock, K. *Feeding the Planet: Environmental Protection through Sustainable Agriculture.* London: Haus, 2009.

Henry, L. *Population: Analysis and Models.* London: Arnold, 1976.

Livi-Bacci, M. *A Concise History of World Population.* Oxford: Blackwell, 1997.

———. *Population and Nutrition: An Essay on European Demographic History.* Cambridge: Cambridge University Press, 1991.

Lovelock, J. *The Revenge of Gaia: Why the Earth Is Fighting Back—and How We Can Still Save Humanity.* New York: Basic Books, 2006.

McNeill, W. L. *Population and Politics since 1750.* Charlottesville: University of Virginia Press, 1990.

Meadows, D. H., D. L. Meadows, and J. Randers. *Beyond the Limits: Confronting Global Collapse, Envisioning a Sustainable Future.* White River Junction, VT: Chelsea Green, 1992.

Meadows, D. L. *The Limits to Growth.* New York: Universe Books, 1972.

Mesarović, M., and E. Pestel. *Mankind on a Turning Point.* New York: Dutton, 1974.

Post, J. D. *Food Shortage, Climate Variability, and Epidemic Disease in Pre-Industrial Europe: The Mortality Peak in the Early 1740s.* Ithaca, NY: Cornell University Press, 1985.

Roberts, P. *The End of Food: The Coming Crisis in the World Food Industry.* London: Bloomsbury, 2008.

Shiva, V. *Stolen Harvest: The Hijacking of the Global Food Supply.* Cambridge, MA: South End Press, 2000.

———. *The Violence of the Green Revolution: Third World Agriculture, Ecology and Politics.* London: Zed Books, 1991.

Simon, J. L., 1990. *Population Matters: People, Resources, Environment, and Immigration.* New Brunswick, NJ: Transaction, 1990.

Spengler, J. J. *Facing Zero Population Growth: Reactions and Interpretations, Past and Present.* Durham, NC: Duke University Press, 1978.

Stein, Z., M. Susser, G. Saenger, and F. Marolla. *Famine and Development: The Dutch Hunger Winter of 1944–1945.* New York: Oxford University Press, 1975.

Teitelbaum, M. S., and J. M. Winter. *The Fear of Population Decline.* Orlando, FL: Academic Press, 1985.

Wilkinson, R. G. *The Poverty of Progress.* London: Methuen, 1973.

Woodham-Smith, C. *The Great Hunger: Ireland, 1845–1849.* London: Hamilton, 1962.

Wrigley, E. A., and R. Schofield. *The Population History of England, 1541–1871: A Reconstruction*. Cambridge: Cambridge University Press, 1989.

Chapter 16

Boserup, E. *The Conditions of Agricultural Growth: The Economics of Agrarian Change under Population Pressure*. Chicago: Aldine, 1965.

Boyden, S. *Western Civilisation in Biological Perspective*. Oxford: Oxford University Press, 1987.

Braungart, M., and W. McDonough. *Cradle to Cradle: Remaking the Way We Make Things*. New York: North Point, 2007.

Catton, W. R. *Overshoot: The Ecological Basis of Revolutionary Change*. Urbana: University of Illinois Press, 1980.

Cohen, M. N. *The Food Crisis in Prehistory: Overpopulation and the Origins of Agriculture*. New Haven, CT: Yale University Press.

Davis, M. *Late Victorian Holocausts: El Niño Famines and the Making of the Third World*. London: Verso, 2002.

Gore, A. *Earth in the Balance: Ecology and the Human Spirit*. New York: Plume, 1993.

Gould, J. D. *Economic Growth in History: Survey and Analysis*. London: Methuen, 1972.

Hardin, G. *Living within Limits: Ecology, Economics, and Population Taboos*. New York: Oxford University Press, 1993.

McMichael, A. J. *Planetary Overload: Global Environmental Change and the Health of the Human Species*. Cambridge: Cambridge University Press, 1993.

Sachs, J. *Common Wealth: Economics for a Crowded Planet*. London: Penguin, 2008.

Spooner, B., ed. *Population Growth: Anthropological Implications*. Cambridge, MA: MIT Press, 1972.

Williams, R. J. P., and J. J. R. Fraústo da Silva. *The Chemistry of Evolution: The Development of Our Ecosystem*. Amsterdam: Elsevier, 2006.

Chapter 17

Abrahamson, M. *Global Cities*. New York: Oxford University Press, 2004.

Browder, J. O., and B. J. Godfrey. *Rainforest Cities: Urbanization, Development, and Globalization of the Brazilian Amazon*. New York: Columbia University Press, 1997.

Brugmann, J. *Welcome to the Urban Revolution*. New York: Harper Collins, 2009.

Castells, M. *The Informational City: Information Technology, Economic Restructuring, and the Urban-Regional Process*. Oxford: Blackwell, 1989.

Clark, P., and P. Slack. *English Towns in Transition: 1500–1700*. London: Oxford University Press, 1976.

Cronon, W. *Nature's Metropolis: Chicago and the Great West*. New York: Norton, 1991.

De Vries, J. *European Urbanization. 1500–1800*. Cambridge, MA: Harvard University Press, 1984.

Herber, L. *Crisis in Our Cities*. Englewood Cliffs, NJ: Prentice Hall, 1965.

Hohenberg, P. M., and L. H. Lees. *The Making of Urban Europe, 1000–1950*. Cambridge, MA: Harvard University Press, 1985.

Neuwirth, R. *Shadow Cities: A Billion Squatters, A New Urban World*. New York: Routledge, 2005.

Reich, R. B. *The Work of Nations: Preparing Ourselves for 21st Century Capitalism*. New York: Knopf, 1991.

Rykwert, J. *The Seduction of Place: The History and Future of the City*. Oxford: Oxford University Press, 2000.

Sassen, S. *The Global City: New York, London, Tokyo*. 2nd ed. Princeton, NJ: Princeton University Press, 2001.

Chapter 18

Cohen, M. N. *The Food Crisis in Prehistory: Overpopulation and the Origins of Agriculture*. New Haven, CT: Yale University Press, 1977.

Headrick, D. R. *The Tools of Empire: Technology and European Imperialism in the Nineteenth Century*. New York: Oxford University Press, 1981.

Kane, H. "The Hour of Departure: Forces that Create Refugees and Migrants." WorldWatch Paper 125. Washington, DC: WorldWatch Institute, 1995.

Sassen, S. *Guests and Aliens*. New York: New Press, 1999.

Spooner, B., ed. *Population Growth: Anthropological Implications*. Cambridge, MA: MIT Press, 1972.

Chapter 19

Barry, J. M. *The Great Influenza: The Epic Story of the Deadliest Plague in History*. London: Penguin, 2004.

Garrett, L. *Betrayal of Trust: The Collapse of Global Public Health*. New York: Hyperion, 2001.

———. *The Coming Plague: Newly Emerging Diseases in a World Out of Balance*. New York: Penguin, 1995.

Hays, J. N. *The Burdens of Disease: Epidemics and Human Response in Western History*. New Brunswick, NJ: Rutgers University Press, 1998.

Hopkins, D. R. *Princes and Peasants: Smallpox in History*. Chicago: University of Chicago Press, 1983.

Hubbert, W. T., W. F. McCulloch, and P. R. Schurrenberger, eds. *Diseases Transmitted from Animals to Man*. Springfield, IL: Thomas, 1975.

Kaufmann, S. *The New Plagues: Pandemics and Poverty in a Globalized World*. London: Haus, 2008.

Livi-Bacci, M. *Population and Nutrition: An Essay on European Demographic History*. Cambridge: Cambridge University Press, 1991.

McNeill, W. H. *Plagues and People*. Oxford: Blackwell, 1976.

Sallares, R. *Malaria and Rome: A History of Malaria in Ancient Italy*. Oxford: Oxford University Press, 2002.

Walters, M. J. *Six Modern Plagues and How We Are Causing Them*. San Francisco: Island Press, 2003.

Zinsser, H. *Rats, Lice, and History*. Boston: Little Brown, 1963.

Chapter 20

Meadows, D. H. *Thinking in Systems: A Primer*. White River Junction, VT: Chelsea Green, 2008.

McNeill, J. R., and W. H. McNeill. *The Human Web: A bird's Eye View of World History*. New York: Norton, 2003.

Post, J. D. *Food Shortage, Climate Variability, and Epidemic Disease in Pre-Industrial Europe: The Mortality Peak in the Early 1740s*. Ithaca, NY: Cornell University Press, 1985.

Chapter 21

Beniger, J. R. *The Control Revolution: Technological and Economic Origins of the Information Society*. Cambridge, MA: Harvard University Press, 1986.

Reich, R.B. *Supercapitalism: The Transformation of Business, Democracy, and Everyday Life*. New York: Knopf, 2007.

——. *The Work of Nations: Preparing Ourselves for 21ˢᵗ Century Capitalism.* New York: Knopf, 1991.

Sassen, S. *The Global City: New York, London, Tokyo.* 2nd ed. Princeton, NJ: Princeton University Press, 2001.

——, ed. *Global Networks: Linked Cities.* New York: Routledge, 2002.

——. *The Mobility of Labor and Capital: A Study in International Investment and Labor Flow.* Cambridge: Cambridge University Press, 1988.

——. *Territory, Authority, Rights: From Medieval to Global Assemblages.* Princeton, NJ: Princeton University Press, 2006.

Zweig, S. *The World of Yesterday.* London: Cassell, 1943.

Chapter 22

Beniger, J. R. *The Control Revolution: Technological and Economic Origins of the Information Society.* Cambridge, MA: Harvard University Press, 1986.

Boulding, K. E. *The Meaning of the 20ᵗʰ Century: The Great Transition.* New York: Harper and Row, 1964.

Castells, M. *The Informational City: Information Technology, Economic Restructuring, and the Urban-Regional Process.* Oxford: Blackwell, 1989.

Georgescu-Roegen, N. *The Entropy Law and the Economic Process.* Cambridge, MA: Harvard University Press, 1971.

Kindleberger, C. P. *Manias, Panics, and Crashes: A History of Financial Crises.* New York: Wiley, 1978.

Kooijman, S. A. L. M. *Dynamic Energy Budget Theory for Metabolic Organization.* 3rd ed. Cambridge: Cambridge University Press, 2010.

Meadows, D. H. *Thinking in Systems: A Primer.* White River Junction, VT: Chelsea Green, 2008.

Odum, H. T. *Environment, Power, and Society.* New York: Wiley, 1971.

Smil, V. *Energy in Nature and Society: General Energetics of Complex Systems.* Cambridge, MA: MIT Press, 2008.

Vaitheeswaran, V. V. *Power to the People: How the Coming Energy Revolution Will Transform an Industry, Change Our Lives, and Maybe Even Save the Planet.* New York: Farrar, Straus and Giroux, 2005.

Wiener, N. *The Human Use of Human Beings: Cybernetics and Society.* Boston: Houghton Mifflin, 1950.

Wrigley, E. A. *Energy and the English Industrial Revolution.* Cambridge: Cambridge University Press, 2010.

Chapter 23

Cipolla, C. M., ed. *The Economic Decline of Empires.* London: Methuen, 1970.

Diamond, J. *Collapse.* New York: North Point, 2004.

Douthwaite, R. J. *The Growth Illusion: How Economic Growth Has Enriched the Few, Impoverished the Many, and Endangered the Planet.* Tulsa, OK: Council Oak Books, 1993.

Ehrlich, P. R. *The End of Affluence: A Blueprint for Your Future.* New York: Ballantine Books, 1974.

Friedman, T. L. *Hot, Flat, and Crowded: Why We Need a Green Revolution, and How It Can Renew America.* New York: Farrar, Straus and Giroux, 2008.

Friedman, T. L. *The World is Flat: A Brief History of the Twenty-First Century.* New York: Farrar, Straus and Giroux, 2005.

Graham, F. D. *Exchange, Prices, and Production in Hyper-Inflation Germany, 1920–1923.* Princeton, NJ: Princeton University Press, 1930.

Haller, S. F. *Apocalypse Soon? Wagering on Warnings of Global Catastrophe.* Montreal: McGill-Queen's University Press, 2002.

Hobsbawm, E. *Globalisation, Democracy and Terrorism.* London: Abacus, 2007.

Huntington, S. *The Clash of Civilizations and the Remaking of World Order.* New York: Simon and Schuster, 1996.

Kaplan, R. D. *The Ends of the Earth: A Journey to the Frontiers of Anarchy.* New York: Random House, 1996.

Karplus, W. J. *The Heavens Are Falling: The Scientific Prediction of Catastrophes in Our Time.* New York: Plenum Press, 1992.

Kennedy, P. *Preparing for the Twenty-First Century.* New York: Random House, 1993.

Klare, M. T. *Resource Wars: The New Landscape of Global Conflict.* New York: Henry Holt, 2002.

Mandelbrot, B. B. *The Fractal Geometry of Nature.* New York: Freeman, 1977.

Meadows, D. H., D. L. Meadows, and J. Randers. *Beyond the Limits: Confronting Global Collapse, Envisioning a Sustainable future.* White River Junction, VT: Chelsea Green, 1992.

Pearce, F. *Peoplequake: Mass Migration, Ageing Nations and the Coming Population Crash.* London: Transworld, 2010.

Reich, R. B. *Supercapitalism: The Transformation of Business, Democracy, and Everyday Life.* New York: Knopf, 2007.

Sassen, S. *Losing Control? Sovereignty in an Age of Globalization*. New York: Columbia University Press, 1996.

Stiglitz, J. *The Roaring Nineties: A New History of the World's Most Prosperous Decade*. New York: Norton, 2003.

Tainter, J. A. *The Collapse of Complex Societies*. Cambridge: Cambridge University Press, 1990.

Teitelbaum, M. S., and J. M. Winter. *The Fear of Population Decline*. Orlando, FL: Academic Press, 1985.

Whyte, I. *World without End? Environmental Disaster and the Collapse of Empires*. London: I. B. Tauris, 2008.

Youngquist, W. L. *GeoDestinies: The Inevitable Control of Earth Resources over Nations and Individuals*. Portland, OR: National Book Co., 1997.

Chapter 24

Albert, M. *Capitalism against Capitalism*. Chicester, UK: Wiley, 1992.

Archer, D. *The Long Thaw: How Humans Are Changing the Next 100,000 Years of Earth's Climate*. Princeton, NJ: Princeton University Press, 2008.

Ayres, R. U. *Turning Point: An End to the Growth Paradigm*. London: Earthscan, 1998.

Beder, S. *Global Spin: The Corporate Assault on Environmentalism*. Totnes, UK: Green Books, 2002.

Bramwell, A. *The Fading of the Greens: The Decline of Environmental Politics in the West*. New Haven, CT: Yale University Press, 1994.

Braudel, F. *Afterthoughts on Material Civilization and Capitalism*. Baltimore, MD: Johns Hopkins University Press, 1977.

Brown, L. R. *Eco-Economy: Building an Economy for the Earth*. New York: Norton, 2001.

Cipolla, C. M. *The Economic History of World Population*. 7th ed. Harmondsworth, UK: Penguin, 1978.

Cole, H. S. D., C. Freeman, M. Jahoda, and K. L. R. Pavitt. *Models of Doom: A Critique of the Limits to Growth*. New York: Universe Books, 1973.

Dilworth, C. *Too Smart for Our Own Good: The Ecological Predicament of Humankind*. Cambridge: Cambridge University Press, 2009.

Eatwell, J., and L. Taylor. *Global Finance at Risk: The Case for International Regulation*. New York: New Press, 2000.

Edgerton, D. *The Shock of the Old: Technology and Global History since 1900*. Oxford: Oxford University Press, 2007.

Galbraith, J. H. *The New Industrial State*. Boston: Houghton Mifflin, 1967.

Garrett, L. *Betrayal of Trust: The Collapse of Global Public Health*. New York: Hyperion, 2001.

Gould, J. D. *Economic Growth in History: Survey and Analysis*. London: Methuen, 1972.

Grove, R. H. *Green Imperialism: Colonial Expansion, Tropical Island Edens, and the Origins of Environmentalism, 1600–1860*. Cambridge: Cambridge University Press, 1996.

Hall, P. A., and D. Soskice, eds. *Varieties of Capitalism: The Institutional Foundations of Competitive Advantage*. Oxford: Oxford University Press, 2001.

Held, D., ed. *A Globalising World? Culture, Economics, Politics*. 2nd ed. London: Routledge, 2004.

Kunstler, J. H. *The Long Emergency: Surviving Converging Catastrophes of the 21st Century*. London: Atlantic Books, 2005.

Lomborg, B., ed. *Global Crises, Global Solutions*. Cambridge: Cambridge University Press, 2004.

———, ed. *Smart Solutions to Climate Change: Comparing Costs and Benefits*. Cambridge: Cambridge University Press, 2010.

Lovelock, J. *The Revenge of Gaia: Why the Earth is Fighting Back—and How We Can Still Save Humanity*. New York: Basic Books, 2006.

Martin, J. *The Meaning of the 21st Century: A Vital Blueprint for Assuring Our Future*. London: Transworld, 2006.

McNeill, W. H. *A World History*. New York: Oxford University Press, 1967.

Meadows, D. L. *The Limits to Growth*. New York: Universe Books, 1972.

Nordhaus, W. D. *A Question of Balance: Weighing the Options on Global Warming Policies*. New Haven, CT: Yale University Press, 2009.

Oreskes, N., and E. Conway. *Merchants of Doubt. How a Handful of Scientists Obscured the Truth on Issues from Tobacco Smoke to Global Warming*. London: Bloomsbury, 2010.

Pearce, F. *The Climate Files: The Battle for the Truth about Global Warming*. New York: Random House, 2010.

Reich, R. B. *Supercapitalism: The Transformation of Business, Democracy, and Everyday Life*. New York: Knopf, 2007.

Rifkin, J. *The Empathic Civilization: The Race to Global Consciousness in a World in Crisis*. Cambridge: Polity Press, 2009.

Roberts, P. *The End of Food: The Coming Crisis in the World Food Industry*. London: Bloomsbury, 2008.

Sachs, J. *Common Wealth: Economics for a Crowded Planet*. London: Penguin, 2008.

Silver, C. S., and R. S. DeFries. *One Earth, One Future: Our Changing Global Environment*. Washington, DC: National Academy Press, 1990.

Simmons, I.G. *Changing the Face of the Earth: Culture, Environment, History*. Oxford: Blackwell, 1989.

Smil, V. *Global Catastrophes and Trends: The Next Fifty Years*. Cambridge, MA: MIT Press, 2008.

Stiglitz, J. *Globalization and Its Discontents*. New York: Norton, 2002.

Walker, G., and D. King. *The Hot Topic: How to Tackle Global Warming and Still Keep the Lights On*. London: Bloomsbury, 2008.

Weisman, A. *The World without Us*. London: Virgin Books, 2007.

Welzer, H. *Klimakriege. Wofür im 21. Jahrhundert getötet wird*. Jena: Fischer Verlag, 2008.

Williams, M. *Deforesting the Earth: From Prehistory to Global Crisis, an Abridgment*. Chicago: University of Chicago Press, 2006.

Worster, D. *The Wealth of Nature: Environmental History and the Ecological Imagination*. New York: Oxford University Press, 1994.

Zalasiewicz, J. *The Earth After Us: What Legacy Will Humanity Leave in the Rocks?* Oxford: Oxford University Press, 2008.